Harley Flanders

Scientific Pascal
second edition

1996

Birkhäuser

Boston • Basel • Berlin

Harley Flanders
Department of Mathematics
University of Michigan
Ann Arbor, MI 48109

Library of Congress Cataloging-in-Publication Data

Flanders, Harley
    Scientific Pascal / Harley Flanders. -- 2nd edition.
        p.   cm.
    Includes index.
    ISBN 0-8176-3760-5 (Boston : acid-free). -- ISBN 3-7643-3760-5
(Basel : acid-free)
    1. Pascal (Computer program language.)   I. Title.
  III. Series.
QA6.73P2F55   1995   95-35243
005.13'3--dc20   CIP

Printed on acid-free paper

Birkhäuser

© 1996 Birkhäuser Boston
Scientific Pascal, First Edition, 1984
published by Reston Publishing Company, Inc.

ISBN 0-8176-3760-5
ISBN 3-7643-3760-5
Typeset by the author using the program *The Publisher.*
Printed and bound by Quinn-Woodbine, Woodbine, NJ.
Printed in the U.S.A.

9 8 7 6 5 4 3 2 1

Contents

Preface

Guide to this Book

My main objective is to teach programming in Pascal to people in the hard sciences and technology, who don't have much patience with the standard textbooks with their lengthy, pedantic approach, and their many examples of no interest to scientists and engineers. Another objective is to present many both interesting and useful algorithms and programs. A secondary objective is to explain how to cope with various features of the PC hardware.

Pascal really is a wonderful programming language. It is easy to learn and to remember, and it has unrivalled clarity. You get serious results in short order.

How should you read this book? Maybe backwards is the answer. If you are just starting with the *Borland Pascal* package, you must begin with Appendix 1, *The Borland Pascal Package*. If you are a Pascal user already, still you should skim over Appendix 1. Appendix 2, *On Programming*, has material on saving programming time and on debugging that might be useful for reference. Chapter 1, *Introduction to Pascal*, will hardly be read by the experienced Pascal programmer (unless he or she has not used units). Chapter 2, *Programming Basics*, begins to sample deeper waters, and I hope everyone will find something interesting there. Chapter 3, *Files, Records, Pointers*, is the final chapter to concentrate on the Pascal programming language; the remaining chapters concentrate on various areas of application. Throughout I have tried to include complete programs and units to illustrate the text. All program code printed in this text and much more is included on the disk that accompanies this book.

References are located in the text where quoted, and all authors are indexed. Note that "CACM" means "Collected Algorithms of the ACM". Items in this collection can be downloaded from Netlib or Simtel.

The following paragraphs are extracted from the preface to the first edition. Some still valid, some rather dated, but fun (at least for me) to read:

> The first purpose of this book is to teach the programming language Pascal to those whose primary use of the computer is for scientific applications. The second purpose is to teach the skill of tackling a computing problem, developing an algorithm for solving it, and writing a program that does the job. The third purpose is to present a number of important programs in usable form.

Why Pascal? First, because it has built-in clarity. A well-written Pascal program can be read easily because of the way the language handles subroutines, links between subroutines, and repetitive processes. Second, the language is widely used. Virtually every large and middle sized (mini) computer has Pascal software, and more and more microcomputers include Pascal. What is more, Pascal is rapidly becoming the first language taught in introductory computer science/systems courses because the structured programming discipline of Pascal engenders better programming technique in general than do, say, BASIC or FORTRAN. A third reason is Pascal's broad flexibility for data types. Finally, the recursive nature of Pascal is especially attractive.

The algorithmic and recursive language ALGOL was developed in the 1950s and 60s by a group of computer scientists. Pascal, an offshoot of ALGOL, was invented by Niklaus Wirth in the early 70s. Compared to ALGOL, Pascal is simpler and clearer. It has much better data handling capability and built-in I/O procedures (input/output), which ALGOL lacked.

I believe one learns most efficiently by studying examples and by working exercises. Students learn calculus by working through hundreds of solved examples and by solving thousands of exercises. I think the same principle applies to learning a programming language. A student should see solved examples on each topic and work on hundreds of exercises.

The examples and exercises, culled from many sources, are probably the bulk of this otherwise modest text. At first they are brief, single programs. Needless to say, organizing a large program presents a rather different type of problem than implementing a single short algorithm. Pascal is particularly suited to the decomposition of a large task into manageable components.

To guarantee accuracy, we have reproduced computer-generated printout of all our programs directly from disk files (after running successfully) without the intervention of typists, copy editors, or typesetters. Yet, there may still be some errors lurking, and the author will be grateful for having errors brought to his attention, large or small.

I have tried to find challenging problems for the examples and exercises. Finding an algorithm that will solve a given problem in a reasonable time with available computer memory can be a difficult task. This is part of the solution. The other part is translating the algorithm into a clear Pascal (or other language) program. I hope this text proves useful for developing these problem-solving skills.

I have set several standards for myself in this book. One is to include only programs that have actually been tested on a computer. A second is to print all programs in an uncluttered, clear format.

History of this Book

Allow me to recount how I started in scientific programming. My work was mostly in pure mathematics until the mid 70s, when I became interested in some problems in electric circuit theory. I needed extensive computations, and as no one would do them for me, I decided to bite the bullet and learn programming. The Tel Aviv University mathematics department had a portable teletype machine that I had to wheel into my office and strap a telephone onto in order to use the CDC main frame. I gradually learned BASIC, and on the advice of my very wise colleague Amir Pnueli, ALGOL; a little later I started on Pascal. On a visit to the Univ. of Pennsylvania I had a private lesson in algorithms from my old friend Albert Nijenhuis, on a CompuCorp Scientist programmable calculator. At Florida Atlantic University, the mathematics department chair, Frank Hadlock, set up an Apple II lab, and I got well into the p-system, and so one thing led to another ...

I wrote *Scientific Pascal*, 1/e, on an Apple *///* computer with 128K and an external hard disk, advanced equipment in its day. Its 5M hard disk held my word processor, *Apple Writer ///*, my p-System *UCSD Apple Pascal* package, all programs for the text and many others, the text of a calculus book I was writing, and a bunch of software packages, like *Advanced VisiCalc*. (My current spreadsheet, Borland's *Quattro Pro for Windows*, fills about 14M by itself.) Times have changed.

The p-System was an adventure in itself: an operating system within the operating system. It was first available for Apple II+ computers (remember them: 4–64K, two 180K floppy drives, ROM Applesoft-BASIC), later for Apple *///*'s, with a compiler option that allowed compiling for a II on a *///*.

Shortly after IBM introduced the PC, two PC versions of Pascal became available, a p-System based *UCSD Pascal*, which crashed and hung with great regularity, and *IBM Pascal*, a form of Standard Pascal. It had the remarkable property of not supporting communication with another computer via the RS232 port!

My *MicroCalc* program for teaching calculus was programmed on the Apple *///*; I decided around 1980 to port it to the PC, about or just before the time that Borland's *Turbo Pascal* became available. Getting an Apple II or *///* and a PC to talk to each other was a real challenge. I lived then in Boca Raton, the birthplace of the PC; I knew some knowledgeable IBM'ers, but none could tell me even how to cable the machines together. It turned out to be one of the great programming and communicating challenges of my career, but I am happy to report that eventually I got the code transported (without retyping anything). I also broke the code of how disks were written on the two platforms, and I learned to write disks for either system on the PC. The Rosetta stone was the one unique feature of *UCSD Pascal* that I regret not having in *Borland Pascal*: the very low level disk procedures "UnitRead" and "UnitWrite".

UCSD Pascal disappeared for a time; then some outfit revived it for the PC, complete with the p-System operating system within the DOS operating system, guaranteeing slow execution. I think the p-System is now gone for good.

Borland's *Turbo Pascal* went through 6 versions before changing to *Borland Pascal 7.0*. It is a remarkable program in many ways, and one of my objects is to teach the use of the package. Please refer to Appendix 1 for installation and use of *Borland Pascal*.

I decided about 12 years ago to get out of the book writing business until the publishers caught up with technology, and have been waiting patiently until now. I rejected TeX, LaTeX, etc on the grounds that composing in these was not fit work for this human; I needed a WYSIWYG package. I did a great deal of testing on a MS-windows based product called *Scientific Word*, a derivative of the old *T3* package. In time it will be suitable for writing a book, but at the moment it seems adequate only for articles.

My big problem for this edition was how to get Pascal programs into the manuscript. Obviously the programs had to be developed in the programming development environment, so they could be thoroughly tested. They needed to be formatted decently before being inserted into the book manuscript, so there would be no possibility of changes or misprints creeping in.

In particular there had to be a great deal of indentation, which word processing packages (ditto TeX and LaTeX) dislike intensely, as they know much better than you or I what is good for us!

The publishing package I used, on a UNIX workstation (first a CompuAdd SS-1+, which I outgrew, then a Tatung MicroComp 5) is *The Publisher*, from ArborText, Ann Arbor, MI. It is both very good, and very strict. I can only put extra space, like   this between words with some difficulty, and actually writing like this:

```
program hello_world;

  const  hw = 'Hello World!';

  begin
  WriteLn(hw);
  end.
```

is like pulling teeth. But... *The Publisher* has a "Computer" mode, which uses an unjustified font and allows computer text with all of its (publishing point of view) unspeakable word spacing and indentation. So, I developed programs in the *Borland Pascal* "integrated environment" on a DTK 486DX33 (which I also outgrew) or CCS Pentium100. After each program was completed, I ran its source code through a formatting program, whose task was to surround each "reserved word" of the Pascal language with *The Publisher's* codes for bold face, and to uniformize the capitalization of terms from *Borland Pascal's* libraries (e.g, change `writeln` to `WriteLn`). The next step was to copy the programs onto a DOS diskette, which fortunately can be read by my UNIX machine(s), and to change from the DOS to the UNIX line-ending convention en route. Finally I imported these files into the text manuscript.

But there is a pitfall. The IBM character set includes many useful symbols besides the usual characters from ASCII 32 to ASCII 127. For instance ASCII 228 is Σ, ASCII 243 is $\leq$, and so on. These and boxing characters are extremely useful in output, and sometimes in program comments, but *The Publisher* forbids me to use these characters in its computer mode, so I had to compromise. But the disk listings include them of course.

The first edition of *Scientific Pascal* was published by Reston Publishing Company, Inc. in 1984. Reston was an independent subsidiary of Prentice-Hall, Inc. As soon as my book was published, Prentice-Hall decided to close down Reston. The marketing department at Prentice-Hall wasn't much interested in Reston's titles, so *Scientific Pascal*, with other titles I forget, was kept secret (the story of my life). It took years before Prentice-Hall would even admit that all programs in the book were available on disk.

Updates

Corrections and additions for both the text and the disk programs will be available at three public sites. First, by ftp (file transfer protocol) from

`ftp.math.lsa.umich.edu`

This is a UNIX site, so remember that UNIX is case-sensitive. You must change directory:

```
cd pub/flanders/scipas
```

Download the ASCII file "READ.ME" for latest information on the other files. Any file with a name "*.exe" is a self-extracting archive file that must be downloaded in binary mode, then run on an MS-DOS machine. All other files are ASCII files, and must be downloaded in ASCII mode.

Suppose that your system prompt is #. For logging onto a remote ftp site, your name is "anonymous" and your password is your full account name. A typical dialog (with responses omitted) after you are logged onto your own system:

```
#ftp <Enter>
ftp>open math.lsa.umich.edu <Enter>
Name: anonymous <Enter>
Password: jqpublic@math.erewhon.edu <Enter>
ftp>cd pub <Enter>
ftp>cd flanders <Enter>
ftp>cd scipas <Enter>
ftp>ascii <Enter>
ftp>get READ.ME <Enter>
ftp>binary <Enter>
ftp>get update1.exe <Enter>
ftp>quit <Enter>
```

If you have a windows based telnet server, you simply open the remote site ftp.math.lsa.edu with the login "anonymous" and password as indicated. Be sure to tell your system that you are opening a UNIX site. Once the remote site is open, double click on subdirectory icons to descend to "scipas", set the transfer mode to ASCII or binary as required, and drag file icons from the remote window to your local window.

The other sites are Birkhaüser Boston's WWW (World Wide Web) page and FTP site. The web address is

```
http://www.birkhauser.com
```

You may search the site for "Flanders" or click your way through "Our books", "Mathematics" (or "Computer Science"), then "Flanders". The FTP address is

```
ftp.birkhauser.com
```

Log on as "anonymous" and use you full email address as the password. Then change directory (cd) to "pub". The next subdirectory may be "mathematics", "cs", "computer-science", or "books", and the final subdirectory is "flanders".

Acknowledgments

I take pleasure in expressing my gratitude to Edwin F. Beschler of Birkhaüser Boston, with whom I have worked since the early sixties, for his encouragement and belief in this project, to my very able production editor, Ann Kostant, and to reviewer Richard W. Nau of Carleton College. I owe special thanks to my reviewer (and mathematical grandson) George D. Parker of Southern Illinois University, who went through every word and every line of code in the manuscript and suggested very many useful improvements.

Harley Flanders
Ann Arbor, MI
Sept, 1995

Scientific Pascal

second edition

Chapter 1 Introduction to Pascal

Section 1. The Pascal Programming Language

The Pascal Language, in the *Borland Pascal* dialect that we shall use, consists of about 70 **reserved words** and symbols.

Frequently used Pascal reserved words:

Declarations of Pascal programs and library units

program   **unit**

Declaration headings in a program or library unit

uses   **const**   **type**   **var**
procedure   **function**

Constructors for user-defined variable types

array   **record...end**   **file...of**   **string**

Words that bracket a sequence of program statements

begin...end

Constructs for controlling program flow

if...then...else   **case...of...end**
for...to...do   **for...downto...do**
repeat...until   **while...do**

Declaration for the public and the private parts of library units

interface   **implementation**

Arithmetic operations

div   **mod**   **shl**   **shr**

Logical operators

not   **and**   **or**

Alphabetical list:

and	array	begin	case
const	div	do	downto
else	end	file	for
function	if	implementation	interface
mod	not	of	or
procedure	program	record	repeat
shl	shr	string	then
to	type	until	var
unit	uses	while	

These reserved words will get us well started. Perhaps later we'll introduce a few others.

Frequently used Pascal symbols and symbol-pairs:

```
Assignment of variable to expression:    :=
Statement separator, declaration end:    ;
Brackets for expressions:                ( ... )
Punctuation in variable declarations
   and in write parameters:              :

Ellipses for lists:                      ..
Binary operators:                        +   -   *   /
Logical equality, inequality:            =   <
Less than, greater than:                 <   >

Less (Greater) than or equal to:         <=  >=
Constant and type declarations:          =
Program or unit end:                     .
String delimiter pair:                   '   '

Comment bracket pairs:                   {  }   (*  *)
Decimal part separator                   .
Separator for items in a list:           ,
Brackets for array domains/components:   [ ... ]
```

With this vocabulary we can actually do nothing at all! There is no input or output. These missing features, and much more, are supplied by library units. (The disk files "res_word.dat" and "symbols.dat" contain the material displayed above.)

A particular library unit, "System" is automatically linked into every program. It contains I/O subroutines, like "Read" and "Write", variable types, like "Integer" and "Real", and much more. Other library units, like "Graph" and "DOS", must be invoked explicitly in the program code.

Two forms of code files can be compiled in Pascal, the **program** and the **unit**. A program is something that can be run. A unit (library unit) contains definitions and subroutines that can be invoked by one or many programs (or by other units). When a program (or unit) is compiled, exactly what is needed—nothing more—is linked from any units that are invoked.

Section 2.   Simple Programs

We can learn Pascal quickly by examining simple examples of units and programs, starting with some programs.

Program 1.2.1

```
program  sum;    { 06/10/94 }

    var   n, s: Word;

    begin
    s := 0;    { Must initialize variables. }
    for  n := 1  to  99  do  s := s + n*(n + 1);
    WriteLn('1*2 + 2*3 + ... + 99*100 = ', s);
    Write('Press <Enter>: ');
    ReadLn;
    end.
```

Let us go through this program line-by-line, with some details postponed. The first line, **program** sum; is a declaration; in this case it declares a program named sum. The date is in comment brackets {}, hence not part of the program.

The second line, **var**   n, s: Word; is a declaration of two variables, n and s, of type "Word", an unsigned integer type with domain [0, 65535].

The action in the program takes place between the third line, **begin** and the closing line **end**.

The fourth line s := 0; initializes the variable s (the partial sum) to 0, and the comment reminds us that variables are not initialized automatically.

The fifth line executes a loop 99 times; each time s is increased by the product $n(n + 1)$, where the counter n goes from 1 to 99.

The next line outputs the result to the screen. First a literal string is written. (Single quotes '...' delimit a literal string.) It is followed by the computed value of s. The last two letters in "WriteLn" mean that the cursor moves to the beginning of the next line (carriage-return/line-feed) after everything is written.

The next line prompts the user to press "Enter" when he is ready, and the line ReadLn waits for the "Enter" key to be pressed; then the program ends. Please compile and run the little program "sum"; you will find it on your disk.

Remark 1   The word **program** is in bold face to emphasize that it is a *reserved word*. Throughout this book, all reserved words in programs will be printed in bold face. That makes the programs easier to read. (Of course, when programs are typed into a file with a text editor, there is no bold face.)

Remark 2   The word sum is a user-defined *identifier*. Identifiers can be any words you please *except* the (sacred) reserved words.

Remark 3   An *identifier* is any sequence of letters, digits, and the underscore character "_". The first character must be a letter or underscore. Pascal is case-insensitive. That is, "Sum", "sum", "SUM", etc are all the same identifier.

Remark 4   Note the punctuation in the program. Semicolons both terminate declarations and separate statements in the action part of a program. The comma is always used to separate items in a list, in this case in lines 2 and 6. The colon separates the list of variables being declared from their type Word. Programs, as sentences, end in periods.

Remark 5   There are two assignments : = in the program. In Pascal, the equal sign by itself is never used for assignment; it has another use. Products must have the binary operator * written between their factors. The math expression ax^2 in a Pascal program is written a*x*x.

Let us look at another program, almost as brief, except for comments:

Program 1.2.2

```
program  cube_sum;    { 06/14/94 }

   var   n0, n: Word;
   { Type Word is a 2-byte unsigned }
           t,            { integer, domain [0, 65535]. }
      an, sum: Real;
              { Type Real is 6-byte, with + or - values
                in domain [2.9E-39, 1.7E38] and 11-12
                significant digits accuracy. }

   begin    { Program  action starts here }
   { Prompt for input }
   Write('Number of terms  n0 = ');
   { Read keyboard input }
   ReadLn(n0);
   sum := 0.0;    { Always initialize variables! }
   for  n := n0  downto  1  do
      begin             { Add smallest to largest to }
      t := 1.0/n;    { decrease roundoff error.    }
      an := t*t*t;    { t^3; Pascal lacks exponents }
      sum := sum + an;
      end;
```

```
WriteLn('S = Sum 1/n^3  from  n = 1  to  n = ', n0);
WriteLn('S =', sum);
{ Prompt for program to end }
Write('Press <Enter>: ');   ReadLn;
end.   { cube_sum }
```

Remark The program uses the types `Real` and `Word`, and the procedures `Write`, `ReadLn`, and `WriteLn` from the library unit "System", which is automatically linked into the program when it is compiled. Note that the **for** loop goes backwards because of the **downto** instead of **to**. The program's purpose is shown in its first output line.

Program 1.2.3

```
program  linear_diophantine_equation;   { 06/10/94 }

    var  x, y: Integer;

    begin
    WriteLn('All integer solutions of  3x - 7y = 1,');
    WriteLn('  -50 <= x <= 50, -25 <= y <= 25:');
    for  x := -50  to  50  do
        for  y := -25  to  25  do
            if  3*x - 7*y = 1  then
                WriteLn('(x, y) = (', x, ', ', y, ')');
    Write('Press <Enter>: ');
    ReadLn;
    end.   { linear_diophantine_equation }
```

This program introduces us to new elements. There is a nested (double) **for** loop, and the statement that it controls is a conditional statement. The `WriteLn` statement only executes if a condition is satisfied. The condition is `3*x - 7*y = 1`, which shows us how the = sign is used in Pascal: only in *Boolean* expressions, i.e, expressions that evaluate to "True" or "False".

The output line

```
WriteLn('(x, y) = (', x, ', ', y, ')');
```

is interesting; four separate items are written to screen, and they are in a list separated, of course, by commas. The sequence of characters ', ' is a literal string consisting of a comma followed by a space (delimited by single quotes). The non-string items are the second x and the second y.

Besides the date of creation comment at the top, there is a comment on the final **end** line, repeating the name of the program that has just ended. This is not needed in such a brief program, but having comments that remind us of where we are and what we are doing is quite important when we come to longer programs. What this particular program does should be clear from reading it, but please do test it.

Program 1.2.4

```pascal
program  partial_sum;     { 06/10/94 }
   { Sum(ln(Sum(ln j))
       i         j             2 <= j <= i <= 500 }

   var    i, j: Word;
          t, sum: Real;

   begin
   sum := 0.0;    { Must initialize }
   for  i := 2  to  500  do
      begin
      t := 0.0;
      for  j := 2  to  i  do  t := t + Ln(j);
      sum := sum + Ln(t);
      end;
   WriteLn('Sum = ', sum:6:5);
   Write('Press <Enter>: ');
   ReadLn;
   end.    { partial_sum }
```

This program computes a double sum of real numbers. Here "Word" is an unsigned 2–byte integer, domain [0, 65535]. "Real" is one of several available types of real number; it occupies 6 bytes and has domain plus or minus [2.9×10^{-39}, 1.7×10^{38}]. "Ln" is the natural logarithm function, built into the "System" unit, hence automatically available to any program. The comment under the program declaration tells what the program does. Run it if you are interested in the answer.

Here is another quite similar program, which you will understand without my help.

Program 1.2.5

```pascal
program  double_sum;     { 06/10/94 }
   { Sum 10/(2i + 3j) }

   var    i, j: Word;
          sum: Real;
```

```
begin
  sum := 0.0;
  for  i := 1  to  100  do
     for  j := 1  to  100  do
         sum := sum + 10.0/(2.0*i + 3.0*j);
  WriteLn('F(i, j) = 10/(2i + 3j)');
  WriteLn;
  WriteLn('  100     100');
  WriteLn('Sigma Sigma F(i, j) =', sum);
  WriteLn('  1         1');
  Write('Press <Enter>: ');   ReadLn;
  end.    { double_sum }
```

Exercises

Write a program to test the given assertion ("Pi" is a built-in constant):

1. $\Sigma_1^\infty 1/n^2 = \pi^2/6$
2. $\Sigma_1^\infty 1/n^4 = \pi^4/90$
3. $\Sigma_1^\infty (-1)^{n-1}/n^2 = \pi^2/12$
4. $\Sigma_1^\infty (-1)^{n-1}/n^4 = 7\pi^4/720$
5. $\Sigma_1^\infty 1/(2n+1)^2 = \pi^2/8$
6. Evaluate $\Sigma_{i=1}^{100} \Sigma_{j=1}^{i} [1/(i+j)]$

Section 3.   More Fairly Simple Programs

We program another Diophantine equation. It introduces us to the LongInt type, and the program's output should not be obvious before running it. Study the **repeat...until** loop carefully; ditto the **if...then...else** control statement.

Program 1.3.1

```
program  cubic_diophantine_equation;    { 06/10/94 }
  { Find all integer solutions of  x^3 = y^2 + 63
    that can be computed with LongInt. }

  var  x, y,
       z, w,
          n: LongInt;
  { MaxLongint = 2^31 - 1 = 2,147,483,647   2.15*10^9
       - MaxLongint - 1 <= n <= MaxLongint }
```

```
begin
{ First find largest n with n + 63 <= MaxLongint }
n := MaxLongint - 63;
n := Trunc(sqrt(n));    { Trunc(t) = floor(t) = [t] }
x := 0;
WriteLn('All integer solutions of x^3 = y^2 + 63,');
WriteLn('with 1 <= y <= ', n, ':');
for  y := 1  to  n  do
   begin
   z := y*y + 63;
   repeat
      Inc(x);   w := x*x*x;    { Inc(x):   x := x + 1 }
   until  w >= z;
   if  w = z  then
      WriteLn('(x, y) = (', x, ', ', y, ')')
   else  Dec(x);    { Dec(x) means   x := x - 1 }
   end;
Write('Done: Press <Enter>: ');   ReadLn;
end.    { cubic_diophantine_equation }
```

Just as we can compute sums, so can we compute products. Note the System unit Boolean-valued function "Odd" in the following program. The program also introduces the reserved word **const**, which allows us to introduce constants up-front. which can then be changed easily. The program also gives us more practice with a **for...to...do** loop.

Program 1.3.2

```
{$N+}
program  infinite_products;    { 05/25/94 }
{ Compute partial products for 3 convergent products }

   const   m = 5000;

   type   Real = Extended;

   var   p1, p2, p3,
                x: Real;
                n: Word;

   begin    { infinite_products }
   p1 := 1.0;   p2 := 1.0;   p3 := 1.0;
```

```
for  n := 2  to  m  do
   begin
   x := Sqr(1.0/n);
   p1 := p1*(1.0 + x);   p2 := p2*(1.0 - x);
   if  Odd(n)  then  p3 := p3*(1.0 - x)
   else  p3 := p3*(1.0 + x);
   end;
WriteLn('Products for  2 <= n <= ', m, ':');
WriteLn;
WriteLn('P (1 + 1/n)          =', p1);
WriteLn;
WriteLn('P (1 - 1/n)          =', p2);
WriteLn;
WriteLn('P (1 + (-1)^n /n) =', p3);
WriteLn;
Write('Press <Enter>: ');   ReadLn;
end.    { infinite_products }
```

Program "test_sum" introduces the reserved word **uses**, which is the way to link in library units. We need the subroutine "ClrScr" for clearing the screen, and it is found, with others, in the library unit "CRT". Note the finesse of assigning an integer (Word actually) variable to a real variable before computing its 4*th* power, to avoid overflow.

Program 1.3.3

```
program  test_sum;   { 06/14/94 }
   { Compares forward and backward summation of
     sums of decreasing terms. }

   uses  CRT;   { Screen procedures unit, e.g, ClrScr }

   var  x, y,
        s, t: Real;
           k: Word;

   begin
   ClrScr;
   WriteLn(' 1/n^5, 1001 to 10000');
   s := 0.0;
   for  k := 1001  to  10000  do
      begin
      x := k;   { sqr(sqr(k))  might overflow! }
      s := s + 1.0/(x*sqr(sqr(x)));
      end;
   t := 0.0;
```

```pascal
for  k := 10000  downto  1001  do
   begin
   x := k;
   t := t + 1.0/(x*sqr(sqr(x)));
   end;
WriteLn('Forward sum   = ', s);
WriteLn('Backward sum  = ', t);
WriteLn('Difference    = ', s - t);
WriteLn;
WriteLn(' 1/n^5, 1 to 1000');
s := 0.0;
for  k := 1  to  1000  do
   begin
   x := k;
   s := s + 1.0/(x*sqr(sqr(x)));
   end;
t := 0.0;
for  k := 1000  downto  1  do
   begin
   x := k;
   t := t + 1.0/(x*sqr(sqr(x)));
   end;
WriteLn('Forward sum   = ', s);
WriteLn('Backward sum  = ', t);
WriteLn('Difference    = ', s - t);
WriteLn;
WriteLn(' 1/2^n, 1 to 500');
s := 0.0;
x := 1.0;
for  k := 1  to  500  do
   begin  s := s + x;  x := 0.5*x;  end;
t := 0.0;
for  k := 500  downto  1  do
   begin  x := 2.0*x;  t := t + x;  end;
WriteLn('Forward sum   = ', s);
WriteLn('Backward sum  = ', t);
WriteLn('Difference    = ', s - t);
WriteLn;  Write('Press <Enter>: ');  ReadLn;
end.    { test_sum }
```

I suggest here that you reread all of the programs to date, and make sure that you understand everything in them before continuing. Experiment with the code, by making changes and running the results.

Section 4.   Subroutines—Functions

Now we look at some programs that are modularized, that is, parts of the programs are organized into subroutines. There are two big advantages in doing this. First, the same code can be used many times, without retyping it. Second, it is much easier to write programs in small pieces than all at once, easier to test and debug the programs, and easier to read them later.

We are getting to more serious material. From now on, programs that do real arithmetic will use the mathematical coprocessor (80x87, automatically included in DX processors like the 486DX). The coprocessor's native real data type is "Extended". Details of the differences between "Real" and "Extended" fill the long comment after **type**. Extended" is a mouthful, so the beginning of each program will redefine the data type "Real" to be "Extended". We instruct the compiler to compile for the coprocessor by a *compiler directive* at the beginning of the program: {$N+}. Note that this is *not* a comment. The construction {$...} with the $ sign immediately after the opening { always means a compiler directive.

Remark You may omit the compiler option {$N+} if instead you choose Options—Compiler—8087/80287 in the *Borland Pascal* "Integrated Development Environment" (IDE). If you actually do not have a mathematical coprocessor, you can choose "Emulation", but then do not expect speed. (Compiling for the 80x87 is the IDE default if one is present.)

The main point to study in program "test_agm" is the definition of the real-valued function of two real variables "AGM". Note in particular the line (marked critical) at the end of its definition.

The **while** loop executes repeatedly until an inequality fails, and there is no way to know in advance how many times it will execute. A **for** loop would be hopeless in a situation like this. That is why Pascal has **while** and **repeat** loops.

AGM is called twice, near the end of "test_agm", within "WriteLn" calls. Of course it could be tested many more times.

Program 1.4.1

```
{$N+}    {"Compiler directive" N+: compile for 80x87 }
program  test_agm;    { 06/14/94 }
   { Purpose: test the AGM function }

   { The "System" unit is automatically linked.  This
     program uses its types and procedures:
       Real  Extended  Sqrt  WriteLn  Write  ReadLn }
```

```
type   Real = Extended;
   { type Extended is the native type of the 80x87
     family of mathematical coprocessors.  The
     storage size is 10  bytes, the decimal accuracy
     19-20  significant digits, and the domain:
     [3.4E-4932, 1.1E4932], positive and negative.
     (This type declaration redefines type Real,
     a 6 bit type with 11-12 significant digits
     and domain [2.9E-39, 1.7E38], positive and
     negative.) }

function   AGM(x, y: Real): Real;
   { Gauss's arithmetic-geometric mean.
     Assumed: x > 0.0, y > 0.0 }

   const   eps = 1.0E-100;

   var   z: Real;    { A local variable (x and y
                        are also local) }

   begin
   while   Abs(y - x) >= eps   do
      begin
      z := x;
      x := Sqrt(x*y);     { geometric mean }
      y := 0.5*(z + y);   { arithmetic mean }
      end;    { while }
   AGM := y;    { This line is critical! }
   end;    { agm }

begin    { test_agm }
WriteLn('Gauss arithmetic-geometric means:');
WriteLn('AGM(1.0, 2.0)        =', agm(1.0, 2.0));
WriteLn('AGM(1.0, Sqrt(2.0)) =',
        agm(1.0, Sqrt(2.0)));
Write('Press <Enter>: ');   ReadLn;
end.    { test_agm }
```

Remark The **function** AGM has its own list of declarations: the **const** eps and the **var** z. These quantities are "local" to AGM, and are unknown outside of AGM.

The next example, also with a **function** declaration, uses the "LongInt" integer type, discussed in Program 1.3.1 above. Note its local variables n and z.

Program 1.4.2

```
program  test_largest_power;   { 02/22/95 }

   function  largest_power(a, b: LongInt): Word;
      { Solution  n  of  b^n <= a < b^(n+1)
        Assumes a >= 1,  b > 1 }

      var  n: Word;
           x: LongInt;

      begin
      x := b;
      n := 0;
      while  x <= a  do
         begin  x := b*x;   Inc(n);   end;
      largest_power := n;    { The critical line }
      end;    { largest_power }

   begin    { test_largest_power }
   WriteLn('3^n <= 10000 < 3^(n+1)');
   WriteLn('n = ', largest_power(10000, 3));
   Write('Press  <Enter>: ');   ReadLn;
   end.
```

Our next example has two subroutines, and introduces the System unit subroutines "Randomize" and "Random".

Program 1.4.3

```
{$N-}   { Ignore the 80x87 }
program  monte_carlo;   { 06/10/94 }
   { Integrates x^2 over [0, 1], Monte Carlo method. }

   const  shots = 50000;

   function  F(x: Real): Real;

      begin  F := x*x  end;

   function  integral: Real;

      var  j, hits: Word;
           x, y: Real;
```

```
begin
Randomize;    { Initialize the pseudo-random number
                  generator }
hits := 0;    { Must initialize variables }

for  j := 1  to  shots  do
   begin
   x := Random;  y := Random;
   if  y <= F(x)  then  Inc(hits);
   end;
integral := hits/shots;
end;    { integral }

begin    { monte_carlo }
WriteLn('Monte Carlo method, ', shots, ' samples:');
WriteLn;
WriteLn('1');
WriteLn('I x^2 dx =', integral);
WriteLn('0');
WriteLn;
Write('Press <Enter>: ');
ReadLn;
end.    { monte_carlo }
```

Program "monte_carlo" has a fault; it depends on the particular function defined in the program. Suppose we want to define a Simpson's rule subroutine that integrates *any* function. We must somehow name the class of real-valued functions of one real variable, then define the Simpson rule integration subroutine to use any function of that class. The final example of this section introduces the idea of a subroutine with a subroutine as one of its calling parameters—not yet for Simpson's rule—but for finding a zero of a function by bisection.

Program 1.4.4

```
{$N+}
program  bisection;    { 06/14/94 }

   type              Real = Extended;
        real_function = function(x: Real): Real;

   function  G(x: Real): Real;  far;
   { "far" is really a compiler directive, required
     for subroutines passed as parameters to other
     subroutines.  Do it; ignore technical details. }

      begin  G := Sqr(x) - 2.0;  end;
```

```
function   zero(F: real_function; x, y: Real): Real;

   const   eps = 1.0e-12;

   var            mid,
           fx, fy, fm: Real;

   begin   { zero }
   fx := F(x);   fy := F(y);
   if   fx*fy > 0.0   then   Halt;
   repeat
      if   Abs(fx) < eps   then
         begin   zero := x;   Exit;   end
      else   if   Abs(fy) < eps   then
         begin   zero := y;   Exit;   end
      else
         begin
         mid := 0.5*(x + y);   fm := F(mid);
         if   fx*fm <= 0   then
            begin   y := mid;   fy := fm;   end
         else
            begin   x := mid;   fx := fm;   end;
         end
   until   false;
   end;   { zero }

begin   { bisection }
WriteLn('G(', zero(G, 1.0, 2.0):10:10, ') = 0');
{ First 10: right justify in at least 10 spaces.
  Second 10: fixed point, 10 decimal places. }
ReadLn;
end.
```

Remark 1 The function "zero" returns a zero of any (continuous) function with opposite signs at the endpoints of the input interval. If the opposite sign condition fails, the "Halt" procedure, while not graceful, halts the whole program.

Remark 2 The function G(x) is defined to test "zero"; it does not enter into the definition of "zero" at all. The example is worthy of careful study.

Exercises

Write a program for each problem:

1. Find all lattice points (x, y) that satisfy $x^2 - 4xy + y^2 \leq 100$.
2. Test the proposition $x^y + y^x > 1$ for positive reals x and y.
3. Suppose $F(n)$ is an integer-valued function defined over the positive integers. Write a **function** for $G(n) = \Sigma_d F(d)$ summed over the divisors of n.

Section 5. Files: I/O

A computer is a machine that takes input, processes it, and produces output. The mechanism for input and output is the *file*. When a Pascal program begins, two files are automatically opened, with the special names "Input" and "Output". The file "Input" handles input of characters from the keyboard. The file "Output" handles output to the video screen. We never refer to these files explicitly; they are defaults. The input procedures "Read" and "ReadLn", unless qualified (as will be explained in time) read your keyboard input. Similarly, the output procedures "Write" and "WriteLn", without qualification, send output to the screen.

There are other input devices that we commonly use: the mouse, disk drives, the serial port etc, and there are other output devices: the printer, disk drives, the serial port, etc. Each of these can be addressed in a Pascal program by defining files. In assigning files to devices, the usual DOS names are available: 'prn', 'lpt2', 'com1', 'c:\my_file.dat', etc.

Let's take a quick look at a program for printing a table of values, then modify it to write the table to a disk file:

Program 1.5.1

```
program  table_of_values;    { 06/14/94 }

    var            x: Real;
                   k: Word;

    function  F(x: Real): Real;

        begin  F := 5.0*x/(1.0 + x*x);  end;

    begin   { table_of_values }
    x := 0.0;
    { Write header lines }
    WriteLn('Table of values of F(x) = 5x/(1 + x^2)');
    WriteLn;
    WriteLn('x':10, 'F(x)':20);
    WriteLn;
    for  k := 0  to  50  do
        begin  { Write the lines of the table }
        WriteLn(x:10:4, F(x):20:10);
        x := x + 0.1;
        if  k mod 10 = 9  then  ReadLn;
        end;    { for }
    WriteLn;
    Write('Press <Enter>: ');  ReadLn;
    end.    { table_of_values }
```

The tabulated function is not particularly interesting, however, run this program to see what it does. Note that it pauses (without warning) every 10 lines until you press <Enter>. To make the pause, we compute the remainder of the line counter k divided by 10. This is done by the **mod** operator. There is a corresponding **div** operator for the quotient in integer division. In general, if m and n are positive integers, then

$$m = (m \ \mathbf{div} \ n)n + (m \ \mathbf{mod} \ n) \qquad 0 \le m \ \mathbf{mod} \ n < n$$

(Do not assume anything for negative arguments.) Note that **div** and **mod** are Pascal reserved words.

Now suppose our interest in this particular table is so great that we want it in a disk file for later insertion in a manuscript on useless rational functions. The following program does this job.

Program 1.5.2

```
program  file_of_values;    { 06/14/94 }

   var              x: Real;
                    k: Word;
          out_file: Text;
          { "Text" is a file of lines of text }

   function  F(x: Real): Real;

      begin  F := 5.0*x/(1.0 + x*x);   end;

   begin    { file_of_values }
   { Link out_file to a disk file; name <= 8 chars,
     ext <= 4 chars (.xyz) }
   Assign(out_file, 'c:\my_table.dat');
   { Open out_file for writing }
   Rewrite(out_file);
   x := 0.0;
   { Write header lines }
   WriteLn(out_file,
           'Table of values of F(x) = 5x/(1 + x^2)');
   WriteLn(out_file);
   WriteLn(out_file, 'x':10, 'F(x)':20);
   WriteLn(out_file);
   for  k := 0  to  50  do
      begin  { Write the lines of the table }
      WriteLn(out_file, x:10:4, F(x):20:10);
      x := x + 0.1;
      if  k mod 10 = 9  then  WriteLn(out_file);
```

```
    end;   { for }
  { Close out_file, making it permanent on the disk }
  Close(out_file);
  end.   { file_of_values }
```

The first important word is "Text"; it is the type of a file of characters organized into lines, the kind of file produced by a vanilla word processor. The System unit procedure "Assign" assigns the file we have named "out_file" to the disk file with the ad hoc name "my_table.dat" in the root directory of C:. (Of course you may change its path before running the program.) At this point the file is just sitting there. The "Rewrite" procedure opens the file for writing, and (don't forget this!) clears out its contents if the disk file "c:\my_table" existed previously. Later the "Close" procedure closes the file and makes it permanent in the disk directory.

The only other change from the previous program "table_of_values" is adding a blank line every 10 lines instead of prompting for <Enter>. Note carefully how "WriteLn" is used when its output is to be directed to a file other than the default "Output" = screen. In general, "WriteLn" has format

```
WriteLn(param_1, param_2, . . . , param_n);
```

with any number of "write parameters". If the first parameter is a file name, then the other parameters are written to that file; otherwise all parameters are written to the screen.

It is important to know the mechanism using a disk file as input to a program. So let us read the file "my_table.dat" that was created by the previous program, and write it to the printer.

Program 1.5.3

```
program  get_file;   { 06/21/94 }

  const  FF = #12;   { ASCII 12: form feed }

  var   in_file,
        out_file: Text;
        k, count: Word;
             ss: string[80];
```

```
begin
  Assign(in_file, 'c:\my_table.dat');
  { Open in_file for reading: }
  Reset(in_file);
  Assign(out_file, 'prn');       { MS-DOS printer name }
  { Open out_file for writing: }
  Rewrite(out_file);
  count := 0;    { No of lines read }
  for  k := 1  to  3  do  WriteLn(out_file);
  for  k := 1  to  4  do
     begin     { Process table heading }
     ReadLn(in_file, ss);
     WriteLn(out_file, ss);
     end;
  count := 0;    { No of table lines on current page }
  while  not Eof(in_file)  do
     begin
     ReadLn(in_file, ss);
     WriteLn(out_file, ss);
     Inc(count);
     if  count = 44  then
        begin
        count := 0;
        Write(out_file, FF);    { New page }
        for  k := 1  to  4  do  WriteLn(out_file);
        end;                    { 4 blank lines }
     end;    { while }
  Write(out_file, FF);    { Finish final page }
  Close(out_file);
  Close(in_file);
end.    { get_file }
```

The most important thing to note in this program is its **while** loop. This is the proper way to read a disk file. It allows for the extreme case that the file is empty; then the loop is not executed at all. The System unit function "Eof" means "end of file". So if we are not yet at the end of the file "in_file", then the statements in the loop are executed.

The reserved word **string** builds string variables. the declaration of ss as a variable of type **string**[80] means that ss is a string of up to 80 characters. Without a qualification [n], the default length of a string of 255 characters. Thus

```
var  ss: string;
```

declares ss as a string of up to 255 characters.

Remark As a matter of course, we shall "Close" every file opened in programs. Actually, *Borland Pascal* automatically closes all open files when a program ends. I think it is good practice to close each file as you are done with it. This protects disk files from certain kinds of crashes.

Exercises

1. Display a table of squares of the integers from 0 to 999. Format the table in rows of 10 squares per row.
2. (Cont.) The table is more than a screen full, so save it as a file, or perhaps print it.
3. Write a Pascal program to draw the Pascal triangle (binomial coefficients). Try for about 16 rows starting at row 0.
4. Print a table of 186 pseudo-random whole numbers in [0, 9999]. They should be in 10 columns of 18 entries each, with the last column short.
5. Test by a program, then prove that $(m \bmod a) \bmod b = m \bmod (a * b)$

Section 6.   Units

Our first small unit will define hyperbolic functions. Any program can have a **uses** clause accessing the three functions in this unit. Note that Pascal has the exponential function "Exp" in its System unit, but no hyperbolic functions.

Program 1.6.1

```
{$N+}
unit  hyp_fun;    { 06/21/94 }
   { The name of this file must be "hyp_fun.pas".
     It will compile to "hyp_fun.tpp" or "hyp_fun.tpu",
     which a program that "uses" this unit can find.
     (.tpp for protected mode, .tpu for real mode. }

interface    { public declarations }

   type  Real = Extended;

   function  sinh(x: Real): Real;
   function  cosh(x: Real): Real;
   function  tanh(x: Real): Real;

implementation    { private part }

   var  t: Real;

   function  sinh(x: Real): Real;
```

```
    begin
    t := Exp(x);
    sinh := 0.5*(t - 1.0/t);
    end;   { sinh }

function  cosh(x: Real): Real;

    begin
    t := Exp(x);
    cosh := 0.5*(t + 1.0/t);
    end;   { cosh }

function  tanh(x: Real): Real;

    begin
    t := Exp(2.0*x);
    tanh := (t - 1.0) / (t + 1.0);
    end;   { tanh }

end.   { hyp_fun }
```

The reserved words **interface** and **implementation** are critical. Each unit has sections headed by these words, with no punctuation. The **interface** section contains declarations of constants, types, variables, and subroutines that are available to the public. The **implementation** section contains the program code for the subroutines. It may also contain local declarations, like the **var** t. (I wrote the unit this way just to make the point; it would be safer programming practice to declare a local variable t for each of the three functions.)

Each unit starts with the reserved word **unit** and ends with **end** followed by a period. This **end** need not have a matching **begin**, although you can put one in just before the **end** if you please. (Later, when we include initialization code for a unit, there must be a **begin**.)

The next program uses the unit "hyp_fun"; use your calculator (low tech) to check its output.

Program 1.6.2

```
{$N+}
program  test_hyperbolic_funs;   { 06/21/94 }

    uses  CRT, hyp_fun;
```

```
begin
ClrScr;
WriteLn('sinh( 0.5) =', sinh(0.5));
WriteLn('cosh(-2.3) =', cosh(-2.3));
WriteLn('tanh( 1.7) =', tanh(1.7));
Write('Press <Enter>: ');  ReadLn;
end.
```

Remark If you compile a file XYZ.PAS whose heading is **unit XYZ**. When unit XYZ is compiled, the result is XYZ.TPU (real mode target) or XYZ.TPP (protected mode target)..

Section 7. Program Flow

Eventually I must state the rules of Pascal more precisely, and here is a good place to start, after digesting many easy examples.

Pascal is a block-structured programming language. In contrast, programming languages like BASIC, FORTRAN, and Assembler are line-structured. In Pascal, the line does not exist as a unit, so far as the language is concerned. Of course, we write our programs in lines for clarity, but the lines may all be run together; the program will still run in exactly the same way. Each "block" does a task, and the program structure governs the flow from task to task.

Each programmer has his style of spacing and indenting. I am a strong believer in lots of white space, and indentation that helps the reader know the level of subordination of each line of the program. Computer programs carry much information in a brief space, like mathematical proofs, and good formatting can help greatly. (I personally find the "standard" formatting of C programs bad.) Always keep in mind that it is very unlikely your program will work correctly the first time, and that you may have to spend more than half of your programming time debugging your work. (This applies to me too.) Attention to good formatting, careful choice of identifier names, and reasonable documentation in your programs can improve your productivity greatly.

Programs

A Pascal program is a sequence of characters, built according to rules, that can be run in a computer programmed to accept Pascal programs. A brief example:

```
program  ABC;
   begin  end;
```

It is the absolute minimum in Pascal programs. It does nothing, yet it compiles without error, and it runs. We can learn something by analyzing its parts. First is the reserved word **program**, with which Pascal programs start. Next is an identifier "ABC". Identifiers may have any number of characters, but *Borland Pascal* only(!) recognizes the first 63 characters of an identifier, far more than ever needed. Recall that an identifier must begin with a letter or underscore _ and then continue with letters, digits, and underscores.

Taken together, the phrase "**program** ABC;" is a declaration, in this case, the declaration of a program named ABC. Each declaration must close with a semicolon. Therefore the first line of any program must contain

program `<identifier>;`

I must tell a little secret here about *Borland (Turbo) Pascal*: The opening program declaration line may be omitted altogether! You will notice that the program name never is actually used in the program, and is completely independent of the file name for the program if it is saved on disk. (In contrast, a unit must declare itself, and its disk file must have the corresponding name.)

Next we come to **begin   end.** These words are like parentheses, only for statements rather than for mathematical expressions. To each **begin** in a program there must correspond an **end**. (The converse is not true: reserved words **case** and **record** require a following **end**.)

Each program must have a part that does something, technically called the program body. Each subroutine declaration (**procedure** or **function**) has its own body.

The least allowable action part of the program is a compound statement

begin `S1; S2; ... ;   Sn;` **end**

where S1,..., Sn are statements. The chain of statements in program ABC above happens to be empty (like the empty set) so the action of program ABC is like that of many committee meetings — do nothing.

The body of each program must be followed by a period, a signal that here is the final stopping point of the program. Thus the minimal possible program is

program `rose;` **begin**   **end.**

where "rose" can be any other identifier.

Here are more examples of programs that accomplish nothing. Consecutive statements are separated by a semicolon, and between every two consecutive semicolons is the empty statement. (Note that extra begin end pairs and extra semicolons are allowed.)

program `DE;` **begin** `; ; ; ; ;` **end.**
program `FG;` **begin** **begin** `; ` **end;** `;` **end.**

Statements

Each program or subroutine body contains a sequence of statements, each of which accomplishes one molecule of the program. We classify statements; first is the assignment statement:

```
variable := expression;
```

Expressions are made up of constants, variables, function evaluations and operators. First the expression is evaluated, then the computed value is stored in the portion of computer memory reserved for the variable. Of course the value of the expression must be compatible with the type of variable on the left. In this regard, anything reasonable is acceptable. Some integer type values can be assigned to some other integer type variables; any integer type expression can be assigned to any real type variable. Example:

```
y := F(3*x) / 5.0*sin(x);
```

Procedure calls

A procedure is called (invoked) simply by writing its name if it has no parameters or its name with expressions (including variables) substituted for its parameters if it has parameters; this is then a call for action which might very well involve quite a bit of action. Examples:

```
WriteLn;    Read(in_file, x);
```

Compound Statement

A "compound statement" is simply a statement of the form

```
begin  S1;  S2;  ... Sn;  end;
```

where Si are statements. Semicolons in Pascal, to be precise, are statement *separators*, not statement terminators. So the semicolon following Sn is unnecessary. However, I always put in semicolons in such cases because it makes program revision (adding additional statements for instance) easier.

Structured Statements

Structured statements are made from special reserved words, Boolean expressions, and other statements. Each involves one or more test, explicitly or implicitly. (The **goto** statement, in a class by itself, is out of fashion, and I will not use it.)

if....then statement

An **if then** statement has the form

```
if   expr_1   then   statement_1;
```

where expr_1 is a Boolean expression.
We can nest if then statements. Since

```
if   expr_2   then   statement_2;
```

is a statement, it can replace statement_1 above:

```
if   expr_1   then   if   expr_2   then   statement_2;
```

If you have set the compiler option (as you certainly should) "short-circuit Boolean evaluation", then the statement is equivalent to the simpler

```
if   expr_1 and expr_2   then   statement_2;
```

if....then....else statement

The form is

```
if   expr   then   statement_1   else   statement_2;
```

Don't make the mistake of writing

```
if   expr   then   statement_1   else;   statement_2;
```

because the first semicolon completes the **if then else** clause (doing nothing in the **else** case); then statement_2 is executed. A good rule: *Never write* **else;** (Illegal semicolon: **if ... then ... ; else ...** . However, see **case** statement below.)

Ambiguous Case

The statement

```
if   expr_1   then
     if   expr_2   then   statement_2   else   statement_1;
```

has two possible interpretations. First interpretation:

```
if   expr_1   then
   begin
   if   expr_2   then   statement_2   else   statement_1;
   end;
```

Second interpretation:

```
if  expr_1  then
   begin  if  expr_2  then  statement_2  end
else  statement_1;
```

Pascal chooses the first: each **else** matches the closest preceding **if**. If you require the second interpretation, then insert statement brackets **begin end**. Indeed, whenever in doubt, use **begin end**, like () in equations, to make clear what is subordinate to what.

Nested if....then....else statements

Consider the statement

```
if  expr_1  then  statement_1
else  if  expr_2  then  statement_2
else  if  expr_3  then  statement_3
.  .  .  .  .  .  .  .  .  .  .  .  .  .  .
else  if  expr_n  then  statement_n;
```

First expr_1 is evaluated. If True then statement_1 is executed and we are done. If expr_1 evaluates to False, then expr_2 is evaluated. If True then statement_2 is executed; if False then expr_3 is evaluated, etc.

Assuming the expr_i's are independent, that is, evaluating the sequence of expr_i's in any order leads to the same result for each, then it is wise to arrange the order so that the most probable expr_i is first, the next most probable second, etc. This will decrease execution time, especially if the nested statement appears in a loop that is traversed many times.

case....of....end statement

The example above involved several alternatives. Sometimes this type of situation is better handled by the Pascal **case** statement. An example will show the idea:

We wish to process an integer N according to its residue class modulo 17. If $N = 0 \pmod{17}$ then N := 0. If $N = 1$ or 6 $\pmod{17}$ then N := −N. If $N = 2, 3,$ or 5 $\pmod{17}$ then N := 2*N. If $N = 4 \pmod{17}$ then N := 3*N. In all other cases, N := 5*N.

```
case  N mod 17  of
        0:  N := 0;
     1, 6:  N := -N;
  2, 3, 5:  N := 2*N;
```

```
        4:  N := 3*N;
    else  N := 5*N;
end;
```

Note carefully the points of this construction. First we have the three reserved words **case of end**. Between **case** and **of** is an expression (in the example, N mod 17) that evaluates to an integer, which possibly appears in one of the four "case label lists" that appear to the left of the colons. Each statement after a colon is separated from the next list of case labels by a semicolon. The **else** clause, covering all other possible cases, is optional.

Variable types called scalar types (discussed later) including integers are allowed, but not real types.

while....do statement (loop)

The structure is

```
while  expr  do  statement_0;
```

As usual, expr is a Boolean expression. When the **while** statement is entered. the Boolean expression expr is evaluated. If True, then statement_0 is executed, and expr is tested again, etc, until expr evaluates to False. If expr happens to evaluate to False the first time, then statement_0 is not executed at all. Note the particular case

```
while  True  do  statement_1;
```

which repeats statement_1 forever. Shortly we'll see how to escape from such an infinite loop.

repeat....until statement (loop)

The structure is

```
repeat  statement_0;  until  expr;
```

On entry, "statement_0" is executed; then the Boolean expression "expr" is evaluated. As long as expr evaluates false, the process is repeated. Of course statement_0 can be any statement, in particular a *compound statement*

```
begin
statement_1; statement_2; ...; statement_n;
end;
```

There is a special rule: in this situation the **begin** and **end** may be omitted (but need not be). Thus the general **repeat ... until** statement has the form (semicolon after `statement_n` optional)

```
repeat
    statement_1;
    ...;
    statement_n;
until  expr;
```

Note the particular case

```
repeat
    statement_1; statement_2; ...; statement_n;
until  False;
```

which loops forever.

In a **while....do** loop, the test comes first; in a **repeat....until** loop the test comes last.

for....to....do and for....downto....do statements (loops)

The form of the first is

```
for  J := expr_1  to  expr_2  do  statement_2;
```

Here J, called the *control variable* of the **for** loop, is any identifier that has been declared as an integer (scalar) type variable. This control variable must be *local* to the block in which it is used. For instance, if the **for** loop occurs in the body of a procedure declaration, then the control variable J must be declared in the declarations local to that procedure. (There is one exception to this assertion: the control variable can be global, at the outermost level, i.e, outside of all subroutines. It is bad practice to use this possibility

When this **for** statement is executed, first the integer-valued expression expr_1 is evaluated; then expr_2 is evaluated. These values remain constant throughout the execution of the **for** loop. Thus the **for** loop is equivalent to

```
J := expr_1;   K := expr_2;
while  J <= K  do
    begin
    statement_2;  inc(J);
    end;
```

provided that each execution of statement_2 does not change J or K.

There are three important points to remember:

(1) The **for** statement

```
for  J := expr_1  to  expr_2  do  statement_3;
```

is definitely not equivalent to

```
begin
J := expr_1;
while  J <= expr_2  do
   begin  statement_3;  inc(J);  end;
end;
```

because expr_2 may change with each execution of statement_3 in the **while** loop.

(2) This is an important rule of Pascal: "statement_0" in

```
for  J := expr_1  to  expr_2  do  statement_0;
```

is not allowed to change the control variable J. For example

```
sum := 0;
for  k := 1  to  100  do
   begin  sum := sum + Sqr(k);  k := k + 2;  end;
```

is an attempt to sum n^2 over the integers $3k + 1$ from 1 to 100. It is wrong because the control variable k is changed in the compound statement controlled by k. Correct is either

```
sum := 0;
for  k := 0  to  33  do  sum := sum + Sqr(3*k + 1);
```

or even better

```
sum := 0;  k := 1;
repeat
   sum := sum + Sqr(k);  k := k + 3;
until  k > 100;
```

(3) After the execution of a **for** loop the control variable is undetermined.

The **for....downto...do** version of the **for** loop is just like the **for....to....do** loop except that the control variable marches down, not up. Its format is

```
for  J := expr_1  downto  expr_2  do  statement_0;
```

Use a **for** loop when you know exactly how many times a statement is to be executed, else use a **repeat** or **while** loop. Use **repeat** if you want the enclosed (compound) statement executed at least once; use **while** if you want the test first, before the enclosed statement is executed at all.

Sometimes it is convenient to test for a possible exit in the middle of the loop, rather than at the start or the finish. This is handled by the *Borland Pascal* System unit procedure "Break", which breaks its *innermost* enclosing loop, whether **for**, **while**, or **repeat**. For instance

```
while  true  do
   begin
   statement_1;
   if  expr  then  Break;
   statement_2;
   end;
```

Exercises

1. What is the difference between the statements?
```
   begin
   if  X = 1  then  statement_1;
   if  X = 2  then  statement_2;
   end;
```
and
```
   if  X = 1  then  statement_1
   else  if  X = 2  then  statement_2;
```

2. In a program preparing a table in a disk file, the following loop occurs. The idea of course is to write a blank line after every ten lines. Can you make it more efficient; 1000 long divisions seems extravagant.
```
   for  j := 1  to  1000  do
      begin
      Line(j);
      if  j mod 10 = 0  then  WriteLn(FF);
      end;
```
In the remaining exercises, assume the declarations
```
   var  i, j, k, m, n: Word;
            a, b, c: LongInt;
          s, x, y, z: Real;
```
The following program bodies have no documentation to explain their actions. Even though quite short, it is not obvious what each accomplishes. Reflect on this point; then explain what each program body does.

3.
```
    begin
    n := 0;  x := 1.0;
    repeat  Inc(n);  x := 2.0*x;  until  x > 1.0e5;
    WriteLn(n, '  ', x);  ReadLn;
    end.
```
4.
```
    begin
    s := 0.0;  n := 1;
    while  n <= 100  do
       begin
       x := 3.0*n + 2.0;  x := 1.0/x;
       s := s + x;  Inc(n);
       end;
    WriteLn(n, '  ', s);  ReadLn;
    end.
```
5.
```
    begin
    s := 0;  n := 0;
    while  n < 100  do
       begin
       Inc(n);  x := 3*n + 2;
       x := 1.0/x;  s := s + x;
       end;
    WriteLn(n, '  ', s);  ReadLn;
    end.
```
6.
```
    begin
    n := 25;  a := 1;  b := 1;
    for  j := 3  to  n  do
       begin  c := b;  b := a + b;  a := c;  end;
    WriteLn('F_', n, ' = ', b);  ReadLn;
    end.
```
7.
```
    begin  n := 25;  a := 1;  b := 1;
    for  j := 2  to  n  do
       begin  b := a + b;  a := b;  end;
    WriteLn('F(', n - 1, ') = ', b);  ReadLn;
    end.
```
8.
```
    begin
    x := Pi;  y := 0.0;
    for  n := 1  to  20  do  y := y*x + n;
    WriteLn('G(', x, ') = ', y);  ReadLn;
    end.
```
9.
```
    begin
    c := 2;  s := 0;
    for  n := 1  to  99  do
       begin
       c := 6 - c;  y := n/(100 + n);  s := s + c*y
       end;    { for }
    s := s + 0.5;
    WriteLn('I = ', s/300.0);  ReadLn;
    end.
```

10.
```
        begin
        for  k := 1  to  100  do
           begin
           j := 2 + k*k;   m := 1;   n := 1;
           while  n <= j  do
              begin
              if  n = j  then
                 WriteLn('(', k, ', ', m, ')');
              Inc(m);   n := m*m*m;
              end;    { while }
           end;    { for }
        ReadLn;
        end.
```

Chapter 2 Programming Basics

Section 1. Procedures

Our subroutines so far have been of type **function**; these subroutines have a result, always an integer type or a real type (but there are limited other possible results). Type **procedure** subroutines can do all kinds of action, including computing one or more results. The following short program introduces the syntax of **procedure**, and also introduces **array** types.

Program 2.1.1

```
program  for_loop_demo;    { 06/22/94 }

   const   m = 25;   n = 30;

   type   matrix = array[1..m, 1..n] of Real;

   var   A, B: matrix;

   procedure  assign_matrix(var  A: matrix);
     { Make a double for loop into a single for loop. }

      var   i, j, k: Word;

      begin
      i := 1;  j := 1;
      for k := 1  to  m*n  do
         begin
         A[i, j] := 1.0/(i + j);
         Inc(j);
         if  j > n  then
            begin  Inc(i);  j := 1;  end;
         end;    { for  k }
      end;    { assign_matrix }

   begin   { for_loop_demo }
   assign_matrix(B);
   A := B;
   { Test }
   WriteLn('A[17, 23] =', A[17, 23]);
   WriteLn('A[23, 17] =', A[23, 17]);
   ReadLn;
   end.    { for_loop_demo }
```

Note the reserved word **array**, used to construct variable types. An array can have any number of dimensions; the "matrix" type defined above has two dimensions.

Also note the **const** declaration near the top of "for_loop_demo". Such declarations are valuable because if you want to change m say to another value, you need do it only in this one line, not throughout the program. As we'll see later, **const** declarations allow some limited calculation. Here m and n are not at all like variables; no memory is reserved for them. Instead, their values are substituted for each occurrence at compile time.

The main thing going in "for_loop_demo" is the procedure "assign_matrix". It has one parameter A declared as a **var** parameter. This means that "assign_matrix" must be called with a declared variable (say) B of type "matrix" substituted for A, and B will be changed as a result of calling "assign_matrix". (B must be declared outside of "assign_matrix".) In other words, parameter A is both input and output to the procedure.

The body of "for_loop_demo" calls "assign_matrix" and then has a few statements. Do notice the assignment A := B. This essentially copies the area of memory occupied by the variable B into that occupied by A. In other words, it is like an assembler MOV call to move data wholesale. Don't be tempted to use **for** loops for such copying of whole arrays!

Procedures may have unexpected *side effects* if you use them to modify non-local variables. The following brief program illustrates how this can happen. The function f does what it is supposed to do, but sneaks in a change in the value of the global variable m. Run the program; even better, "play computer" first and predict its output.

Program 2.1.2

```
program side_effect;    { 06/24/94 }

   var  m: Integer;

   function  f(n: Integer): Integer;

      begin  f := 3*sqr(n);   m := m + 6;   end;

   begin    { side_effect }
   m := 2;
   WriteLn('Commutative Law:');
   WriteLn('   m * f(4) = f(4) * m');
   WriteLn('   ', m, ' * ', f(4),
          '   = ', f(4), ' * ', m);
   Write('Press <Enter>: ');   ReadLn;
   end.
```

Remark It is time to clarify three standard procedures: "Break", "Exit", and "Halt". The Break procedure is used in a loop to stop further execution of that loop. Remember that it only breaks *one* loop, the innermost one. The "Exit" procedure stops execution of a **procedure** (**function**). Again, it only stops the execution of one subroutine, the innermost one in which it is called. (This is particularly important when it is used in procedures nested recursively.) The "Halt" procedure stops execution of a program—without qualification.

It is useful to report the running time of programs or parts of programs. The computer on-board clock measures to hundredths of a second, and we can use it for timing. The DOS unit has a procedure GetTime for reading the clock; it allows us to write the following unit timer. Read this code carefully, as it contains a lot of structure you will use repeatedly. Note that there are three public procedures declared. The main point of the first two, whose implementations are pretty simple, is to spare you from remembering the syntax of "GetTime". Note that these two procedures have deliberate side effects; they alter variables that are global to the **implementation** section of the unit, not local to themselves.

Program 2.1.3

```
unit  timer;    { 02/22/95 }

interface

    procedure  get_time0;    { Read start time }
    procedure  get_time1;    { Read stop time }
    procedure  put_time;     { Output routine }

implementation

    uses  DOS;    { for procedure  GetTime }

    var  hour0, min0, sec0, hund0,
         hour1, min1, sec1, hund1,
            hour, min, sec, hund: Word;

    procedure  get_time0;    { Start time }

       begin
       GetTime(hour0, min0, sec0, hund0);
       end;

    procedure  get_time1;    { Stop time }
```

```
begin
GetTime(hour1, min1, sec1, hund1);
end;    { get_time1 }

procedure  put_time;

    procedure  format(hour, min, sec, hund: Word);

        procedure  print(w: Word);

            begin
            if  w < 10  then  Write('0');
            Write(w);
            end;

        begin   { format }
        print(hour);  Write(':');
        print(min);  Write(':');
        print(sec);   Write('.');   print(hund);
        WriteLn;
        end;    { format }

    procedure  compute_elapsed_time;

        begin
        { Account for midnight oil and borrows }
        if  hour1 < hour0  then  Inc(hour1, 24);
        { Now  hour0 + 23 >= hour1 >= hour0 }
        if  hour1 > hour0  then
            begin  Dec(hour1);  Inc(min1, 60);  end;
        { Now  hour1 >= hour0,  min1 >= min0 }
        if  min1 > min0  then
            begin  Dec(min1);  Inc(sec1, 60);  end;
        { Now  hour1>=hour0, min1>=min0, sec1>=sec0 }
        if  sec1 > sec0  then
            begin  Dec(sec1);  Inc(hund1, 100);  end;
        { Now  hour1 >= hour0, ..., hund1 >= hund0 }
        hour := hour1 - hour0;   min := min1 - min0;
        sec := sec1 - sec0;  hund := hund1 - hund0;
        if  hund >= 100  then
            begin  Dec(hund, 100);  Inc(sec);  end;
        if  sec >= 60  then
            begin  Dec(sec, 60);   Inc(min);  end;
        if  min >= 60  then
            begin  Dec(min, 60);  Inc(hour);  end;
        end;    { compute_elapsed_time }
```

```
begin    { put_time }
{ Display results }
Write('Stop time:      ');
format(hour1, min1, sec1, hund1);
Write('Start time:     ');
format(hour0, min0, sec0, hund0);
compute_elapsed_time;
Write('Elapsed time:  ');
format(hour, min, sec, hund);
end;     { put_time }
```

end.   { unit timer }

The procedure "put_time" makes an attractively formatted output. Note that it has two sub-procedures, and one of these in turn has a sub-procedure itself. Having these sub-procedures avoids duplication of code. The sub-procedure "format" has local variables with the same names as variables declared outside of it; they are different.

Remark 1 Some further illustrations of the differences between *value* and **var** parameters of a routine are in order. Suppose we declare

```
type   polynomial = array[0..5000] of Extended;
```

We propose to code a procedure whose input is polynomials F and G, and whose output is their sum H = F + G. We have two choices for its declaration:

```
(1)    procedure   sum(F, G: polynomial;
                       var  H: polynomial);

(2)    procedure   sum(var  F, G, H: polynomial);
```

Suppose we use declaration (1). Then we are allowed to call "sum" with expressions substituted for F and G, but only an (external) variable substituted for H. But it is quite impossible that we would substitute anything for F and G except variables (external to "sum"), even though F and G are input only. What is more, declaration (1) implies that memory is reserved for the local variables F and G of "sum" (100,000 bytes) and copies are made of the external variables substituted for F and G in a call of "sum", which costs time. Conclusion: declaration (2) wins in both space and time.

Remark 2 However, there is another kind of parameter for procedures and functions, a **const** parameter that is even better. If we used it in "sum" above we would write

(3)   **procedure**   sum(**const**   F, G: polynomial;
                   var   H: polynomial);

Copies are *not* made for **const** parameters, so they are efficient in space. They are not local variables to the procedure, i.e., the procedure body *cannot* change their values. This provides some security against accidental assignment of parameters. Borland asserts that **const** parameters are also more efficient in time than **var** parameters for strings, arrays, records, etc.

Remark 3 There is an important rule about **var** parameters: The actual variables substituted for several **var** parameters of a routine *must* be distinct.

Exercises

Use the unit "timer" for tests:

1. Compare the speed of **for to do**, **while do**, and **repeat until** loops.
2. Compare the speeds of addition for Integer, Word, Longint, and Extended types.
3. Compare the speeds of multiplication for Integer, Word, Longint, and Extended types.
4. Compare the speeds of **div** for Integer, Word, Longint types, and division (/) for Extended type.
5. The operators **div** and **mod** compute the quotient and the remainder for integer division. Each is a division operation, slower than multiplication. If you have already computed the quotient by m **div** n, then by a multiplication and a subtraction you can compute the remainder. Is this a good idea?
6. Compare the speeds of inc(n) and   n := n + 1 for a variable n of various integer types. Also check out inc(n, 3) which increases n by 3, etc.
7. Define a vector of say 1000 reals and two variables V and W of that type. Compare the timing of the data move done by V := W with that done element-by-element with a **for** loop.
8. The (at most) max nonzero elements of a sparce m × n real matrix A are stored in three vectors, row_index[ ] and col_index[ ] vectors of words, value[ ] a vector of reals, so that

    A[row_index[k], col_index[k]] = value[k]

Define a function F(j, k) that returns A[j, k]. (Note the thrift in storage.)

9. In a graphing program that depends on speed, how should we program a **for** loop from x0 to x1 when we do not know in advance which limit is smaller? Two possibilities:

```
if  x1 < x0  then
    begin  x := x0;   x0 := x1;   x1 := x;   end;
for  x := x0  to  x1  do   act;


if  x0 <= x1  then
    for  x := x0  to  x1  do   act
else  for  x := x1  to  x0  do   act;
```

I conjecture that one is 4% faster than the other. Which?

10. I received a trial program package (Microsoft FORTRAN actually) whose instructions said that it would be valid until 12:00 midnight, Jan. 30, 1995. Look up Pascal's "GetDate" procedure and make one of your programs user-unfriendly in this way.

Section 2.   Graphics

As a first introduction to computer graphics, we'll start a unit on graphics, and apply it to draw a few things. Basic to all graphics is Borland's unit "Graph". Our unit "graphs" (so far) simply spares you from some of the details involved in opening and closing graphs.

Program 2.2.1

```
unit  graphs;    { 06/22/94 }

interface

    procedure  open_graph;
    procedure  close_graph;

implementation

    uses  Graph;

    procedure  open_graph;

        var  graph_device, graph_mode: Integer;
```

```
    begin
    graph_device := Detect;
    InitGraph(graph_device, graph_mode, '');
    if  GraphResult <> 0  then
        begin
        WriteLn('Can''t open graph');
        ReadLn;
        Halt;
        end;
    end;

procedure  close_graph;

    begin
    CloseGraph;
    if  GraphResult <> 0  then
        begin
        WriteLn('Can''t close graph');
        ReadLn;
        Halt;
        end;
    end;


end.
```

Remark This is a practical warning. Sometimes while you are developing and testing programs, a program will crash while the screen is in graphics mode, and you seem to be stuck there. Just keep handy some program that works correctly going in and out of graphics mode, and run it. Don't panic and hit the <Alt-Ctrl-Del> button! (DOS's "Mode CO80" does the trick too!)

The next program will give you a simple test of graphics, and familiarize you with some of the Graph unit's procedures for colors and lines. It uses the Delay procedure from the CRT unit, whose parameter is milliseconds. (Note that "Delay" may be inaccurate for some processors, such as a 486DX2. You can use the unit "timer" to check if Delay(10000) is about 10 sec, or much shorter, and calibrate your programs accordingly.)

Program 2.2.2

```
program  test_graphs;    { 06/22/94 }
    { The "Borland graphics interface" file
      for your monitor, e.g, egavga.bgi, must be
      in the same directory as this file. }

    uses  CRT, Graph, graphs;
```

```
begin
open_graph;
SetBkColor(Brown);    { background color }
SetColor(LightCyan);
{ Draw principal diagonal }
Line(0, 0, GetMaxX, GetMaxY);
{ Upper left: (0, 0).
  Lower right: (GetMaxX, GetMaxY). }
Delay(1000);
SetColor(Yellow);
{ Draw opposite diagonal }
Line(0, GetMaxY, GetMaxX, 0);
Delay(1000);
SetColor(LightGreen);
Circle(GetMaxX div 2, GetMaxY div 2, GetMaxY div 3);
{        centerX,           centerY,          radius }
Delay(1500);
close_graph;
end.
```

The next program shows how really easy it is to produce an elaborate figure, in this case, a complete regular polygon. Actually the sides of the polygon are omitted to make it interesting; all diagonals are drawn. You should play around with this program.

Program 2.2.3

```
program  polygon;    { 06/28/94 }

  uses  CRT, Graph, graphs;

  const        max = 21;
         recip_max = 1.0/max;
             twopi = 2.0*Pi;

  var  cx, cy: Integer;    { center: (cx, cy) }
         x, y: array[1..max] of Integer;
         { j-th vertex: (x[j], y[j]) }
       color,
      radius: Word;
         j, k: Word;
```

```
begin
open_graph;
cx := GetMaxX div 2;
cy := GetMaxY div 2;
radius := cy - 10;
for  j := 1  to   max  do
   begin
   x[j] := cx
             + Round(radius*Cos(twopi*j*recip_max));
   y[j] := cy
             + Round(radius*Sin(twopi*j*recip_max));
   end;
color := 13;
for  j := 2  to   max  do
   begin
   SetColor(color);
   for  k := 1  to  j - 2  do
      Line(x[j], y[j], x[k], y[k]);
   Dec(color);
   if  color = 10  then  color := 13;
   end;
SetColor(White);
OutTextXY(10, 10, 'Press <Enter>: ');   ReadLn;
close_graph;
end.    { polygon }
```

There are some interesting points in "polygon". First note the computations done in the **const** declarations. Next, we use two arrays to hold the coordinates of the polygon's vertices. This is not the most natural way to handle lists of points, we'll learn better later with **record** types.

Again, note that the upper left corner of the graphics screen has coordinates (0, 0) and the functions GetMaxX and GetMaxY return the lower right corner coordinates, (639, 479) for VGA high resolution mode.

The procedure OutTextXY is one of the ways to write text to the graphics screen. The upper left corner of the text starts at (10, 10) in this case.

Section 3.   Char

The type "Char" is used for declaring variables of character type. The actual characters are Chr(0) to Chr(255), that is, ASCII character 0 to ASCII character 255. So the **function** "Chr" simply maps bytes to characters. Another way to do this mapping is by *type casting*. Any two types that have the same size may be type cast into each other. For instance, suppose types type1 and type2 have the same size. Suppose x is a variable of type type1 and y is a variable of type type2. Finally, suppose that x has been assigned. Then the *type-cast*

    y := type2(x)

simply interprets the data stored in x as a quantity of type type2, referenced by y.

In particular, if N is of type Byte and ch of type Char, then

    ch := Char(N);   N := Byte(ch);

are valid assignments. The function "Ord", which applies to scalar types (Boolean, Char, etc) is yet another possibility:

    N := Ord(ch);   N := Byte(ch);

are the same:

    #65 = 'A';   Ord(#65) = 65;   Byte('A') = 65;
    Chr(65) = 'A';   Char(65) = 'A';

are all True.

Two characters can be compared by <, >=, etc, and the ordering is according to their orders.

There is a function "UpCase" that converts lower case letters to caps, and does nothing to other characters:

    UpCase('p') = 'P'
    UpCase('B') = 'B'
    UpCase('+') = '+'

You can refer to a control character by its ASCII number or by its Ctrl-sequence. For instance, suppose we have declared

    var   ch: Char;

Then

    ch := chr(7);   ch := #7;   ch := ^G;

all assign to ch the same character. Here ^G denotes Ctrl-G (BEL).

Among the 32 control characters, the ones you may need are

```
#7  =  ^G  =  Chr( 7)  (BEL),    #10  =  ^J  =  Chr(10)  (LF),
#12  =  ^L  =  Chr(12)  (FF),     #13  =  ^M  =  Chr(13)  (CR),
#27  =  ^[  =  Chr(27)  (ESC)
```

The disk program "chr0_127" prints out all characters numbers 0 ... 127 with their ASCII numbers. It prints two versions, the first with the mnemonics ETX, SOH, etc (from teletype days) for control characters, the second with the IBM characters (little faces, etc) for these controls. (This program is for HP printers, and uses control sequences from the HP PCL printer language. The printer control strings are all inserted as **const** declarations, hence easily changed for other printers.)

IBM characters numbers 127 - 255 contain foreign letters, boxing symbols, and mathematical symbols, and are often useful in programs. You can insert them directly into your programs. For instance to insert Chr(227) = π, hold down the Alt key, type 227 *on the numeric keypad*, then release Alt. Or you can enter #227. My program listings on disk all have the actual characters, but for technical reasons, I have to use the # method for program listings in this text. Sorry about that.

The disk program "chr128_" prints a table of these characters. The disk program "boxes" prints a nice display of the boxing characters with their numbers, in a way that shows you how to make all sorts of ruled boxes: single, double, and mixed. It also contains a table of the 16 screen colors. (These programs are also for HP printers.)

The disk program "ascii" prints several pages of these tables, slightly differently; play with it. Finally, the data file "ascii.dat", ready for printing, contains several pages of these tables, including the full names of the control characters (in case you want to know that ETB stands for "End of Text Block", etc). You might use a word processor to spread some of the lines vertically somewhat. This is easy in *PCWrite* for instance.

Keyboard Characters

Generally, when you type a character, you expect to see what you typed on the screen at the cursor. But sometimes you just want an action taken, not necessarily with echoing your keypress to the screen. Let's explore this topic a bit. The following little program "testread" uses "ReadLn" to read a single character and report what it has read on the next line. But when you enter the character, you see it on the screen, before you press <Enter>. Change the first "ReadLn" to "Read" in "testread" and run it again. This time just press 'a', then hold down 'b'. You will see lots of b's on the screen and learn something about the keyboard buffer, which holds 127 characters (I think). When you can enter no more b's, press <Enter>. The program will report that it has read 'a'.

Program 2.3.1

```
program testread;   { 07/09/94 }

   var  ch:  Char;

   begin
   Write('Enter a character: ');  ReadLn(ch);
   WriteLn('Character read: ', ch);
   Write('Press <Space>: ');  ReadLn;
   end.
```

Is it possible to read a key that is typed without its being echoed to the screen? Yes; the CRT unit contains a function "ReadKey" that does just this. It does not advance the cursor, so it is possible to write in place of the character entered any other character. The following program "test_readkey" does just that, changing each lower case letter entered to a cap. (You could map each key to another character, thus making a secret code for kids to play with.)

Program 2.3.2

```
program test_readkey;   { 07/09/94 }

   uses  CRT;

   var  ch: Char;

   begin
   WriteLn('Enter lower case characters, z to quit:');
   repeat
     ch := ReadKey;  Write(Upcase(ch));
   until ch = 'z';
   end.
```

Run the "testread" above and press the "F1" key for your character. Nothing happens, nor does anything happen with keys like "PgUp" and "↓". How does a program read these keys? The "ReadKey" function provides a method for input of these keys also.

To understand its working, you must first know something about how the keyboard works. When you press a key, two codes are returned, a *character code* and a *scan code* (also called "extended code"). For ordinary characters, the character code is the character, and you can ignore the scan code. When "ReadKey" reads a key like "F1", it returns character code 0 (corresponding to #0). In this case, a *second* call of "ReadKey" returns the scan code. So all you need is a table of what scan key is returned by each non-character key, and you are in business. The disk file scancode.dat is such a table, but you can make your own with the following program "scan_codes". I suggest that you spend some time testing it; try every key combination that you can think of (*not* "Ctrl-PrtScr" or "Alt-Ctrl-Del") and finally "Ctrl-Break" for the hardware interrupt that halts the program.

Program 2.3.3

```
program  scan_codes;   { 07/09/94 }

uses  CRT;

var   ch: Char;

begin
WriteLn(
'Program for character and scan (extended) codes.');
WriteLn('Press <Ctrl-Break> to halt program.');
repeat
  Write('Next key: ');
  ch := ReadKey;  WriteLn;
  if  ch <> #0  then
     WriteLn('Regular key, Ord(ch) = ', Ord(ch))
  else
     begin
     Write('Character returned: #0.   ');
     ch := ReadKey;
     WriteLn('Scan code: ', Ord(ch));
     end;
  WriteLn;
until   false;
end.
```

You will find that nothing happens with certain keys, like F11, Ctrl-UpArrow, etc. These indeed have character and scan codes, but the DOS function that *Borland Pascal* uses for ReadKey is tuned to the PC Keyboard, not the Enhanced Keyboard. Stay tuned for more on this subject.

Exercises

1. Suppose the System unit lacked an "UpCase" function. Program a function. Hint: Ord('A') = Ord('a') − 32.
2. Program a function "LowCase". Hint: Ord('a') = Ord('A') + 32.
3. Count the caps in a random character matrix.

Section 4.   Strings

The reserved word **string** is used to build string types, literally, string of characters. For example, to declare a variable SS that is to be assigned by

```
SS := 'abcd';
```

and never reassigned to a longer string of characters, you would declare

```
var  SS: string[4];
```

The maximum string length is 255; the reserved word **string** by itself has this
length by default (automatically). Thus **var**   TT: **string**   declares TT to
be a string of up to 255 characters.

We can declare type *names* for strings, for instance

```
type  str80 = string[80]
```

Then the type name str80 may be used to declare variables and to declare
parameters in subroutines:

```
procedure  print(UU: str80);
```

(Declarations like

```
procedure  print(UU: string[80]);
```

are illegal. The type declared for a subroutine parameter must be the *name* of a
type, not a structure building up a type.)

The declaration

```
var  SS: string[80];
```

reserves 81 bytes of memory to hold SS. Byte 0 contains the actual length of SS,
after SS has been assigned. For instance, the assignment

```
SS := 'abcdefgh';
```

results in

```
SS[0] = #8 = Chr(8), SS[1] = 'a', ... , SS[8] = 'h'.
```

A single byte can have 256 different values, #0 to #255, which explains the upper limit of 255 on string length. When you declare (say)

```
var   SS: string[50];
```

you *must assign* SS before you do anything with it. Not initializing strings is a frequent cause of difficult-to-find programming bugs. Thus the assignment

```
SS[8]  :=  'H';
```

is an error if SS itself has not been assigned. Something like

```
SS  :=  'abcdefghij0123456789abcdefghij0123456789';
```

does the job. Now SS is a string of length 40, and the assignment above to SS[8] results in

```
SS  :=  'abcdefgHij0123456789abcdefghij0123456789';
```

This particular string could have length as high as 50, and we might concatenate another string to make it so:

```
SS  :=  SS + 'ABCDEFGHIJ';
```

This introduces our first operation on strings, "+". For instance

```
'Scientific ' + 'Pascal' = 'Scientific Pascal';
```

A string constant, meaning single quotes delimiting several characters, *must* be on one line. Simply use "+" to build up a long string over several lines, like

```
SS  :=  '1111122222333334444455555666667777788888'
            + '99999000001111122222 etc';
```

If you want to include control characters in strings, no problem, just use the # or ^ notation and juxtaposition. For instance

```
Writeln(^G^G^G^G^G'Warning!');
```

rings the bell 5 times and prints the warning. If you want to skip a line after printing something,

```
Writeln('something'^J);  or  Writeln('something'#10);
```

will put in an extra line feed, with exactly the same result as the two statements.

```
Writeln('something');  Writeln;
```

By the way, to put the single quote character ' itself in a string, just put in two consecutive single quotes:

```
SS := 'Jack and Jill''s pail';
```

This also applies to the single character

```
Chr(39) = '''';
```

Characters may be added to strings; in a sense they act like strings of one character. Thus

```
SS := 'Jack and Jill' + '''' + 's pail';
```

is the same as the previous assignment of SS.

The function "Length" returns the length of a string. Here is a practical application of "Length". Some vanilla word processors (e.g, *PCWrite*) tend to leave many spaces at the ends of lines, swelling file size. The program "remove_trailing_blanks" rewrites such a text file, with the extra spaces removed. The procedure "initialize_files" uses some DOS features to make the output have the same name as the input file. Don't worry about this or the unit "textfile" now.

Program 2.4.1

```
program remove_trailing_blanks;   { 06/30/94 }
   { Removes space characters at the ends of lines
     in a text file. }

   uses  DOS, textfile;

   var  ss: string;

   procedure  initialize_files;

      begin . . . end;
```

```
begin    { remove_trailing_blanks }
initialize_files;
while   not Eof(in_file)   do
   begin
   ReadLn(in_file, ss);
   while  ss[Length(ss)] = ' '   do
      Delete(ss, Length(ss), 1);
   WriteLn(out_file, ss);
   end;
Close(out_file);
Close(in_file);
end.
```

Concentrate on the string procedure "Delete". To delete n characters from a string variable TT, starting at index i, the call is

```
Delete(TT, i, n);
```

For instance:

```
TT := '1234xyz567890';  Delete(TT, 5, 3);
```

results in TT = '1234567890'.

There is a corresponding "Insert" procedure that inserts one string into another string. To insert ss into string variable TT, starting at index i, the call is

```
Insert(ss, TT, i);
```

For instance

```
TT := '1234567890';  Insert('xyz', TT, 6);
```

results in TT = '12345xyz67890'.

There is a function "Copy" for copying substrings from strings. To assign a string variable TT to the n characters from string SS, starting at index i, the call is

```
TT := Copy(SS, i, n);
```

For instance,

```
TT := Copy('1234567xyzw890', 8, 6);
```

results in TT = 'xyzw89'.

Another useful function is "Pos" that finds the starting position of a substring in a string. For instance

```
Pos('F(x)', 'Let F(x) = 2x') = 5
```

"Pos" returns 0 if the substring does not occur.

There are two other string procedures, "Str" and "Val" for converting numbers to strings and vice versa. They are best discussed later. Finally, two strings can be compared by >, <=, etc; the ordering is lexicographic.

Exercises

1. Write a procedure whose input is a string T and whose output is the string reversed.
2. Write a program that removes all indentation from a text file.
3. Write a program to double space a single spaced text file.
4. Write a program to do the opposite.
5. A text file is to be sent out for translation to another language. It is convenient for the translator to have two copies of each line, so he can keep sight of the whole English line while replacing the other line with his translation. Write a program that takes for its input a text file, and produces a text file with each line of the original twice.
6. (Continued) Write the inverse program that removes the original English lines; then test your work.
7. Write a program that adds line numbers, like { 32 } to the end of each line of a source code file *.pas. Assume the lines are sufficiently short, and that there are no multi-line comments (why?).
8. The same, except add line number to the beginning of each line.
9. "Pos" finds the first occurrence of a substring in a string. How would you find the second or the third occurrence?

Section 5. Arrays

The Pascal reserved word **array** is used to form structured types such as vectors, matrices, and tensors in general. For instance

```
var  space_time: array[1..4] of Real;
```

reserves storage for 4 reals, the components of "space_time". They are referenced by

```
space_time[1] ... space_time[4]
```

More generally, 1..4 can be replaced by any (finite) scalar type. For instance

```
const  m = 9;
type   index = 0..m;
```

```
var   v: array[index] of Real;
```

Equivalent: the sequence of declarations

```
type    index = 0..9;
         vector = array[index] of Real;
var   v: vector;
```

Note that the declarations

```
var   A: array[1..6, 1..10] of Word;
```

```
var   A: array[1..6] of array[ 1..10] of Word;
```

are equivalent, and A[3, 7] = A[3][7] in either case.

We declare an array type, give it a name, and then declare variables of that type by name. A general array type declaration has the form

```
type   tensor = array[type_1, type_2, ..., type_n]
                 of base_type
```

Each type_j is a finite scalar type and base_type is any type whatever. The number of bytes a variable of this "tensor" type occupies is computed as a product by the standard "SizeOf" function, which returns a LongInt:

```
SizeOf(tensor)
    = SizeOf(type_1)*...*SizeOf(type_n)
        *SizeOf(base_type)
```

Pascal has a limit of 64K = 65536 bytes for any variable, so 64K is an upper bound for SizeOf(tensor). For instance

```
var   A: array[1..100, 1..100] of Extended;
```

reserves 10,000*8 = 80,000 bytes for A, which is over the legal limit. Ditto for the tensor T declared by

```
var   T: array[1..10, 1..10, 1..10, 1..10] of Real;
```

(Later we'll learn ways to overcome the 64K var-size barrier.)

If A and B are array variables of the same type, then A := B is a valid assignment, and makes A a copy of B.

The time has arrived for examples. We set up a unit "matrices" for dealing with square real matrices; first we examine its interface:

Program 2.5.1

```
unit  matrices;    { 07/02/94 }

interface

   const  max = 8;

   type    Str4 = string[4];
           matrix = array[1..max, 1..max] of Real;

   procedure  random_matrix(bound: Real;
                                 var  A: matrix);
       { |A[j, k]| <= bound }
   procedure  display(var  A: matrix;  name: Str4);
   procedure  product(var  A, B, C: matrix);
   function   trace(var  A: matrix): Real;
   function   sup_norm(var  A: matrix): Real;
   function   L2_norm(var  A: matrix): Real;

implementation

   uses  CRT;    { for GotoXY }
```

Let's first implement the easier of the subroutines, starting with "trace" and "product":

```
   function  trace(var A: matrix): Real;

      var  j: Word;
           t: Real;

      begin  t := 0;
      for  j := 1  to  max  do  t := t + A[j, j];
      trace := t;
      end;

   procedure  product(var  A, B, C: matrix);

      var  i, j, k: Integer;
```

```
begin
for  i := 1  to   max  do
   for  k := 1  to   max  do
      begin
      C[i, k] := 0.0;
      for  j := 1  to   max  do
         C[i, k] := C[i, k] + A[i, j]*B[j, k];
      end;
end;
```

These two are pretty routine. The standard formula for matrix multiplication translates almost literally into Pascal. Note, however, that each element C[i, k] of the product must be initialized to 0 before adding to it products of elements. "Random_matrix" is also routine for us by now:

```
procedure   random_matrix(bound: Real;
                                 var  A: matrix);

   var  j, k: Word;

   begin
   Randomize;
   for j := 1  to   max  do
      for  k := 1  to   max  do
         A[j, k] := bound*(2.0*Random - 1.0);
   end;   { random_matrix }
```

The two norm functions have little new for us really. Remember that "Sqr" is a function for squaring real or integer type expressions, and "Sqrt" is for taking the square root of a nonnegative real type expression. Perhaps the absolute value function "Abs" is new. It applies to any expression of an integer or a real type and returns a number of the same type.

```
function   sup_norm(var  A: matrix): Real;

   var  i, j: Word;
        s, t: Real;

   begin
   s := 0.0;
   for  i := 1  to   max  do
      for  j := 1  to   max  do
         begin
         t := Abs(A[i, j]);
         if  t > s  then  s := t;
```

```
          end;
    sup_norm := s;
    end;

function  L2_norm(var  A: matrix): Real;

    var  i, j: Word;
           s: Real;

    begin
    s := 0.0;
    for  i := 1  to  max  do
       for  j := 1  to  max  do
          s := s + Sqr(A[i, j]);
    L2_norm := Sqrt(s);
    end;
```

Remark A comment on the functions "Sqr", "Sqrt", and "Abs" is in order. They are "overloaded" in the sense that they can be applied to expressions of several different types. Pascal does not offer us easy ways of overloading our own subroutines or operators, alas.

This brings us to the remaining procedure to implement, "display". Its purpose is to make a nice display of a small matrix with the name of the matrix and matrix brackets. The brackets are pieced together from IBM boxing characters, located within ASCII 179—218. I am sorry that I cannot show them in the printed listing, but if you will bring up matrices.pas on your computer, they will show plainly.

```
procedure  display(var  A: matrix;  name: Str4);

    var  j, k, l: Word;

    begin
    GotoXY(1, 1 + (max + 1) div 2);
    Write(name, ' = ');
    l := Length(name) + 4;
    GotoXY(l, 1);  Write(#218);
    GotoXY(l + 8*max + 1, 1);  Write(#191);
    for  j := 1  to  max  do
       begin
       GotoXY(l, 1 + j);
       Write(#179);
       for  k := 1  to  max  do
          Write(A[j, k]:8:2);
       Write(#179);
       end;
```

```
   GotoXY(1, max + 2);   Write(#192);
   GotoXY(1 + 8*max + 1, max + 2);   WriteLn(#217);
   end;   { display }

end.   { unit  matrices }
```

I hope I typed in the right numbers: #179 is a vertical line, #191 is an upper right corner, etc. Note the use of the GotoXY(x, y) procedure (CRT unit) for positioning the cursor. The text screen has upper left corner (1, 1) and lower right corner (80, 25). Thus increasing x moves *right*, increasing y moves *down*.

A common programming tool is to use the elements of one array as indices for another array, a form of indirect addressing. The following program supplies an example. A 20×20 matrix A of "Word" in the interval [1, 120] is given. The problem is to print a list, in increasing order, of those "Word"s that actually appear in A, with their frequencies. Program "frequency_list" omits the main body, which is on your disk. Its procedure "find_frequencies" uses the indirect address technique.

Program 2.5.2

```
program  frequency_list;    { 07/02/94 }

  const  no_rows = 20;
          no_cols = 20;
         elem_max = 120;

  type   range = 1..elem_max;
         matrix = array[1..no_rows, 1..no_cols]
                    of range;

  var            A: matrix;
            list,
       frequency: array[range] of Word;
     word_count: Word;

  procedure  find_frequencies;

    var   j, k: Word;

    begin
    for  k := 1  to  elem_max  do
      begin
      frequency[k] := 0;  list[k] := 0;
      end;
```

```
for  j := 1  to  no_rows  do
   for  k := 1  to  no_cols  do
      Inc( frequency[ A[j, k] ] );
word_count := 1;
for  k := 1  to  elem_max  do
   begin
   if  frequency[k] > 0  then
      begin
      list[word_count] := k;
      Inc(word_count);
      end;
   end;
Dec(word_count);
{ Now list[1..word_count] contains the words
  in A, in increasing order, and the frequency of
  list[k] equals frequency[ list[k] ] }
end;    { find_elements }

begin . . . end.    { frequency_list }
```

A polynomial (of one variable) can be described by the array of its coefficients, indexed by degree. Consider polynomial multiplication: $H(x) = F(x)G(x)$, where degrees are bounded, and terms of the product beyond the max degree are chopped off. The following code is natural for a computer program, but a mathematician would compute each H[k] separately.

Program 2.5.3

```
program  poly_prod;    { 07/03/94 }

   const  max = 5000;

   type  polynomial = array[0..max] of Real;

   procedure  product(var  F, G, H: polynomial);

      var  j, k: Word;

      begin
      for  k := 0  to  max  do  H[k] := 0.0;
      for  j := 0  to  max  do
         for  k := 0  to  max - j  do
            H[j + k] := H[j + k] + F[j]*G[k];
      end;    { product }

begin . . . end.
```

Exercises

1. It is often the case that variables declared earlier can be accessed faster than those declared later. Use the unit "timer" from Chapter 1 to test this: Have the declarations of (say) two Word variables separated by a declaration of an array of 1000 reals. Then time accessing the simple variables.

2. Each element of an array is like a variable. But it takes longer to access an array element than a simple variable because its offset from the first element of the array must be computed. Use the unit "timer" from Chapter 1 to test this, keeping in mind the results of the previous exercise.

3. Write a program to test the routines in the unit "matrices".

4. Write a function that tests a square matrix of Integer for symmetry.

5. Assume

```
const   max = 10000;
type    vector = array[1..max] of Integer;
```

   Write a procedure with output the largest component of a vector, where it first occurs, last occurs, and frequency of occurrence.

6. Start with

```
const   max_deg = 30;
type    polynomial
             = array[-1..max_deg] of Integer;
var  F, G, H: polynomial;
```

   (Here F[-1] = deg F, etc.) Set up procedures "sum" and "product" for integer polynomials of degree <= max_deg = 30. Assume you will only multiply polynomials whose product has degree at most 30. (The zero polynomial should be assigned degree -1.)

7. Given: a sequence $x_1 < x_2 < \cdots < x_n$ of reals where, say, n = 1000, and a real y. Problem: to locate k so that $x_k \leq y < x_{k+1}$. (Assume $x_1 \leq y < x_n$.) Define a function that returns k.

8. Complete the program "poly_prod" in order to test "product". Of course you should change max to a small value.

9. By a *D-polynomial* we mean a truncated Dirichlet series with integer coefficients:

$$\sum_{j=1}^{n} \frac{a_j}{j^s} \qquad \text{where} \quad a_1 = 1$$

   Set up suitable declarations and a procedure for multiplying D-polynomials.

10. Some commercial programs put a small twirling windmill on the screen while doing a long calculation. Try to write a program what has such a windmill, using the four characters —, /, \. |.

11. Use IBM boxing characters to construct a large rectangular box on the screen with gaps, which is rotating slowly, while a message flashes inside the box.

Section 6. Sets

The reserved word **set** is used to build structured types. The base type of a set can be any scalar (finite) type of at most 256 elements. Thus natural set declarations are

```
var   S: set of Char;
      B: set of Byte;
```

Set constants are written with [ ] notation and lists of elements. Examples:

```
const   alphas = ['A'..'Z', 'a'..'z'];
        empty = [];
        digits = [0..9];
```

The latter is a set of bytes, very different from the set

```
['0'..'9']
```

of characters. The reserved word **in** is used for set membership:

```
if  ch in alphas   then ...
```

The negation of "ch **in** alphas" is not "ch **not in** alphas" which is illegal, but rather

```
not (ch in alphas)
```

If S and T are set constants or set variables, then S + T is their union, S*T is their intersection, S–T is their difference. The relational operators = (equal), <> (not equal), <= (is a subset), >= (is a superset) apply to sets, but proper inclusion < and > do not apply. There is not much finite set theory you can do with set types, particularly because of their size limit, but they are useful in programming. For instance, it is much easier and clearer to write

```
if  ch in alphas   then ...
```

than it is to write

```
if  ( (ch >= 'A') and (ch <= 'Z') ) or
    ( (ch >= 'a') and (ch <= 'z') )   then ...
```

or even

```
if   ( (UpCase(ch) >= 'A')
     and ( (UpCase(ch) <= 'Z') ) ) then ...
```

To add an element to a set, you have a choice of adding a singleton set or using the procedure "Include". The following code builds the set of all even numbered caps, ['A', 'C', ..., 'Y']:

```
var    S: set of 'A'..'Z';
       ch: Char;

begin
S := [];   ch := 'A';
repeat
    S := S + [ch];   Inc(ch);   Inc(ch);
until  ch >= 'Z';
```

An alternative to S := S + [ch]; is Include(S, ch); Borland claims that "Include" generates the more efficient code. Maybe. There is also a corresponding procedure "Exclude".

The unit "sets" contains routines for sets of characters, and you can easily modify them to suit your own needs. We first look at its interface:

Program 2.6.1

```
unit   sets;     { 07/03/94 }
   { Routines for sets of characters }

interface

   type   char_set = set of Char;
          Str3 = string[3];

   function   cardinal(var   S: char_set): Word;
   function   smallest_member(var   S: char_set): Char;
   function   largest_member(var   S: char_set): Char;
   procedure  print(name: Str3;   S: char_set);
      { Prints elements in increasing order }
   procedure  lex_relation(S, T: char_set);
       { Returns S > T, S = T, T > S according to the
         lexicographic ordering of the sets }
   procedure  successor(var   S: char_set);
       { Replaces S by its lexicographic successor. }
```

```
procedure  random_subset(first, last: Char;
                          var  S: char_set);
   { random subset of first..last; all elements
           equally likely }
procedure  random_k_subset(first, last: Char;
                           k: Word;
                           var  S: char_set);
procedure  all_subsets(S: char_set);
   { Lists all subsets of S }
procedure  all_subsets2(S: char_set);
   { List of subsets in which each subset differs
     from the previous subset in one element only }
```

The purpose of each routine should be clear from its name and the comments. Most of their implementations will be left to exercises, but we shall look at several that offer particular points of interest, starting with "cardinal".

implementation

```
var  no_bits: array[Byte] of Word;
     { Number of bits in each byte }

function  cardinal(var  S: char_set): Word;

   type  bits32 = array[0..31] of Byte;

   var  a:bits32;
        n: Word;
        j: Byte;

   begin
   n := 0;
   a := bits32(S);    { type cast }
   for  j := 0  to  31  do
      n := n + no_bits[a[j]];
   cardinal := n;
   end;    { cardinal }
. . . . . . . . . . .
procedure  init_no_bytes;

   var  j, k, m: Byte;
```

```
begin
for k := 0  to   255   do
   begin
   no_bits[k] := 0;
   m := k;
   for  j := 0  to   7  do
      begin
      if  Odd(m)  then  Inc(no_bits[k]);
      m := m shr 1;    { fast division by 2 }
      end;
   end;    { for k }
end;    { init_no_bytes }

begin
init_no_bytes
end.    { sets }
```

This code uses the way sets are stored in memory, as arrays of bytes. Each bit represents one element of the set. In our case, **set of** Char requires 256 bits, i.e, 32 bytes to hold all possible such sets. So to count the elements of a set, we count the bits that are 1. The type cast of a set into an array of 32 bytes gives us direct access to the bits.

The array "no_bits" is computed once and for all by the procedure "init_no_bytes", which is invoked in the *initialization* section of the unit.

The procedure "random_k_subset" is a really interesting algorithm. Recall that the problem is to choose a random set of exactly k elements in the interval first..last of characters. To illustrate the algorithm: suppose, for example, you want a random set S of 3 elements from [4..10]. First, choose a random element i from [4..8] and assign S := [i]. Second, choose a random j from [4..9]. If j is not in S then assign S := S + [j]; otherwise S := S + [9]. Last, choose a random k from [4..10]. If k is not in S, then S := S + [k]; otherwise S := S + [10];

```
procedure  random_k_subset(first, last: Char;
                            k: Word;
                            var  S: char_set);
   { Attributed to R W Floyd }

   var   ch, stop: Char;
               j: Byte;
```

```
begin
if  k > Ord(last) - Ord(first)  then
    begin  S := [first..last];  Exit;  end;
Randomize;
stop := last;
for  j := 1  to  k - 1  do  Dec(stop);
{ #[stop..last] = k }
j := Ord(stop) - Ord(first) + 1;
{ j = #[first..stop] }
S := [];
while  stop <= last  do
    begin
    { Choose a random ch in [first..stop] }
    ch := Chr(Ord(first) + Random(j));
    if  ch in S  then  S := S + [stop]
    else  S := S + [ch];
    Inc(stop);  Inc(j);
    end;    { while }
end;    { random_k_subset }
```

I'll leave a prosaic algorithm for "all_subsets" for you to work out. The problem of listing the subsets of S in such a way that each subset in the list is obtained from the previous subset by either adding or deleting one element is rather difficult, and you should appreciate the following neat algorithm. Note that deciding which next element to add or delete depends on the parity of the cardinality of the subset.

```
procedure  all_subsets2(S: char_set);
    { List of subsets in which each subset differs
      from the previous subset in one element only. }

    var   even: Boolean;
        { even number of elements of T }
            T: char_set;
        firstS,
         lastS: Char;

    procedure  next_subset;

        var  ch0, ch1: Char;
```

```
        begin
        if  even  then  ch1 := firstS
        else
            begin
            { ch0 := first element of T }
            ch0 := firstS;
            while  not  (ch0 in T)  do  Inc(ch0);
            { ch1 := first element of S beyond ch0 }
            ch1 := Succ(ch0);
            while  not  (ch1 in S)  do  Inc(ch1);
            end;
        if  ch1 in T  then  T := T - [ch1]
        else  T := T + [ch1];
        even := not even;
        end;    { next_subset }

    begin    { all_subsets2 }
    firstS := smallest_member(S);
    lastS := largest_member(S);
    T := [];
    even := true;
    repeat
        print('', T);
        next_subset;
    until  T = [lastS];
    print('', T);
    end;    { all_subsets2 }
```

Exercises

1. Implement "smallest_member" from the unit "sets".
2. Implement "largest_member" from the unit "sets".
3. Implement "print" from the unit "sets".
4. Implement "lex_relation" from the unit "sets".
5. Implement "successor" from the unit "sets".
6. Implement "random_subset" from the unit "sets".
7. Implement "all_subsets" from the unit "sets".
8. Write a program for testing all of the procedures in unit "sets".
9. Prove that the procedure "random_k_subset" is correct; each element of the base set has equal probability of being chosen for the random k-element subset.

Section 7. Integers and Booleans

We have done much with *structured* types constructed with **array**, **set**, and **string**. We now get back to the main building blocks, numbers, and their operations.

Integers

Pascal offers us several types of integers. The signed integer types are *ShortInt*, *Integer*, and *LongInt*. The unsigned integer types are *Byte* and *Word*.

Given that variables are normally stored on word (two byte) boundaries, why ever use the one-byte types ShortInt and Byte? The answer is, only in large arrays, provided you can live with their limited domains: a variable of type ShortInt has domain [-128, 127] and a variable of type Byte has domain [0, 255]. Consider this:

```
type   short_1000 = array[1..1000] of ShortInt;
        int_1000 = array[1..1000] of Integer
```

You can easily test the assertions

```
SizeOf(short_1000) = 1000
SizeOf(int_1000) = 2000
```

I hope the point is clear; the same comparison holds for an array of Byte *vs* an array of Word. Types Integer and Word are two-byte types; their domains are

$$[-32,768, \ 32767] = [-2^{15}, \ 2^{15} - 1]$$
$$= [-MaxInt - 1, \ MaxInt]$$

and [0, 65,535] respectively. As noted several times already, LongInt is a 4–byte type, with domain

$$[-MaxLongint - 1, \ MaxLongint]$$
$$= [-2^{31}, \ 2^{31} - 1]$$
$$= [-2,147,483,648, \ 2,147,483,647]$$

That Integer and LongInt have domains that go down further on the negative than on the positive side is a consequence of the PC family being "two's complement" machines.

There are many operations on integers.   The basic binary operations are +, -, *.   When you use these (and other) operators on operands of different integer types, there will be type conversions as are necessary.   Some precautions are in order.   For example, assume the declarations

```
var   a, b: Word;   x: Integer;
```

and the assignments

```
a := 3;   b := 5;
x := a - b;
```

This will crash! Why is that? The point is that the expression a - b will be evaluated *before* the assignment to x is made, and this expression is a Word–valued expression; but a Word cannot be negative. However,

```
x := - b + a;
```

is legal! This seems a freak to me, and a much safer way to do the subtraction is

```
x := a;   x := x - b;
```

We have already worked with the two division operators **div** and **mod**. As noted, the first operand in

```
m div n   and   m mod n
```

should be nonnegative and the second operand positive. You're on your own if you want to divide integers with negative operand(s). However, the one thing you can be sure of, no matter what the signs, is

```
m = (m div n)*n + (m mod n)
```

There are two unary operators (functions) on integers in the System unit: Abs, Sqr. Also there is the Boolean-valued function Odd with the obvious meaning.

The operators **shl** (shift left) and **shr** (shift right), when applied to unsigned integers are the same as multiplying or dividing by powers of 2. For example, suppose n is a Word variable. Then n is represented by 16 bits, low bit on the right, high bit on the left. The shift operators move these bits right or left so many places, discarding the bits moved out, and placing zeros in the emptied slots. For instance (not Pascal)

```
11111111 shl 3 = 11111000
11111111 shr 3 = 00011111
```

For a Word n,

```
n shl 3 = 8*n and n shr 3 = n div 8
```

It is risky to use these operators on signed integers because the high bit is the sign bit. Any shift right puts a 0 there, making the sign +, and any shift left has unexpected sign. Play with this a little.

The operators `shl` and `shr` use fast machine level op codes. The Pascal compiler produces optimal code using these shift operators for integer multiplication (or `div`) by (constant) powers of 2. So, for example, you will not produce a faster program by substituting `x := y shl 5` for `x := 32*y`.

There are four "logical" operators that can be applied to integers: **not, and, or**, and **xor**. These apply bitwise in the usual way. First **not** is a unary operator; it changes each bit 0 to 1 and each bit 1 to 0. **And** and **or** work the usual way. These examples show all possibilities:

```
         not 11110000 = 00001111
11110000 and 10101010 = 10100000
11110000 or  10101010 = 11110101
```

Xor is not as common as the other operators. It can be expressed in terms of them as

```
a xor b = (a and not b) or (not a and b)
```

Example:

```
11110000 xor 10101010 = 01011010
```

Use of these operators on integers in scientific programming is rare.

It is sometimes useful to enter Byte or Word variables in hex (hexadecimal). The digits are '0'..'9' and 'A'..'F', lower case acceptable. Put a $ before the entry: $FF = 255 for instance.

The purpose of the following program is to convert a decimal integer to hex. It actually handles integers with up to about 200 digits, strings of course, as no integer type handles more than 6 or 7 digits. Program "dec2hex" provides lots of exercise in working with strings and with integers, and it uses much of what we have studied to date, so it should be worth your while to study and run the program. Note its use of the standard procedure "Val" for converting strings that look like numbers to numbers.

Program 2.7.1

```
{$S+,R+}
program Dec2hex;    { 07/05/94 }
   { Input: a string of up to about 200 decimal digits,
     possibly with commas.
     Output: the number in hexadecimal. }

   uses   CRT;   { for KeyPressed function }
```

```pascal
const   h15 = '0123456789ABCDEF';

var     hold_in,
        dec_str,
        hex_str: string;
     hex_digits: array[0..15] of Char;
            ch: Char;
             k: Integer;

procedure  initialize;

    var     j: Word;
          temp: string[Length(h15)];

    begin
    temp := h15;
    { Note that h15 itself, a constant,
      may not be used in the next line. }
    for  j := 0  to  15  do
       hex_digits[j] := temp[j + 1];
    end;

procedure  get_decimal(var  dec_str: string);

    var  k: Word;

    begin
    repeat
       WriteLn('Enter a string of decimal digits:');
       ReadLn(dec_str);
       if  dec_str = ''  then  Halt;
       { Delete non-digits }
       for  k := Length(dec_str) downto 1  do
          if  not (dec_str[k] in ['0'..'9'])  then
             Delete(dec_str, k, 1);
       { Delete leading zeros }
       while  (Length(dec_str) > 1) and
              (dec_str[1] = '0')  do
          Delete(dec_str, 1, 1);
       if  Length(dec_str) > 0  then  Exit;
       WriteLn('Input unacceptable; try again.');
       WriteLn;
    until  false;
    end;    { get_decimal }
```

```
procedure  convert(var  dec_str, hex_str: string);

    var  remainder: Integer;

    procedure  divide_by_16;
    { The string dec_str of decimal digits is divided
      by 16.  The remainder converts later to the
      next hex digit.  The quotient is changed into
      the string dec_str. }

        var      j, k,
              dividend: Word;
                 error: Integer;

        begin    { divide_by_16 }
        { Delete leading zeros }
        remainder := 0;    { 0 <= remainder <= 15}
        for  j := 1  to  Length(dec_str)  do
           begin
           if  KeyPressed  then  Halt;
           Val(dec_str[j], k, error);
           { Val converts '4' to 4, etc }
           dividend := 10*remainder + k;
           { 0 <= dividend <= 159 }
           remainder := dividend mod 16;
           { 0 <= remainder <= 15 }
           dec_str[j] := hex_digits[dividend div 16];
           { '0'..'9' }
           end;
        Delete(dec_str, 1, 1);    { leading '0' }
        end;    { divide_by_16 }

    begin    { convert }
    hex_str := '';
    while  Length(dec_str) > 0 do
       begin
       if  KeyPressed  then  Halt;
       divide_by_16;
       hex_str := hex_digits[remainder] + hex_str;
       end;
    end;    { convert }
```

```
begin    { dec2hex }
initialize;
repeat
   get_decimal(dec_str);
   hold_in := dec_str;
   convert(dec_str, hex_str);
   k := Length(hold_in) - 2;
   while  k > 1  do
      begin
      Insert(',', hold_in, k);
      Dec(k, 3);
      end;
   Write('Decimal ', hold_in, ' = $');
   while  Length(hex_str) mod 4 <> 0  do
      hex_str := '0' + hex_str;
   k := Length(hex_str) - 3;
   while  k > 1  do
      begin
      Insert(' ', hex_str, k);
      Dec(k, 4);
      end;
   WriteLn(hex_str);  WriteLn;
   WriteLn('<Enter>: Continue; <Esc>: Quit: ');
   ch := ReadKey;
   if  ch = #27  then  Break;
   WriteLn;
until  false;
end.    { dec2hex }
```

Booleans

There is not much to say about the type *Boolean*. A Boolean occupies a byte of memory, though in practice, a single Boolean might very well use up two bytes because of the usual word boundaries for variables. The Boolean operations are

not and or xor

and act as was explained for integers. Of course, for Booleans, only the single low order bit means anything as a Boolean expression evaluates to one of the two values: "True" or "False".

Xor can be used to simplify Boolean expressions in tests. Consider

```
if   ( new_str and not Odd(m) )
     or ( not new_str and Odd(m) )   then   ...
```

This contorted test can be replaced by

```
if   new_str xor Odd(m)   then   ...
```

There is something important to say about the order of evaluation of Boolean expressions. When N. Wirth first introduced Pascal, the standard was always to evaluate first E1, second E2 in expressions

```
E1 and E2     E1 or E2
```

then evaluate the **and** or **or** on the resulting values. This led to complicated programming devices to get around the fact that half of the time the result is determined by the evaluation of the first operand alone. Note that

```
False and E2 = False;   True or E2 = True;
```

no matter to what E2 evaluates.

This may sound abstract, but it is actually very practical, notably when arrays are involved. For instance:

```
var   A: array[1..n] of Real;
. . . . . . . . . . . . . . . .
k := 1;
while (k <= n) and (A[k] < b)   do
    begin ... inc(k);   end;
```

When k reaches n + 1, the first test fails, but then the second test attempts to access A[n + 1], which causes a run-time error.

Borland Pascal offers what is called "short circuit" Boolean evaluation to avoid this, and I recommend that you use it always. It is the default when you start the Integrated Development Environment (IDE); keep it that way. In about 20 years of Pascal programming, I have never once seen a program statement in which the opposite, "complete Boolean evaluation" was more appropriate.

Under "Options – Compiler" is where you find the choice.

Scalar Types

The type Boolean is essentially defined by

```
type  Boolean = (False, True);
```

We are allowed to define variables in this manner. For instance:

```
type prism = (red, orange, yellow, green,
              blue, indigo, violet);
```

This is just a convenient way of associating names to the numbers $0, \ldots, 6$. Indeed,

```
Ord(red) = 0,  Ord(blue) = 4,  Ord(indigo) = 5,
orange < violet,  yellow = Pred(green),
blue = Succ(green),
```

and so on. We could just as well have declared

```
type  prism = 0..6;
```

but without the convenience of the names.

Exercises

1. If there were no function "Odd" how would you write one, say to apply to LongInt type?
2. Write **function** round_up(x: LongInt; n: Word): LongInt; to round x up to the nearest multiple of n.
3. Write a function that returns the number of 1's in the binary expansion of a Word variable.
4. Write a program that displays an input Integer in binary, then checks the result by changing the 16 binary bits back to an Integer.
5. Write a program that displays an input Word in binary, then checks the result by changing the 16 binary bits back to a Word.
6. The 7th grade teacher makes busy-work: add 20 25–digit integers. Some pupils do their own work (serial). But one group of 27 bright pupils decides to work in parallel: 25 each add one column, then pass the carry to the left, etc. Program these two methods, and convince yourself that the sum might be 27 digits. Find how many passes are needed for the parallel crew.
7. Define **function** power(x: Real; n: Integer): Real; that returns x^n. By the way, make it *fast*.
8. Define **function** mod_w(x: Word): Word; that returns $(ax + c) \bmod w$, where a and c are words with constant values and $w = 2^{16}$.
9. (Cont.) Define **function** mod_w1(x: Word): Word; that returns $(ax + c) \bmod (w+1)$.

Section 8.   Reals

As with integers, Pascal offers us several species of reals: *Single* (4 bytes), *Real* (6 bytes), *Double* (8 bytes), *Extended* (10 bytes), and *Comp* (8 bytes). All except Real use the 80x87 coprocessor, and either it is present, or you compile in the emulation {$E} mode. If it is present, it is selected by default unless you specifically request non-coprocessor code generation by the compiler directive {$N–} or by setting that compiler option. Type *Comp* is somewhat off the main line; its purpose is to do (integer) business arithmetic with the speed of the 80x87. The positive parts of the domains of all these types are:

```
Type            Positive Interval       # Sig. digits

Single:         [1.5E-45, 3.4E38]          7 -  8
Real:           [2.9E-39, 1.7E38]         11 - 12
Double:         [5.0E-324, 1.7E308]       15 - 16
Extended:       [3.4E-4932, 1.1E4932]     19 - 20
Comp:           [-2^63 + 1, 2^63 - 1]     19 - 20
```

(Each domain reflects in the origin to a negative interval.)

In addition to the binary operators +, –, *, / there are the following unary operators (functions) in the System unit:

```
Abs   Sqr   Sqrt   Exp   Ln   Sin   Cos   ArcTan
```

Any of these may be applied to integer arguments; the arguments will be converted automatically to reals before the function is applied. The constant Pi is built in, (but not *e*). The functions

```
Round   Trunc   Int   Frac
```

have a different nature. "Round" sends a real to its nearest LongInt. You have to be careful of overflow with it and with "Trunc", which simply chops off the decimal part of a real, leaving a LongInt. "Int" and "Frac", the integer part (rounding towards 0) and fractional part of a real are again *reals*.

Don't try to remember all the details of these functions, some of which you may never use. If you need enormous arrays of reals, and size is more important than accuracy, you might find a use for the Single type; I never have, and I almost always use Extended because of its great accuracy. Also Extended is the native type of the 80x87, so you maximize speed of floating point operations by using Extended type all the time.

I'll put in one program using *Comp* and probably not mention that type again. The program offers me the possibility of introducing a way to handle runtime errors that might interest you.

Its purpose is to compute n!, both as a *LongInt* and as a *Comp*. Please run the program and note its results. First, n = 10 works fine with both types. Next, n = 20 works with *Comp*, but results in a *negative* answer with *LongInt*! How can that be? The answer is in the way overflow is handled by signed integer types; overflow just cycles to the other end of the spectrum.

Finally, n = 25 works with LongInt, giving a very wrong answer (25! = approx. 1.55E25), but crashes Comp. Overflow with real types always results in an fp (floating point) error and crash.

Program 2.8.1

```
{$N+}
program  factorials;    { 07/05/94 }

   var   save_exit: Pointer;
              n: Word;

   procedure   clean_exit;   far;

      begin
      ExitProc := save_exit;
      WriteLn;
      if  ExitCode  <> 0  then
         WriteLn('Runtime error!  ExitCode = ',
                 ExitCode);
      Write('Press <Enter>: ');
      ReadLn;
      end;

   function   n_factorial(n: Word): LongInt;

      var   j: Word;   f: LongInt;

      begin
      f := 1;
      for  j := 2  to  n  do    f := j*f;
      n_factorial := f;
      end;

   function   x_factorial(n: Word): Comp;

      var   j: Word;   f: Comp;
```

```
   begin
   f := 1;
   for  j := 2  to  n  do    f := j*f;
   x_factorial := f;
   end;

begin    { factorials }
save_exit := ExitProc;
ExitProc := @clean_exit;
n := 10;
WriteLn('With LongInt: ', n, '! = ',n_factorial(n));
WriteLn('With Comp:    ', n, '! = ',x_factorial(n));
n := 20;
WriteLn('With LongInt: ', n, '! = ',n_factorial(n));
WriteLn('With Comp:    ', n, '! = ',x_factorial(n));
n := 25;
WriteLn('With LongInt: ', n, '! = ',n_factorial(n));
WriteLn('With Comp:    ', n, '! = ',x_factorial(n));
end.
```

Remark Don't worry now about the details of the runtime error handling.

The binary representation of the various real types follows the IEEE (Institute of Electrical and Electronics Engineers) standard. The following program demonstrates the representation for type *Extended*. Because the function "Random" returns "Real" type, I use a little trick to generate more decimal places. The program also illustrates the use of *type casting* and of the operators **shr** and **and**. (Review "type casting" at the beginning of Section 3.)

Program 2.8.2

```
program  ieee;    { 06/22/94 }
{ To demonstrate:
   1) IEEE format for type Extended.
   2) Type casting.

Extended type:  80 bits:
   79  78.....64  63  62..........0
    s      e       u       m

 If                          then
 0 <= e < 32767              x = (-1)^s*2^(e - 16383)*u.m
 e = 32767  and  m = 0       x = (-1)^s * Inf
 e = 32767  and  m <> 0      x = NaN (Not a Number) }

   uses  CRT;
```

```
const   mask = 127;   { 01111111 }

type   byte_array = array[0..9] of Byte;

var   x, y: Extended;
         b: byte_array;
         k: Word;
         e: Integer;
      s, u: 0..1;
         m: Extended;

begin
ClrScr;
Randomize;
x := 2000.0*Random - 1000.0 + 1.0e-9*Random;
b := byte_array(x);
WriteLn('x = ', x:26);
WriteLn('Bytes of x:');
Write('Byte no.:');
for k := 9 downto 0 do  Write(k:5);  WriteLn;
Write('Value:    ');
for k := 9 downto 0 do  Write(b[k]:5);
WriteLn;  WriteLn;
e := 256*(b[9] and mask) + b[8];
s := b[9] shr 7;
u := b[7] shr 7;
m := 0.0;
for  k := 0 to 6  do  m := m / 256.0 + b[k];
m := (m / 256.0 + (b[7] and mask)) / 128.0;
WriteLn('Exponent       e = ', e);
WriteLn('Sign bit       s = ', s);
WriteLn('Underflow bit  u = ', u);
WriteLn('Mantissa       m =', m);
WriteLn;

y := u + m;
e := e - 16383;
if  e > 0  then
   for  k := 1 to e  do  y := 2.0 * y
else
   for  k := 1 to -e  do  y := y / 2.0;
{ Check sign bit }
if  s = 1  then  y := -y;
WriteLn('x = [(-1)^s]*[2^(e - 16383)]*[u.m]');
WriteLn('(Reconstructed)  x = ', y:26);
WriteLn('(Original)       x = ', x:26);
```

```
ReadLn;
end.    { ieee }
```

Here is a short program along the same line; its task is to display in binary a variable of *any* type. While this seems impossible at first sight, when you realize that *typeless* **var** parameters of procedures are allowed, then the possibility seems more feasible. In the program you may replace "my_type" with any type of your choice. Go for it!

Program 2.8.3

```
{$R+}   { Range checking on }
program  display_binary;   { 06/04/94 }
  { Displays any variable in binary }

  type   my_type = LongInt;    { Example }

  var  x: my_type;

  procedure  binary(var  x;  size: Word);
    { Call with size replaced by SizeOf(x), which
      must be computed outside of "binary" because
      x is an untyped variable. }

    var   hold: array[0..7] of Boolean;
          b: Byte;
       j, k: Word;
       bits: array[1..1024] of Byte;

    begin
    WriteLn('size = ', size);
    Move(x, bits, size);
    for  k := size  downto  1   do
      begin
      b := bits[k];
      for  j := 0  to  7  do
        begin
        hold[j] := Odd(b);
        b := b shr 1;
        end;
      for  j := 7  downto  0  do
        if  hold[j]  then   Write(1)
        else  Write(0);
      end;   { for k }
    end;   { binary }
```

```
begin    { display_binary }
x := $FF00FF0F;    { Example }
binary(x, SizeOf(x));
ReadLn;
end.
```

Remark The standard procedure "Move" is a low level procedure, based on the machine instruction "MOV". In the application above, it moves "size" contiguous bytes, with the source starting at the first byte of the variable "x" and the destination starting at the first byte of the variable "bits". I really meant *low level*; nothing is checked, however, overlap is avoided. Do be careful using "Move".

Exercises

1. Look up the IEEE binary representation for type Double, and write a program like "ieee" above for Double rather than Extended.
2. Look up the IEEE binary representation for type Real. Then write a program for multiplication of two Real's that traps overflow.
3. Write a program for multiplication of two Extended's that traps overflow.

Chapter 3 Advanced Programming

Section 1. Text Files

Let us summarize what we know about a file of type *Text*. It contains a sequence of strings, each terminated by a carriage return CR = #13 and a line feed LF = #10, also possibly by a form feed FF = #12. We declare a variable my_text to be of type text by

```
var  my_text: Text;
```

We attach the variable my_text to a disk file or a device by statements like

```
Assign(my_text, 'D:\scipas\my_file');
Assign(my_text, 'prn');
```

We open my_text for reading by Reset(my_text); and for writing by Rewrite(my_text); and we close my_text by Close(my_text);

We read the value of a variable from my_text. For example if N is of type Word and X of type extended, then

```
Read(my_text, N, X);
```

assigns N and X to the next two items stored in my_text. The file name my_text is omitted if input is from the standard file "Input", meaning the keyboard. In this case we usually use *ReadLn* instead of *Read*, so that a press of <Enter> terminates the input.

The corresponding procedures *Write* and *WriteLn* for output use expressions for their arguments beyond the (optional) first file name. If E represents an integer, real, Boolean, or string valued expression, then

```
WriteLn(E:n);
```

writes E flush right in a field of width n. If n is less than the length of the value of E, it is ignored, e.g,

```
Writeln('12345':3);
```

If E represents a real-valued expression, then two formatting fields may be specified:

```
Write(E:12:5);
```

means write E in fixed, rather than scientific, form, with 5 decimal places, and right justified in a field of width 12.

In Chapter 1, we looked at a very simply unit of text file procedures. The following unit may be of more use to you. Note that it contains precise information in the form of error messages on possible failures of operations on files of type "Text". (The DOS unit function *FExpand* expands a file name into a full path name. See the beginning of Chapter 4 for more on this.)

Program 3.1.1

```
unit  textfile;    { 06/30/94 }
{ A set of text file handling procedures }

interface

    uses  DOS;

    type  message = string[60];

    var  in_file, out_file: Text;
         in_name, out_name: PathStr;
                       dir: DirStr;    { string[67] }
                      name: NameStr;   { string[8]  }
                       ext: ExtStr;    { string[4]  }

    procedure  get_file_name(prompt:  message;
                                  var  name: PathStr);
       { Returns a full path name }
    procedure  test_validity(var  ss: PathStr);
    procedure  open_input(ss: PathStr;
                          var GG: Text);
    procedure  open_output(ss: PathStr;
                           var  GG: Text);

implementation
{$I+}    { I/O checking on }

    procedure  get_file_name(prompt:  message;
                                  var  name: PathStr);
```

```pascal
begin
Write(prompt);
ReadLn(name);
name := FExpand(in_name);
   { Now name is a full path name }
end;

procedure  test_validity(var  ss: PathStr);

   var  n: Integer;

begin
n := IOResult;
{ This call always resets the error flag. }
if  n > 0  then
   begin
   WriteLn('File ', ss, ' cannot be processed.');
   case  n  of  { DOS errors }
      1: ss := 'Non existent DOS function call';
      2: ss := 'File not found';
      3: ss := 'Path not found';
      4: ss :=
         'More than 15 open files per process';
      5: begin
         WriteLn(
          'One of the following errors:');
         WriteLn(' Write to a read only file');
         WriteLn(
          ' Rename a file to existing file');
         WriteLn(' Rewrite a read-only file');
         WriteLn(' Full directory');
         WriteLn(
          ' Erase directory or read-only file');
         WriteLn(' Create duplicate directory');
         ss :=
          ' Read/Write file not open for such';
         end;
       6: ss :=
          'Trashed file; invalid file handle';
       8: ss := 'Not enough memory';
      12: ss := 'Invalid file mode';
      15: ss := 'Invalid drive number';
      16: ss := 'Removing current directory';
      17: ss :=
          'Attempting to rename across drives';
      18: ss := 'No more files';
```

```
            end;
        case   n   of   { I/O errors }
            100: ss :=
                    'Attempt to read past end of ' + ss;
            101: ss := 'The disk is full';
            102: ss := ss + ' has not been assigned';
            103..105: ss := ss + ' is not open';
            106: ss :=
                    'Format error in reading number';
            end;
        WriteLn(ss);
        Write('Press <Enter> to halt the program: ');
        ReadLn;
        Halt;
        end;
    end;

procedure   open_input(ss: PathStr;
                        var   GG: Text);

    begin
    Assign(GG, ss);
    {$I-}    { I/O checking off }
    Reset(GG);
    test_validity(ss);
    {$I+}    { I/O checking back on }
    end;

procedure   open_output(ss: PathStr;
                         var   GG: Text);

    begin
    Assign(GG, ss);
    {$I-}    { I/O checking off }
    Rewrite(GG);
    test_validity(ss);
    {$I+}    { I/O checking back on }
    end;

end.    { unit  textfile }
```

Exercises

1. Write programs to test the unit "textfile", particularly its error messages.

Section 2.   Typed Files

We have used "Text" files quite a bit. Pascal allows other file types. The declaration of a *typed file* is

```
var   GG: file of any_type;
```

where "any_type" is any type whatsoever (except a file type). The procedures "Assign", "Reset", "Rewrite" are the same as for files of type "Text". The procedures *Read* and *Write* are different; the main difference is that each of their parameters must be a *variable*. Expressions and constants are not allowed. ("ReadLn" and "WriteLn" apply *only* to files of type "Text".)

The following example illustrates a file of real numbers side-by-side with a file of type "Text", both used to store the same information, more or less.

Program 3.2.1

```
program  file_of_reals;    { 07/10/94 }

    var        EE: file of Extended;
               TT: Text;
        x, y, z: Extended;
               k: Word;

    begin
    Assign(EE, 'trig_tab.ext');
    Assign(TT, 'trig_tab.txt');
    Rewrite(EE);   Rewrite(TT);
    x := 0.0;
    for  k := 1  to  1000  do
        begin
        y := Sin(x);
        Write(EE, y);
        WriteLn(TT, y);
        x := x + 0.001;
        end;
    Reset(EE);   Reset(TT);
    for  k := 1  to  500  do
        begin
        Read(EE, y);   ReadLn(TT, z);
        end;
    for  k := 1  to  20  do
        begin
        Read(EE, y);   ReadLn(TT, z);
        WriteLn('Error = ', Abs(y - z));
        end;
```

```
Close(EE);  Close(TT);
Write('Press <Enter>: ');  ReadLn;
end.
```

There is much of interest here. First of all, the file EE has size 10,000 bytes, but file TT has 25,000 bytes. How can that be? Well, a variable of type "Extended" occupies 10 bytes, so a **file of** 1,000 "Extended" requires 10,000 bytes. But when you write an "Extended" variable to a text string, that string has 23 characters, including the exponential part 'E+nnnn' or 'E–nnnn'. Add to that the control characters CR LF that end each line when you call WriteLn, you have 25 characters for each Extended written to TT.

So it is thrifty to use **file of** Extended. What about accuracy? As the program shows, there is an error of about 5E-16 near the middle of the files. Obviously writing an Extended to a string of characters loses accuracy, and the program shows us some of this error.

Exercises

1. Predict the outcome of the following program, and explain it:

```
program  file_of_string;   { 07/10/94 }

    type  Str80 = string[80];

    var  FF: file of Str80;
         GG: Text;
         ss: Str80;
          k: Word;

    begin
    Assign(FF, 'strings.str');
    Rewrite(FF);
    Assign(GG, 'strings.txt');
    Rewrite(GG);
    ss := 'abc';
    for  k := 1  to  100  do
       begin
       Write(FF, ss);
       WriteLn(GG, ss);
       end;
    Close(FF);  Close(GG);
    end.
```

2. Let FF be a disk file of reals. Find the second largest number in FF.
3. Let FF be a disk file declared as **file of** Char. Count how many times each control character (#0..#31) appears in FF.
4. Let FF be a disk file declared as **file of** Char. Read FF until either its end, or all characters 'A'..'Z' have been read once, and report the result.

5. Program the following scheme for encrypting a file: Bytes are read from the source 8 at a time. Think of them as the rows of an 8×8 matrix of bits. The 8 columns of this matrix then supply the next 8 encoded bytes for the destination file. Padding of up to 7 bytes has to be added to the final attempt to read 8 bytes, in order to make an even 8. The program should keep track of how many bytes are actually read from the source file, and add this information to one final block of 8 bytes in the output.

6. (Continued) Now write the decoding program, and test that it produces an identical copy of the original source file. (Warning: These programs are protected by international copyright.)

Section 3.   Untyped Files

Suppose we have a disk file created by a Pascal program. Without having the program, we have no way of knowing whether it was created as "Text", as "**file** of Integer", or what. All we see in the disk directory is that it is a file of length so many bytes. Each disk file is just that, a sequence of bytes (sometimes called a *stream*) no matter how it was created.

The purpose of the following program is read a text file created under the Unix operating system and write it as a text file for MS–DOS. (Maybe you received such a file by uuencoded email.) The difference is that Unix ends each line with a single control character LF = #10 while DOS ends each line with two consecutive control characters, CR = #13 and LF. The input file is not in the Pascal format "Text", so we cannot open it as Text. We open it as a "**file** of Char". We read characters one at a time. If we read a LF, then we write CR LF, otherwise, we just write the character read. This program is called at the command line with two parameters, the names of the input and the output files:

```
unix2dos name1 name2 <Enter>
```

The standard function *ParamStr* returns the parameters in order. I have omitted checking the parameters for brevity. (A production program should be far more robust than this.)

Program 3.3.1

```
program  Unix2DOS;    { 04/23/94 }

var     GG, HH: file of Char;
     LF, CR, ch: Char;
```

```
begin
LF = #10;  CR = #13;
Assign(GG, ParamStr(1));  Reset(GG);
Assign(HH, ParamStr(2));  Rewrite(HH);
while  not Eof(GG)  do
   begin
   Read(GG, ch);
   if  ch = LF  then
      begin
      Write(HH, CR);  Write(HH, LF);
      end
   else  Write(HH, ch);
   end;
Close(HH);
Close(GG);
end.    { Unix2dos }
```

This program is fine except that it is very slow, and the reason for its slowness is that it goes to the disk for one character at a time, processes that character, then goes back to the disk.

It is for doing things like this much faster that Pascal has *untyped* files—they come with very fast low-level disk operations. The declaration of an untyped file variable FF is simply

```
var  FF: file;
```

Such a file variable FF is assigned as usual. The procedures *Reset* and *Rewrite* have a different form for an untyped file; in addition to the file variable, there is an (optional) second Word parameter. The calls are

```
Reset(FF, n);  Rewrite(FF, n);
```

Here n represents the size in bytes of individual records in the file, assuming that is known. If you do not include n, then its default value is 128, which might be dangerous to use accidently. I know of no reason to use any value other than 1, so I recommend always using 1.

The procedures for reading from and writing to an untyped file are *BlockRead* and *BlockWrite*. The following rewrite of Unix2DOS uses these procedures. Note that the program marks an area of memory as a *buffer* into which to read, and from which to write. This is done by a **var** declaration; the type of the variable makes no difference whatsoever, only its size matters. Remember, this stuff is *low level*! Look at the program first; I'll discuss it after:

Program 3.3.2

```pascal
program  Unix2DOS;    { 04/23/94 }

  const  LF = #10;
         CR = #13;

  var     GG, HH: file;
             ch,
          buffer: array[1..3048] of Char;
       bytes_in,
      bytes_out,
          j, k: Word;

  begin
  Assign(GG, ParamStr(1));
  Reset(GG, 1);
  Assign(HH, ParamStr(2));
  Rewrite(HH, 1);
  WriteLn('Changing Unix file' , ParamStr(1),
          ' to DOS file ', ParamStr(2));
  WriteLn('Size of ' , ParamStr(1), ': ',
          FileSize(GG), ' bytes');
  repeat
     BlockRead(GG, buffer, 2048, bytes_in);
     for  k := bytes_in  downto  1  do
        begin
        if  buf[k] = LF  then
           begin
           Inc(bytes_in);
           Move(buffer[k], buffer[k + 1],
                bytes_in - k);
           buffer[k] := CR;
           end;
        end;
     BlockWrite(HH, buffer, bytes_in, bytes_out);
  until  (bytes_in = 0) or (bytes_out <> bytes_in);
  Close(HH);  Close(GG);
  end.   { Unix2dos }
```

The variable "buffer" has size 2048 bytes. We try to read 2048 bytes at a time into buffer. As we are going to add some CR's, buffer has to be larger than the number of bytes read. This works fine until the end of the input file GG. Now, "BlockRead" is a low-level procedure, but not so low-spirited as to read past the end of GG. The variable "bytes_in" reports how many bytes are actually read. It is always 2048 until the final time "BlockRead" is called. Note that there are four parameters to "BlockRead", the file variable, the buffer variable, the number of bytes to read, and the number of bytes actually read. "BlockWrite" has similar structures. The final test, (bytes_out <> bytes_in) is a little extra protection against a full (floppy) disk.

Exercises

1. Write a program to copy any file. (Test it on someone else's computer.)
2. There are various "UUEncode" programs around. Their purpose is to change any file to a text file that can be transmitted by email, which only transmits 7–bit characters, not all of them. The XXEncoding scheme uses the 64 characters

```
+ - 0..9 A..Z a..z   (numbered 0..63)
```

only. Three consecutive bytes of the input file can be considered as four consecutive 6–bit words, each of which has 64 possible values, hence corresponds to one of the characters above, according to its order. The input file is read 45 bytes at a time (the final read possibly less). The number of characters actually read (45 except for the final read) is encoded, and becomes the first character (usually "h") of the next line of the output text file. Then the (usually) 45 characters read are encoded to (usually) 60 characters, making up the rest of that line. The final output line may be shorter. Write a UUEncode program. Its first line should be

```
begin xxencode <source file name>
```

and its last two lines should be

```
++
end
```

so the decoding program can know when it ends.
3. (Continued) Write a corresponding "UUDecode" program, and test your work on actual files.

Section 4.   Records

The reserved word **record** allows us to define data types that contain a mix of data of various types. The syntax is roughly

```
type type_name = record
                 <variable declarations>
             end;
```

Let us start with an example. A real monomial in 9 variables is specified by its coefficient, a real number, and by the degrees of each variable, which suggests an array. So we want to lump together into one variable type a real coefficient and an array of degrees:

Program 3.4.1

```pascal
program  monomials;    { 08/04/94 }

   const   max = 9;

   type   term = record
                       coeff: Real;
                        deg: array[1..max] of Word;
                   end;
```

Suppose F is a variable of type term and we wish to refer to its coefficient. The direct way is by *dereferencing* with a point notation, for instance,

```pascal
x := F.coeff;
```

There is another way using the reserved word **with**:

```pascal
with  F   do   x := coeff;
```

After the **do** any statement can follow, in particular a compound statement bracketed by **begin** ... **end**. Let us continue our monomial example with a procedure for printing a term:

```pascal
procedure  print(F: term);

   var   d: 1..max;
         k: Word;

   begin
   with  F   do
      begin
      Write(coeff:2:2);
      for  d := 1  to   max   do
         begin
         k := deg[d];
         if  k > 0   then
            Write('(x', d, '^', k, ')');
         end;
      end;    { with }
   WriteLn;
   end;    { print }
```

To continue the example, let us multiply two monomials:

```
procedure  product(F, G: term;  var  H: term);

    var  d: 1..max;

    begin
    with  H  do
       begin
       coeff := F.coeff*G.coeff;
       for  d := 1  to  max  do
          deg[d] := F.deg[d] + G.deg[d];
       end;
    end;    { product }
```

It may be the case that several statements combined in a **with** statement execute minutely faster than if dereferenced—without the **with**. Anyhow, in this case, the coefficient calculation could have been coded

```
        H.coeff := F.coeff*G.coeff;
```

just as well. It remains to test this work:

```
var  F, G, H: term;
              d: Word;
begin
with  F  do
   begin
   coeff := 3.0;
   for  d := 1  to  max  do  deg[d] := d;
   end;
with  G  do
   begin
   coeff := 5.0;
   for  d := 1  to  max  do  deg[d] := 2*d + 1;
   end;
product(F, G, H);
Write('F   = ');   print(F);
Write('G   = ');   print(G);
Write('F*G = ');   print(H);
ReadLn;
end.    { monomials }
```

Pascal does not have a built-in data type for complex numbers, so we must make one from scratch. We could define a complex number as

```
type  complex = array[1..2] of Real;
```

but it does seem awkward to refer to the real part of z and z[1], etc. A record of two reals appears the better idea here. Unlike the program "monomials" above, which was just a little example, we now embark on a serious project, a unit for complex arithmetic. We begin with the **interface**:

Program 3.4.2

```
{$N+}
unit  complexs;   { 07/12/94 }

interface

type      Real = Extended;
      complex = record   re, im: Real;   end;

const  zero: complex = (re: 0.0;   im: 0.0);
        one: complex = (re: 1.0;   im: 0.0);

procedure  sum(z, w: complex;   var  s: complex);
procedure  difference(z, w: complex;
                      var  s: complex);
procedure  negate(var  w: complex);
procedure  product(z, w: complex;   var  s: complex);
procedure  quotient(z, w: complex;
                    var  s: complex);
function   modulus(z: complex): Real;
procedure  polar_form(z: complex;
                      var  r, theta: Real);
procedure  print_complex(z: complex;  n: Word);
procedure  print_to_file(z: complex;  n: Word;
                         var  out_file: Text);
```

Now we refer to the real and imaginary parts of z as z.re and z.im. Note the *structured typed constants* (zero, one) of record type, and how they are initialized. In general *typed constants* are really variables that are initialized. They can be assigned later just like any variables.

Everything should be obvious in the subroutine declarations, except possibly the n's in the print procedures: they are for formatting. The **implementation**:

```
implementation

procedure  sum(z, w: complex;   var  s: complex);
```

```
begin
  s.re := z.re + w.re;
  s.im := z.im + w.im;
end;

procedure difference(z, w: complex;
                     var s: complex);

  begin
    s.re := z.re - w.re;
    s.im := z.im - w.im;
  end;

procedure negate(var w: complex);

  begin
    w.re := -w.re;
    w.im := -w.im;
  end;

procedure product(z, w: complex; var s: complex);

  begin
    s.re := z.re*w.re - z.im*w.im;
    s.im := z.re*w.im + z.im*w.re;
  end;

procedure quotient(z, w: complex;
                   var s: complex);

  var r: Real;

  begin
  with w do
    begin
    r := Sqr(re) + Sqr(im);
    re := re/r;  w.im := -im/r;
    end;
  product(z, w, s);
  end;

function modulus(z: complex): Real;

  begin
  modulus := Sqrt(Sqr(z.re) + Sqr(z.im));
  end;
```

```
procedure  polar_form(z: complex;
                       var  r, theta: Real);

    begin
    r := modulus(z);
    if  z.re > 0  then   theta := Arctan(z.im/z.re)
    else  if  z.re < 0  then
        theta := Pi + Arctan(z.im/z.re)
    else  if  z.im >= 0  then   theta := 0.5*Pi
    else  theta := -0.5*Pi
    end;     { polar_form }

procedure  print_complex(z: complex;  n: Word);

    begin
    if  z.re < 0.0  then
        Write(' - ',  Abs(z.re):(n + 2):n)
    else  Write(' + ',  z.re:(n + 2):n);
    if  z.im < 0.0  then
        Write(' - ',  Abs(z.im):(n + 2):n)
    else  Write(' + ',  z.im:(n + 2):n);
    Write('*i');
    end;

procedure  print_to_file(z: complex;  n: Word;
                         var  out_file: Text);

    begin
    if  z.re < 0.0  then
        Write(out_file, ' - ',  Abs(z.re):(n + 2):n)
    else  Write(out_file, ' + ',  z.re:(n + 2):n);
    if  z.im < 0.0  then
        Write(out_file, ' - ',  Abs(z.im):(n + 2):n)
    else  Write(out_file, ' + ',  z.im:(n + 2):n);
    Write(out_file, '*i');
    end;

end.    { unit  complexs }
```

We use records a lot, so by all means go back to the beginning of this section and read it again to be sure that the ideas and syntax are clear.

Exercises

1. Write a program to test the unit complexs.
2. Why not name the unit "complexes" or "complex_numbers"?

Section 5.   Variant Parts

The variable declarations between **record** and **end** comprise the *fixed part* of the record declarations. There may also be a *variant part* of a record declaration; only one is allowed, and it must follow the fixed part (if any). The idea of a variant part is thrift: the same memory (or storage) is used for different possible objects. The syntax is like that of a case statement:

```
type   rec = record
                < var declarations>
                case  n: Word  of
                    0: ( <field list> );
                    1: ( <field list> );
                    .......
                    7: ( <field list> );
            end;
```

Only one **end** is used. A *field list* is like any list of **var** declarations, but without the word **var**. For illustration only, I used "Word" for the *tag field type* of the *tag field identifier* n of the record.

Let's look at an example. Suppose we set out to do arithmetic on polynomials with each coefficient either rational or real. We can certainly set up the elementary arithmetic operations on the coefficients; there will be several cases to consider of course. We eventually might have large matrices of these polynomials, so memory size is an issue. This is where variant parts comes in. A real fills 6 bytes; a rational (quotient of longints) takes 8 bytes. We certainly should not use 14 bytes for each coefficient when at most 8 bytes are needed. Here is a little program with both possibilities illustrated and measured:

Program 3.5.1

```
program  variants;    { 08/04/94 }

type   rec1 = record
                r: Real;
                q: record  x, y: LongInt;  end;
            end;

        tag_type = (real_no, rational);
```

```
      rec2 = record
                case  tag: tag_type  of
                    real_no: (r: Real;);
                    rational: (q: record
                                     x, y: LongInt;
                                   end);
              end;

begin
WriteLn('type rec1: ', SizeOf(rec1), ' bytes');
WriteLn('type rec2: ', SizeOf(rec2), ' bytes');
ReadLn;
end.
```

In the second case, tag itself is like another field of the fixed part of rec2. Suppose w is a variable of type rec2. If w.tag = rational, then w.q.y is defined.

There is a variation on variant part that is seldom used; the tag type is there, but not the tag. Then the variant part is called a *free union*. This is used when you require the same space to be interpreted in two different ways. In other words, free unions provide an alternative to type casting. Let us inspect an example of a free union:

Program 3.5.2

```
program  free_union_demo;    { 08/04/94 }

  type   tag_type = (extend, bits);
         vector = array[0..9] of Byte;
         number = record
                    case  tag_type  of
                        extend: (x: Extended);
                          bits: (B: vector);
                    end;

  var   z: number;

  procedure  print_bits(z: number);

    var   j, k: Word;
          bits: array[0..7] of 0..1;

    begin
    for  j := 9  downto  0  do
       begin
       for  k := 7 downto  0   do
           Write((z.B[j] shr k) and 1);
```

```
        if  j = 5  then
            begin
            WriteLn;
            Write('':Length('Binary form: '));
            end;
        end;
    WriteLn;    WriteLn;
    end;    { print_bits }

begin    { free_union_demo }
while   true  do
    begin
    Write('Enter a real number:  ');
    ReadLn(z.x);  WriteLn;
    WriteLn('Decimal form: ', z.x);
    Write('Binary form: ');
    print_bits(z);
    WriteLn('<Ctrl-Break> to Quit');
    end;
end.
```

In this case, the type "tag_type" has two members. The two field lists (each
containing only one variable) each total the same length; that is essential for a
free union. When the variable z is dereferenced as z.x, it is an extended real. But
when it is dereferenced as z.B, it is an array of 10 bytes.

Exercises

1. Find the size of type "number" in "free_union_demo".
2. Compare program "free_union_demo" with "ieee" of Chapter 2.
3. Use a free union of a 5×4 matrix and a 20–term vector to determine Pascal's store
   order for a matrix, row-major or column-major.

Section 6.   Pointers

A *pointer* is, roughly, the address of a variable. Pointers provide a mechanism
for indirect addressing. They also allow us to define data types recursively and
to create variables during the execution of programs, that is, *dynamically* rather
than by static declarations. At first this latter does not sound like a big deal.
After all, whenever you call a procedure, all of its local variables are created. But
they disappear after the procedure finishes its task. With pointers, however, we
create variables that are global and enduring (and sometimes hard to get rid of,
let's add). Since a pointer is a memory address, it is essentially a double word
(segment and offset).

We can do much with pointers: create graph-like data structures, create vectors, polynomials, matrices, etc, whose sizes are not given in advance, but determined during execution, and so on.

The syntax for a pointer type declaration is

type   zilch_ptr = ^zilch;

where "zilch" is a type name.

Here is the one and only place in Pascal in which an identifier may be used before it has been declared. In the declaration above, the type identifier "zilch" need not necessarily have been declared earlier; if not, it must be declared eventually.

The following little program will show common usage of pointers.

Program 3.6.1

```
program  ptr_demo;    { 08/09/94 }
   { Test both in real and protected modes }

   const  max = 200;

   type   vector_ptr = ^vector;
          vector = array[1..max] of LongInt;

   var  VP: vector_ptr;
        i, j: LongInt;
         k: Word;

   begin
   i := MemAvail;
   WriteLn('Before allocation, free: ', i:8, ' bytes');
   New(VP);   { Allocates space in the heap }
   j := MemAvail;
   WriteLn('After allocation, free:  ', j:8, ' bytes');
   for  k := 1  to  max  do  VP^[k] := 3*k;
   WriteLn('VP^[', max div 2, '] = ',
           VP^[max div 2]);
   k := i - j;
   WriteLn('Allocated:                   ', k:8, ' bytes');
   WriteLn(max, ' LongInt''s:         ',
           4*max:8, ' bytes');
   k := k - 4*max;
   WriteLn('Overhead:                    ', k:8, ' bytes');
   Dispose(VP);   { Releases space }
   VP := nil;
   i := MemAvail;
```

```
WriteLn('After release, free:     ', i:8, ' bytes');
ReadLn;
end.   { ptr_demo }
```

The first thing to observe is the standard procedure *New*. When applied to a pointer variable P, *New* allocates memory for a variable of the type that P points to, and assigns P the address of that variable. Then P^ is that new variable. We might think of P as the *pointer* and P^ as the *pointee*.

The first type declaration is "vector_ptr", declared as a pointer to a type "vector", so far unknown. The next line declares "vector" as an array.

Most of the program consists of bookkeeping, to show what happens to available memory. If we omit this for the moment, what remains is

```
begin
New(VP);    { Allocates space in the heap }
for  k := 1  to  max  do  VP^[k] := 3*k;
Dispose(VP);   { Releases space }
VP := nil;
end.   { ptr_demo }
```

We should examine these lines carefully. The "New" call allocates memory, as noted. Then VP^ is the name of an array, and the next line assigns the elements of this array, simply to emphasize the point. The "Dispose" procedure is the inverse of "New", and frees the memory allocated by "New". The following line, assigning VP to **nil** is unnecessary, but is insurance against misuse of the pointer VP later, say by an accidental second call of "Dispose".

One of the rules of life is always to *write a "Dispose" statement as soon as you write a "New" statement.* This is far more serious than the rule to write an **end** for every **begin**, because the compiler will not let you get away without the **end**, but doesn't notice a missing "Dispose", which can cause a crash or worse.

Note the advise in the first comment of the program: to test it both in real and protected modes. (Click on Compile/Target to change Real<—>Protected.) Protected mode has a complicated way of keeping track of what is allocated, and you can find details in the Borland Pascal *Language Guide*'s chapter *Memory Issues*. There is a little overhead. In real mode there is no overhead. In this case, the cause of overhead is explained by how memory is allocated.

First of all, "New" allocates memory in the "heap", not in the 64K data segment of your program. This means that usually much more memory is available for dynamically created variables than for static variables. This is the good news. The bad news (real mode) is that blocks of memory are allocated on 8–byte boundaries (so the heap manager can operate efficiently). Thus, for instance, allocating 500 pointers to Boolean's will use up 4000 bytes, not the 1000 you would guess at first.

Remark The result returned by a **function** can be a pointer.

The program "prt_demo" really has no value other than showing the use of "New" and "Dispose". Now let's move on to a serious use of pointers. We are going to write a program for multiplying rectangular integer matrices whose sizes are not given in advance, but can be specified dynamically. "New" and "Dispose" are not adequate for handling this kind of situation. We need something that can take declarations like

```
type   vector = array[1..20000] of Integer;

var   V: ^vector;
```

and just allocate 10 bytes (5 Integers) to V^ in case we want only a vector of 5 components. The standard procedure "GetMem" handles this and "FreeMem" later releases the memory. The rule of life applies here too: *whenever you write "GetMem", put in the corresponding "FreeMem"*.

We begin the program "matrices"

Program 3.6.2

```
program  matrices;    { 08/08/94 }

type   row_index = array[1..200] of ^row_vector;
      row_vector = array[1..200] of Integer;
         matrix = record
                     no_rows,
                     no_cols: Word;
                    index_ptr: ^row_index;
                  end;
```

We allow for integer matrices up to 200×200, but actually will work with much smaller ones. The actual size of a variable of type "matrix" is a mere 8 bytes: two words and one pointer. You should think of the pointer as pointing to a column vector of row indices. The length of this column is the number of rows in the matrix. Each element of the column is itself a pointer to a row vector. Thus if A is a matrix, then its i, k-th element is

```
A.index_ptr^[i]^[k]
```

Don't read another word until you understand this.

We move on to the **var** declarations of "matrices":

```
var     A, B, C: matrix;
        inds_size,
      elems_size,
            i, k: Word;
            m, n: LongInt;
```

After the declaration of A, B, C, the other variables are for bookkeeping only. (It might be worth your while to strip them out of the program at first.) Later, the main action will assign A and B randomly as 4×5 and 5×3 matrices, and C will be computed as their product. The first two procedures are the crucial ones for creating and destroying an $r \times c$ matrix; here "GetMem" and "FreeMem" are used:

```
procedure   create(var   A: matrix;   r, c: Word);

    var   j, k: Word;

    begin
    with   A   do
      begin
      no_rows := r;   no_cols := c;
      j := no_rows*SizeOf(Pointer);
      Inc(inds_size, j);
      GetMem(index_ptr, j);
      for   j := 1   to   no_rows   do
        begin
        k := no_cols*SizeOf(Integer);
        Inc(elems_size, k);
        GetMem(index_ptr^[j], k);
        for   k := 1   to   no_cols   do
          index_ptr^[j]^[k] := -5 + Random(11);
        end;
      end;   { with }
    end;   { create }

procedure   destroy(var   A: matrix);

    var   j: Word;

    begin
    with   A   do
      begin
      for   j := 1   to   no_rows   do
        FreeMem(index_ptr^[j],
```

```
                  no_cols*SizeOf(Integer));
       FreeMem(index_ptr, no_rows*SizeOf(Pointer));
       index_ptr := nil;
     end;    { with }
  end;    { destroy }
```

There is an important point here: Procedures "GetMem" and "FreeMem" take two parameters, the first a pointer variable, the second a word specifying how many *bytes* of memory to allocate/deallocate. As most types are larger than one byte, it is very good practice *always to use the "SizeOf" function to compute how many bytes are required.*

By the way, did you notice `SizeOf(Pointer)`? The system unit includes the type "Pointer", a pointer type that points to nothing. Perhaps we'll find a use for variables of this type later.

You must understand the procedures "create" and "destroy" (without the bookkeeping distractions) in our program "matrices"; the remaining procedures are pretty routine, one for multiplying two matrices and one for displaying a matrix:

```
procedure  product(var  A, B, C: matrix);

   var  i, j, k: Word;
           sum: Integer;

   begin
   if  A.no_cols <> B.no_rows  then  Halt;
   create(C, A.no_rows, B.no_cols);
   for  i := 1  to  A.no_rows  do
      for  k := 1  to  B.no_cols  do
         begin
         sum := 0;
         for  j := 1  to  A.no_cols  do
            sum := sum + A.index_ptr^[i]^[j]
                           *B.index_ptr^[j]^[k];
         C.index_ptr^[i]^[k] := sum;
         end;
   end;    { product }

procedure  print(var  A: matrix;  ss: string);

   var  i, k: Word;

   begin
   with  A  do
```

```
      begin
      WriteLn(ss);
      for  i := 1  to  no_rows  do
         begin
         for  k := 1  to no_cols  do
            Write(index_ptr^[i]^[k]:4);
         WriteLn;
         end;
      end;     { with }
   end;     { print }
```

Finally, the main action of program "matrices" demonstrates all of this:

```
begin    { matrices }
inds_size := 0;   elems_size := 0;
m := MemAvail;
WriteLn('Before allocation, free: ', m:8, ' bytes');
Randomize;
create(A, 4, 5);
create(B, 5, 3);
product(A, B, C);
n := MemAvail;
WriteLn('After allocation, free:  ', n:8, ' bytes');
WriteLn('Allocated:               ', m-n:8,
        ' bytes');
WriteLn('Pointers:                ',
        inds_size:8, ' bytes');
WriteLn('Matrix elements:         ',
        elems_size:8, ' bytes');
m := m - n - inds_size - elems_size;
WriteLn('Overhead:                ', m:8, ' bytes');
print(A, 'A = ');   print(B, 'B = ');
print(C, 'A*B = ');
destroy(C);
destroy(B);
destroy(A);
m := MemAvail;
WriteLn('After release, free:     ', m:8, ' bytes');
ReadLn;
end.    { matrices }
```

Test "matrices" both in real and in protected modes.

Exercises

1. If A is an m×n matrix, v an m-term row vector, w' an n-term column vector, then vAw' is a scalar. Write a function to output this value.
2. An n×n symmetric matrix does not need n^2 storage elements, only $n(n+1)/2$. Revise the unit "matrices" to handle symmetric matrices. You should include a function for the value of the quadratic form vAv'.

Section 7.   Linear Lists

Records and pointers reach their full bloom with their application to data structures like linear lists and trees. A *linear list*, or just *list*, is a chain of records where each record contains, besides its useful data, a pointer to the next record in the chain. At the head of the list is a pointer, often called "root" that points to the first record in the list. The pointer in the last record in the list is usually set to **nil**, to signal the end of the list. We need operations to insert a record into a list and to delete a record from a list; they require some operations (like plumbing) to connect pointers to their correct targets.

We shall write a unit for handling lists. With (sparce) polynomials in mind, we shall make the data a word and a real, but it will be clear that the procedures in the unit could be adapted to any list. Let us start with the interface:

Program 3.7.1

```
unit   list;    { 08/09/94 }


interface

    type   node_ptr = ^node;
            node = record
                        coefficient: Real;
                            degree: Word;
                                next: node_ptr;
                end;

    var   root: node_ptr;

    procedure   create_list(n: Word);
    procedure   destroy_list;
    procedure   insert_after(var   q, s: node_ptr);
    procedure   insert_before(var   q, s: node_ptr);
    procedure   traverse_insert_before
                    (var q, s: node_ptr);
    procedure   insert_at_end(var   s: node_ptr);
    procedure   remove_after(q: node_ptr);
    procedure   remove(var q: node_ptr);
```

As usual, we first declare the pointer type "node_ptr", then declare "node". The procedure "create" sets up and loads a list of n nodes with "root" at its head. The procedure "destroy" recycles the memory allocated by "create". "Root" doesn't need a size parameter because **nil** signals the end of the list if we traverse it.

implementation

```
procedure  create_list(n: Word);
   { Creates a list of n random nodes,
     each added to the head of the list }

   var  temp: node_ptr;
          k: Word;

   begin
   root := nil;
   for  k := 0  to  n - 1  do
      begin
      temp := root;
      New(root);
      with  root^  do
         begin
         coefficient := 100.0*Random - 50.0;
         degree := k;
         next := temp;
         end;
      end;
   end;    { create_list }

procedure  destroy_list;

   var  temp: node_ptr;

   begin
   while  root <> nil  do
      begin
      temp := root^.next;
      Dispose(root);
      root := temp;
      end;
   end;    { destroy_list }
```

Note how "create" operates; it adds each new node to the head of the list.

We have several insert procedures. The first, "insert_after", inserts an external node s^ into the list immediately after a node q^ pointed to by a node_ptr q. Note its exit conditions: s = **nil**, and q still points to the same node.

```
procedure  insert_after(var  q, s: node_ptr);
     { q points to a node in the list;  s points to a
       node not in the list.  Insert s^ between q^ and
       its successor.  q should be unchanged, }

     begin
     s^.next := q^.next;  q^.next := s;
     s := nil;
     end;
```

The "insert_before" procedure does magic; given q pointing to a node in the list; to insert s^ *before* q^. How on earth to move backwards? Don't; switch the *data* in the records. This code is worthy of careful study:

```
procedure  insert_before(var  q, s: node_ptr);
{ q points to a node in the list;
  s points to a node not in the list.
  To insert s^ between the predecessor of q^ and q^.
     On exit, q points to the new node. }

     var  temp: node_ptr;

     begin
     New(temp);
     { Put contents of q^ into temp^ }
     temp^ := q^;
     { Put contents of s^ into q^ }
     q^ := s^;
     q^.next := temp;
     Dispose(s);  s := nil;
     end;   { insert_before }
```

A more obvious insert method is to traverse the list from its root to locate the node before the insertion point. Of course, with a long list this can be costly, however, we should see how it is programmed:

```
procedure  traverse_insert_before
              (var q, s: node_ptr);
{ q points to a node in the list;
    s points to a node not in the list.
```

To insert s^ between the predecessor of q^ and q^.
On exit, q points to the new node. }

```
    var   run: node_ptr;

    begin
    if  q = root   then
       begin
       s^.next := root;   root := s;   q := s;
       s := nil;
       end
    else
       begin
       run := root;
       while  run^. next <> q  do
          run := run^.next;
       { Now run points to the node before q }
       insert_after(run, s);
       q := run^.next;
       end;
    end;    { traverse_insert_before }
```

Finally, to insert a node at the end of the list, we are forced to traverse the list
first to locate a pointer to its end:

```
    procedure  insert_at_end(var  s: node_ptr);
       { Inserts  s^ at the end of the list }

    var   run: node_ptr;

    begin
    if  root = nil   then   root := s
    else
       begin
       run := root;
       while  run^.next <> nil  do
          run := run^.next;
       { Now run points to the list's final node }
       run^.next := s;
       end;
    s^.next := nil;
    end;    { insert_at_end }
```

This completes our insertion procedures; we have two deletion procedures, one to delete the node after a node, the other to delete the node itself:

```
procedure  remove_after(q: node_ptr);
    { q points to a node in the list.
      To remove the node after q }

    var  temp: node_ptr;

    begin
    temp := q^.next;
    if  temp <> nil   then
       begin
       q^.next := temp^.next;
       Dispose(temp);
       end
    end;

procedure  remove(var q: node_ptr);
    { q points to a node in the list.
      To remove the node q^.
      On exit, q = nil }

    var  temp: node_ptr;

    begin
    temp := q^.next;
    if  temp <> nil   then
       begin  q^ := temp^;  Dispose(temp);  end
    else  if  q = root   then    { q^.next = nil }
       begin  root := nil;  Dispose(q);  end
    else   { q <> root, q^ = final node in the list }
       begin   { traverse the list }
       temp := root;
       while  temp^.next <> q  do
          temp := temp^.next;
       { Now temp^ = immediate predecessor of q^ }
       remove_after(temp);
       end;
    q := nil;
    end;

end.   { unit lists }
```

Exercises

1. Write a program to test all the procedures in the unit "list". (This is really about 8 exercises.)

2. A *double linked list* has a pointer "head" replacing "root" and a pointer "tail" at the other end. Each node has two pointers, one pointing forwards, the other backwards. Write a unit for inserting and removing nodes from a double linked list.

3. If you close up a double linked list, you have a *circular list*. Write a unit for handling circular lists. See if you can splice two circular lists together.

4. Start from scratch, all by yourself, to build confidence in your list skills: write a program that makes a list whose data consists of a key from 32 to 255 and the corresponding ASCII character. There should be a procedure for searching the list for the record with a given key, and printing the character.

5. Add to the unit "list"
    ```
    procedure skip(var  p, q: node_ptr);
    ```

Here p and q are pointers in the list, q a descendent of p. The procedure should clobber all nodes between p and q, leave p pointing to where q was pointing, and push q to its successor.

Section 8.   Sparce Polynomials

Our first example, merely to soften up the reader, is the beginning of a unit to handle sparce complex polynomials. In dealing with polynomials, if you do not know in advance an absolute bound on degrees, there is no way to use static memory, i.e., arrays. A linked list really seems the natural data structure. The following unit "sparpoly" uses the unit "complexs" of Section 4. Everything in the unit should seem pretty routine to you; an exercise later will request a test of the unit.

Program 3.8.1

```
{$N+}
unit  sparpoly;    { 03/10/95 }
   { For sparce complex polynomials }

interface

uses  complexs;

var  out_file: Text;
```

```
type   sparce_poly = ^term;
            term = record
                degree: Word;
                 coeff: complex;
                  next: sparce_poly;
            end;
        { A sparce_poly is a linked list of terms,
          in order of decreasing degree. }

function   degree(P: sparce_poly): Integer;
procedure  insert_term(t: term;  var  P: sparce_poly);
procedure  evaluate(P: sparce_poly;  z0: complex;
                    var  w: complex);

implementation

procedure  initiate(var  r: sparce_poly);

   begin
   New(r);
   r^. next := nil;
   end;

function  degree(P: sparce_poly): Integer;

   begin
   if  P = nil  then  degree := -1
   else  degree := P^.degree;
   end;

procedure  insert_term(t: term;  var  P: sparce_poly);
   { Inserts term t in the sparce_poly  P.  Overwrites
     the previous term of that degree. }

   var  run, q: sparce_poly;

   begin
   if  P = nil  then
      begin  initiate(P);  P^ := t;  end
   else   { P <> nil }
      begin
      run := P;
      while  (run^.degree > t.degree)
             and (run^.next <> nil)  do
          run := run^.next;
      if  run^.degree < t.degree  then
```

```
        begin    { t inserted before  run^ }
        New(q);
        q^ := run^;
        run^ := t;
        run^.next := q;
        end
   else if  run^.degree = t.degree  then
        run^.coeff := t.coeff   { Replace run^.coeff }
   else
   { run^.degree > t.degree  and  run^.next = nil }
        begin    { q^ inserted at end }
        New(q);
        q^ := t;
        q^.next := nil;
        run^.next := q;
        end;
      end;    { P <> nil }
    end;    { insert_term }

procedure  evaluate(P: sparce_poly;   z0: complex;
                    var  w: complex);
{ Computes  P(z0),  where  P  is a sparce complex
  polynomial and  z0  is any complex number.
  Method: Horner's procedure }

   var  gap: Word;
         t: complex;
        run: sparce_poly;

   begin
   if  P = nil  then  w := zero
   else  w := P^.coeff;
   run := P;
   while  run <> nil  do
      begin
      if  run^.next = nil  then
         gap := degree(run)
      else  gap := degree(run) - degree(run^.next);
      { Now multiply  w  by  z0^gap }
      t := z0;
      while  gap > 0  do
         begin
         if  Odd(gap)  then
            begin  product(w, t, w);  Dec(gap);  end
         else
```

```
      begin
      product(t, t, t);  gap := gap shr 1;
      end;
   end;
 run := run^.next;
 if  run <> nil  then  sum(w, run^.coeff, w);
 end;   { while }
end;   { evaluate }

end.   { unit  sparpoly }
```

Exercises

1. Write a program for testing the unit "sparpoly".

Section 9.   Real Polynomials

Now we get down to real work. The following example will be a set of procedures for doing all sorts of manipulations with real polynomials with any number of terms of any degrees. Therefore a list structure is indicated. As above, a polynomial will be a pointer to a term: a record of a degree, a coefficient, and a pointer to the next term. The final term has the pointer **nil**, which serves as a sentinel. Note the gaps; terms with zero coefficient do not appear. Indeed, there is a small positive real constant eps used to purge each term with coefficient satisfying $|c| <$ eps. The polynomial P = **nil** is interpreted as the zero polynomial. We begin the unit "realpoly" with its interface:

Program 3.9.1

```
{$N+}
unit  realpoly;   { 08/10/94 }

interface

   type       Real = Extended;
        polynomial = ^term;
            term = record
                        deg: Integer;
                      coeff: Real;
                       next: polynomial
                  end;
    { A polynomial is a linked list of terms,
      in descending degree from the root. }
```

```pascal
function  degree(p: polynomial): Integer;
function  leading_coeff(p: polynomial): Real;
procedure insert(t: term;  var  p: polynomial;
                  augment: Boolean);
{ Inserts the term  t  into the polynomial p. If a
  term of that degree is already in p, then it is
  replaced (augment false) or added in (true). }
procedure load(var  p: polynomial);
procedure recycle(var  p: polynomial);
procedure purge(var  p: polynomial);
{ Deletes all terms with negligible coefficient }
function  copy_poly(p: polynomial): polynomial;
procedure print(p: polynomial);
function  mult_by_scalar(c: Real;  p: polynomial):
                  polynomial;
function  sum(p, q: polynomial): polynomial;
function  product(p, q: polynomial): polynomial;
function  derivative(p: polynomial): polynomial;
procedure division(f, g: polynomial;
                  var  q, r: polynomial);
function  gcd(f, g: polynomial): polynomial;
function  value(p: polynomial;  c: Real): Real;
procedure evaluate(p: polynomial;  c: Real;
                  var  D0, D1: Real);
procedure eval_der2(p: polynomial;  c: Real;
                  var  D0, D1, D2: Real);
procedure find_zero(f: polynomial);
```

The first two procedures are routine; let's get them out of the way:

implementation

```pascal
function  degree(p: polynomial): Integer;

   begin
   if  p = nil  then  degree := -1
   else  degree := p^.deg;
   end;


function  leading_coeff(p: polynomial): Real;

   begin
   if  p = nil  then  leading_coeff := 0.0
   else  leading_coeff := p^.coeff;
   end;
```

The "insert" procedure is somewhat longer, mainly because there are two cases: add the new term to what is there already, or replace the existing term:

```
procedure  insert(t: term;  var  p: polynomial;
                  augment: Boolean);
   { Inserts the term t into the polynomial p }

   var  run, temp: polynomial;

   begin
   if  p = nil  then
      begin  New(p);  p^ := t;  p^.next := nil;  end
   else   { p <> nil }
      begin
      run := p;
      while  (run^.deg > t.deg)
             and (run^.next <> nil)  do
         run := run^.next;
      if  run^.deg < t.deg  then
         begin   { Insert t before run^ }
         New(temp);
         temp^ := run^;
         run^ := t;
         run^.next := temp;
         end
      else  if  run^.deg = t.deg  then
         begin   { Replace run^.coeff }
         if  augment  then
            run^.coeff := run^.coeff + t.coeff
         else  run^.coeff := t.coeff;
         end
      else
      { run^.deg > t.deg  and  run^.next = nil }
         begin   { t inserted at end }
         New(run^.next);
         run := run^.next;
         run^ := t;
         run^.next := nil;
         end;
      end;    { p <> nil }
   end;    { insert }
```

(You may remember that "Insert" is a standard identifier in Borland Pascal. No matter; there is nothing illegal about redefining identifiers. We can still use the built-in "Insert" after defining our "insert" by referring to it as "system.insert".)

This brings us to two critical procedures: "load" to create a new polynomial and "recycle" to get rid of a polynomial. Let us look at "load":

```pascal
procedure  load(var  p: polynomial);

   var  t: term;  k: Integer;

   begin
   p := nil;
   t. next := nil;
   repeat
      Write
        ('Enter degree of new term (-1 to quit): ');
      ReadLn(k);
      if  k >= 0  then
         begin
         with  t  do
            begin
            deg := k;
            Write('Enter coefficient: ');
            ReadLn(coeff);
            end;   { with }
         insert(t, p, false);
         end;   { if k }
   until  k < 0;
   end;   { load }
```

The implementation of "recycle" is remarkably brief; that is because it uses a really important feature of Pascal that we have not used so far, *recursion*. This means a procedure calling itself. To recycle a polynomial P, first recycle P^.next, then dispose of P:

```pascal
procedure  recycle(var  p: polynomial);

   begin
   if  p <> nil  then  recycle(p^.next);
   Dispose(p);
   p := nil;
   end;   { recycle }
```

This short code ranks high on the careful study list.

The "purge" procedure, which gets rid of terms with negligible coefficients, is pretty humdrum. We know by now how to traverse a linked list looking for something:

```
procedure  purge(var  p: polynomial);
{ Deletes all terms with negligible coefficient }

    const  min_eps = 1.0e-50;

    var  run1, run2: polynomial;

    begin
    if  p <> nil  then
       begin
       run1 := p;  run2 := p^.next;
       while  run2 <> nil  do
          begin    { Purge non-leading terms }
          if  Abs(run2^.coeff) < min_eps  then
             begin
             run1^.next := run2^.next;
             run2^.next := nil;
             Dispose(run2);
             run2 := run1^.next;
             end   { if abs }
          else
             begin
             run1 := run2;  run2 := run2^.next;
             end;
          end;    { while }
       if  Abs(p^.coeff) < min_eps  then
          begin   { Purge leading term }
          run1 := p;  p := p^.next;
          run1^.next := nil;
          Dispose(run1);
          end;
       end;    { if p <> nil }
    end;   { purge }
```

The procedure "copy_poly" for creating a clone of a polynomial is our second example of a recursive call. It is important to understand that when you make a copy of a linked list, you must copy all of its nodes, not just the pointer to the head of the list.

```pascal
function  copy_poly(p: polynomial): polynomial;

    var  s: polynomial;

    begin
    if  p = nil  then   s := nil
    else
        begin
        New(s);
        s^ := p^;
        s^.next := copy_poly(p^.next);
        end;
    copy_poly := s;
    end;    { copy_poly }
```

This brings us back to the mundane, an output procedure:

```pascal
procedure  print(p: polynomial);

    var  run: polynomial;  k: Integer;

    begin
    if  p = nil then   WriteLn(0:2)
    else
        begin
        with  p^  do
            begin
            Write(coeff:5:5);
            if  deg > 0  then   Write(' x^', deg);
            run := next;
            end;    { with }
        k := 4;
        while  run <> nil  do
            begin
            Dec(k);
            if  k = 0   then
                begin
                k := 5;   WriteLn;   Write(' ':15);
                end;
            with  run^  do
                begin
                if  coeff < 0.0  then
```

```
                  Write(' - ', -coeff:5:5)
            else  Write(' + ', coeff:5:5);
            if  deg > 0  then  Write(' x^' , deg);
            run := next;
            end   { with }
         end;   { while }
      WriteLn;
      end   { else }
   end;   { print }
```

Now we come to algebraic operations on polynomials.   The first two,
"mult_by_scalar" and "sum" are straightforward:

```
function  mult_by_scalar(c: Real;  p: polynomial):
                  polynomial;

   var  r, s: polynomial;

   begin
   s := copy_poly(p);  r := s;
   while  r <> nil  do
      with  r^  do
         begin  coeff := c*coeff;  r := next  end;
   purge(s);
   mult_by_scalar := s;
   end;

function  sum(p, q: polynomial): polynomial;

   var  p_run, s: polynomial;
             t: term;

   begin
   s := copy_poly(p);  p_run := q;
   while  p_run <> nil  do
      begin
      t := p_run^;
      t.next := nil;
      insert(t, s, true);
      p_run := p_run^.next;
      end;
   purge(s);
   sum := s;
   end;
```

"Product" is more complicated because the factors must be traversed independently. Multiplication also motivates the use of linked lists rather than arrays for polynomials. No matter how large an array for storing the terms of a polynomial, a product of two polynomials requires more terms.

```
function  product(p, q: polynomial): polynomial;

    var  p_run, q_run, s: polynomial;
                        t: term;

    begin
    s := nil;
    t. next := nil;
    p_run := p;
    while  p_run <> nil  do
        begin
        q_run := q;
        while  q_run <> nil  do
            begin
            t.deg := p_run^.deg + q_run^.deg;
            t.coeff := p_run^.coeff*q_run^.coeff;
            insert(t, s, true);
            q_run := q_run^.next;
            end;    { while  q_run }
        p_run := p_run^.next;
        end;    { while  p_run }
    purge(s);
    product := s;
    end;    { product }
```

Long division of one polynomial by another, producing a quotient and a remainder is another important polynomial operation. It is very easy in this code to make a mistake resulting in dangling pointers (like a land mine ready to explode at any moment). So read it carefully.

```
procedure  division(f, g: polynomial;
                    var  q, r: polynomial);

    var   run, h,
            temp: polynomial;
                t: term;
          gap, d: Word;
               a: Real;
```

```
begin
q := nil;   d := degree(g);
r := copy_poly(f);
while  degree(r) >= d  do
   begin
   gap := degree(r) - d;
   a := leading_coeff(r)/leading_coeff(g);
   { H(x) := - a*(x^gap)*G(x) }
   h := copy_poly(g);   run := h;
   repeat
      with  run^  do
         begin
         deg := gap + deg;
         coeff := -a*coeff;
         run := next;
         end
   until  run = nil;
   { R(x) := R(x) + H(x), lowering degree }
   temp := sum(r, h);
   recycle(r);   recycle(h);
   purge(temp);   r := temp;
   { Add a*x^gap to Q(x) }
   t.deg := gap;   t.coeff := a;
   insert(t, q, false);
   end;   { while }
end;   { division }
```

The Euclidean algorithm for the greatest common divisor goes hand-in-hand with long division:

```
function  gcd(f, g: polynomial): polynomial;

   var  f1, g1, q, r: polynomial;

begin
f1 := copy_poly(f);   g1 := copy_poly(g);
while  g1 <> nil  do
   begin
   division(f1, g1, q, r);
   recycle(q);
   recycle(f1);
   f1 := g1;
   g1 := r;
   end;
gcd := mult_by_scalar(1.0/leading_coeff(f1), f1);
```

```
            recycle(f1);   recycle(g1);
            end;    { gcd }
```

Our final operation is taking the derivative of a polynomial to produce another polynomial:

```
    function  derivative(p: polynomial): polynomial;

        var   run: polynomial;

        begin
        if  degree(p) = 0  then   derivative := nil
        else
            begin
            run := copy_poly(p);
            derivative := run;
            repeat
               with  run^  do
                   begin
                   coeff := deg*coeff;  Dec(deg);
                   if  next = nil  then  Exit;
                   if  next^.deg = 0  then
                       begin
                       Dispose(next);  next := nil;
                       end;
                   run := next;
                   end;    { with }
            until  run = nil;
            end;    { else }
        end;    { derivative }
```

Next we have three procedures for evaluating a polynomial, of increasing complexity. The first merely evaluates F(c) by what is known as the Horner algorithm. The second evaluates this and F'(c), and the third also produces the value of the second derivative F''(c), both by modified Horner. The procedures are efficient and interesting. Note that the symbolic derivatives are not computed first.

```
    function  value(p: polynomial;   c: Real): Real;

        var   i, k: Integer;
                 s: Real;
              run: polynomial;
```

```
begin
if  p = nil  then  s := 0.0
else  s := p^.coeff;
run := p;
while  run <> nil  do
   begin
   if  run^.next = nil  then  k := degree(run)
   else  k := degree(run) - degree(run^.next);
   for  i := k  downto  1  do  s := s*c;
   run := run^.next;
   if  run <> nil  then  s := s + run^.coeff;
   end;   { while }
value := s;
end;   { value }
```

```
procedure  evaluate(p: polynomial;  c: Real;
                    var  D0, D1: Real);

   var  i, gap: Integer;
        run: polynomial;

   begin
   D1 := 0;
   if  p = nil  then
      begin  D0 := 0.0;  Exit;  end;
   D0 := p^.coeff;
   run := p;
   while  run  <>  nil  do
      begin
      if  run^.next = nil  then
         gap := degree(run)
      else  gap := degree(run) - degree(run^.next);
      for  i := gap  downto  1  do
         begin
         D1 := D1*c + D0;
         D0 := D0*c;
         end;
      run := run^.next;
      if  run <> nil  then  D0 := D0 + run^.coeff;
      end;   { while }
   end;   { evaluate }
```

```
procedure  eval_der2(p: polynomial;  c: Real;
                     var  D0, D1, D2: Real);
```

```
var   i, gap: Word;
         run: polynomial;

begin
D1 := 0.0;  D2 := 0.0;
if  p = nil  then
   begin  D0 := 0.0;   Exit;   end;
D0 := p^.coeff;
run := p;
while  run <> nil  do
   begin
   if  run^.next = nil   then
      gap := degree(run)
   else  gap := degree(run) - degree(run^.next);
   for  i := gap  downto  1  do
      begin
      D2 := D2*c + 2.0*D1;
      D1 := D1*c + D0;
      D0 := D0*c;
      end;
   run := run^.next;
   if  run <> nil  then  D0 := D0 + run^.coeff;
   end;    { while }
end;    { eval_der2 }
```

The final procedure of our unit "realpoly" locates a zero of a polynomial by Newton steps. Of course the procedure "evaluate" is used for values $F(c)$ and $F'(c)$.

```
procedure  find_zero(f: polynomial);

var    c1, c2,
     eps, eta,
        v, d: Real;
     max, n: Integer;

begin
WriteLn('Newton''s method');
WriteLn('================');
WriteLn('Two stopping constants, ' +
        'eps and eta, required.');
WriteLn('Search stops if');
WriteLn('a)   |F''(c)| < eta');
WriteLn('b)   |c_n+1 - c_n| < eps');
WriteLn
```

```
       ('Defaults:  eps = 1.0e-10   eta = 1.0E-3: ');
    WriteLn(
      'Enter reals  eta and eps  (0 for defaults):');
    Write('eta = ');  ReadLn(eta);
    Write('eps = ');  ReadLn(eps);
    if  eta <= 0.0  then  eta := 1.0e-3;
    if  eps <= 0.0  then  eps := 1.0e-10;
    Write('Enter max. no. of Newton steps: ');
    ReadLn(max);
    Write('Enter initial approximate zero: ');
    ReadLn(c1);
    n := 0;
    repeat    { until  n >= max }
       evaluate(f, c1, v, d);
       if  Abs(d) < eta  then
          begin
          WriteLn('f''(', c1:5:5, ') = ', d:8,
                  ': process halted.');
          Break;
          end
       else    { abs(d) >= eta }
          begin
          c2 := c1 - v/d;
          if  Abs(c2 - c1) < eps  then
             begin
             Write(
               'After ', n + 1,' iterations,  F(');
             WriteLn(c2:8:8, ') = ',
                     value(f, c2):10:10);
             Break;
             end
          else
             begin  c1 := c2;  Inc(n);  end;
          end;    { else  abs(d) >= eta }
    until  n >= max;
    if  n = max  then
       begin
       WriteLn('Newton failed to converge after ',
               max,' iterations.');
       WriteLn('Last  F(', c2:5:5, ') = ',
       value(f, c2):5:5);
       end;
    end;    { find_zero }

end.   { unit realpoly }
```

Exercises

1. Write a program for testing the unit "realpoly". It should be user-friendly, and it should verify that the memory management has been done correctly.
2. Program "recycle" without recursion.
3. Program "copy_poly" without recursion.
4. Prove that the Horner algorithm in "value" is correct. Hint: Set

$$F_k(x) = a_k + a_{k-1}x + \cdots + a_0 x^k$$

Prove the iteration

$$F_0(x) = a_0 \qquad F_{k+1}(x) = xF_k(x) + a_k$$

5. Prove that the algorithms in "evaluate" and "eval_der2" are correct. Hint: Differentiate the iteration in the previous exercise.

Section 10.   Addressing Memory

This brief section deals with several Borland Pascal operators used to address memory or gather information about memory and addresses of various objects: *Addr, Seg, Ofs, Ptr*, @, @@. Recall some basic DOS facts. Memory is a sequence of bytes. Each byte has an address, given by a segment address and an offset address. A segment is 64K bytes. A typical memory address is $ssss:$oooo. The segment part of the address $ssss is multiplied by $10 (16) to get the starting byte of the segment, and the offset part of the address $oooo is added to that to get the actual address in bytes. So actually, each byte of memory may have many different segment:offset forms of its address. For instance

```
$1000:$0000  =  $0FFF:$0010  :
    $10000          $0FFF0
  +$ 0000         +$ 0010
   -------         -------
    $10000          $10000
```

The range of addreses for (conventional) memory is $0 to $FFFFF, exactly $100000 = 1M.

Each Pascal **var** starts at an address in memory of its lowest byte. The "Addr" function creates a pointer to that address, as does essentially the same @ operator. The "Seg" and "Ofs" functions return words, the segment and offset of a variable. "Ptr" is an inverse to ("Seg", "Ofs"), creating a pointer to the memory location with a given segment:offset. Each unit and the main program have their own data segments, and addresses are relative to these segments. In protected mode, memory from all over the map is mapped by the protected mode manager into conventional segment addresses, and you absolutely do not have to worry about this.

Procedures and functions have entry points, and the operators above apply to these addresses also. (In DOS, a pointer to a procedure is called a *vector*.) The following example illustrates (I hope) all possibilities, including the mysterious @@ operator. We'll list it in pieces, with some discussion between the pieces.

Program 3.10.1

```
program  test_address_ops;    { 08/17/84 }

   type   proc = procedure(var  x: Real);

   var        p: Pointer;
              x: Real;
        my_proc: proc;
          dword: array[0..1] of Word;

   procedure  zilch(var  x: Real); far;

      begin  x := 2*x;   end;
```

The **procedure** type proc is set up so we can apply later tests to the variable my_proc of that type. The **var** x and the **procedure** zilch are for testing addresses.

The first test computes Seg(x) and Ofs(x). Then it assigns the pointer P to Addr(x) and checks that p^ is indeed x. Finally it does the same thing with the operator @ instead of the function Addr.

```
begin
WriteLn(
'Test 1.  Compare @ and Addr on variable  x:');
WriteLn('  (Seg(x), Ofs(x))   = (',
        Seg(x), ', ', Ofs(x), ')');
p := Addr(x);
WriteLn('p := Addr(x):');
WriteLn('  (Seg(p^), Ofs(p^)) = (',
        Seg(p^), ', ', Ofs(p^), ')');
p := @x;
WriteLn('p := @x:');
WriteLn('  (Seg(p^), Ofs(p^)) = (',
        Seg(p^), ', ', Ofs(p^), ')');
WriteLn;
Write('Press <Enter> to continue: ');   ReadLn;
WriteLn;
```

The second test shows the use of the function Ptr:

```
WriteLn(
'Test 2.   Create a pointer to  x  with Ptr:');
p := @x;
WriteLn('  (Seg(x), Ofs(x))    = (',
        Seg(x), ', ', Ofs(x), ')');
p := Ptr(Seg(x), Ofs(x));
WriteLn('p := Ptr(Seg(x), Ofs(x))');
WriteLn('  (Seg(p^), Ofs(p^)) = (',
        Seg(p^), ', ', Ofs(p^), ')');
WriteLn;
Write('Press <Enter> to continue: ');   ReadLn;
WriteLn;
```

Test 3 locates the entry point of procedure "zilch" both by means of Addr and of @. It also shows a third way to compute the entry point: apply Seg and Ofs to zilch directly:

```
WriteLn(
'Test 3.   Compare @ and Addr on procedure  zilch:');
WriteLn(
'The address of a procedure: its entry point.');
p := Addr(zilch);
WriteLn('p := addr(zilch): ' +
        '-- points to the entry point of zilch:');
WriteLn('  (Seg(p^), Ofs(p^)) = (',
        Seg(p^), ', ', Ofs(p^), ')');
p := @zilch;
WriteLn('p := @zilch -- ' +
        'points to the entry point of zilch:');
WriteLn('  (Seg(p^), Ofs(p^)) = (',
        Seg(p^), ', ', Ofs(p^), ')');
WriteLn('The same address is obtained from ' +
        'Seg and Ofs of zilch itself:');
WriteLn('  (Seg(zilch), Ofs(zilch)) = (',
        Seg(zilch), ', ', Ofs(zilch), ')');
WriteLn;
Write('Press <Enter> to continue: ');   ReadLn;
WriteLn;
```

The final test uses the @@ operator. The @ operator on a procedure variable locates its entry point. This entry point address is stored someplace, and @@ creates a pointer to that place.

```
   WriteLn('Test 4  Compare @ and @@ on ' +
                 'procedure variable my_proc:');
   WriteLn('A procedure variable, like any variable,' +
               ' is stored at');
   WriteLn(
'an address, accessed by the @@ operator -- because');
   WriteLn(
'the @ operator yields the procedure''s entry point.');
   my_proc := zilch;
   p := @my_proc;
   WriteLn(
'p := @my_proc -- points to the entry point of');
   WriteLn(
'                  procedure my_proc:');
   WriteLn('  (Seg(p^), Ofs(p^))  = (',
           Seg(p^), ', ', Ofs(p^), ')');
   p := @@my_proc;
   WriteLn(
'p := @@my_proc -- points to the procedure variable');
   WriteLn('                      my_proc --  where ' +
           'the entry point of');
   WriteLn(
'                  the procedure is stored:');
   WriteLn('  (Seg(p^), Ofs(p^))  = (',
           Seg(p^), ', ', Ofs(p^), ')');

   WriteLn('Proof of the pudding: 2 words of p^:');
   p := Ptr(Seg(p^), Ofs(p^));
   Move(p^, dword, 4);
   WriteLn('(high word, low word) = (',
           dword[1], ', ', dword[0], ')');
   WriteLn('Note:  (Seg(p), Ofs(p)) = (',
           Seg(p), ', ', Ofs(p), ')');
   WriteLn;
   WriteLn('For your interest:  SelectorInc = ',
           SelectorInc);
   WriteLn;
   Write('Press <Enter> to exit: ');  ReadLn;
   WriteLn;
   end.  { test_address_ops }
```

Well, as they say: "There it is; enjoy." Actually, I haven't quite told the whole truth, at least not for protected mode. In protected mode, Seg does not return the segment of an object, but something called its "Selector", an index into a table where actual segments are stored. That is part of the way vast amounts of memory can be addressed in protected mode. You should *never* change a segment in a protected mode program, as that will almost surely lead to a fatal error, and often a hung computer. The selectors in a protected mode program always differ by a fixed increment, held by the system unit's variable *SelectorInc*. To increase (decrease) a pointer by 64K, you add to (subtract from) the pointer's selector this amount. Look at "SelectorInc" if you are interested (done above), but don't fool around with it.

Chapter 4 DOS and BIOS

Section 1. The DOS Unit

The DOS unit provides many procedures for direct access to various device drivers, particularly disk drives. We used one of these earlier, "GetTime", to read the system clock. Now we are going to look at some of the procedures for handling disk directories and files.

The DOS unit defines a type *PathStr* as follows:

```
type   PathStr: string[79];
```

The quite useful function *FExpand* expands a file name into a full path name. In the DOS unit, its declaration is

```
function  FExpand(path: PathStr): PathStr;
```

For example:

```
FExpand('my_file.pas') = 'E:\SCIPAS\CH3\MY_FILE.PAS'
FExpand('..\help.exe') = 'F:\FORTRAN\HELP.EXE'
```

Note that the result is in upper case, and that the parent directory reference .. is removed.

In addition to "PathStr", the DOS unit defines other string types:

```
ComStr  = string[127];   { Command line string }
DirStr  = string[67];    { Drive and directory }
NameStr = string[8];     { File name }
ExtStr  = string[4];     { Extension, including '.' }
```

The DOS unit procedure *FSplit* splits a PathStr into its components:

```
procedure   FSplit(path: PathStr;
                    var   dir: DirStr;
                    var   name: NameStr;
                    var   ext: ExtStr);
```

There are two procedures for searching a directory for a file. Each returns a variable of type *SearchRec* with complete directory information about the file:

```
type   SearchRec = record
                       fill: array[1..21] of Byte;
                       attr: Byte;
                       time,
                       size: LongInt;
                       name: string[12];
                   end;
```

The "fill" field is reserved by DOS; don't touch it. The "time" field contains the file date and time, packed into a 4-byte LongInt. If you are interested, the procedure *UnpackTime* will access it, and there are other procedures for setting the time, etc; we'll leave this one alone too. The "size" and "name" fields have their obvious meanings. Size is in bytes of course. "Name" really means an up to 8 character pre-name and an up to 4 character extension, starting with '.'.

The "attr" byte is quite useful, and its bits tell you what kind of directory entry you are dealing with: subdirectory, hidden, archive, etc. Several bits may be set because a file can have several attributes. Meanings of the individual bits are in the following table. The DOS unit also defines constants with values corresponding to these bits set, and we list them also:

```
Bit         Attribute          const
 0          read only          ReadOnly    = $01;
 1          hidden             Hidden      = $02;
 2          system file        SysFile     = $04;
 3          volume name        VolumeID    = $08;
 4          directory          Directory   = $10;
 5          archive bit        Archive     = $20;
                               AnyFile     = $3F;
```

The final entry "AnyFile" is the sum of all the others, i.e, bits 0 – 5 are all set. It is useful for running over all the entries in a directory. (Of course, "VolumeID" will only be found in a root directory.)

Your search of a directory begins with a call to *FindFirst*, so let's look at its DOS unit declaration:

```
procedure  FindFirst(path: string;   attr: Word;
                     var  srec: SearchRec);
```

After the first file in a directory is found, the search continues with repeated calls of *FindNext*:

```
procedure  FindNext(var  srec: SearchRec);
```

Some DOS unit procedures return a nonzero *IOResult* in case of failure; others return a nonzero *DOSError*; it is hard to know which to expect. The **function** "IOResult" is in the "System" unit; the **var** "DOSError" in the "DOS" unit. When IOResult is called, the internal error flag is reset, and further I/O operations can be made. But (with the I– compiler option), if there is an I/O error and "IOResult" is not called, your program will hang at the next I/O operation.

Appendix 1 describes briefly three programs included on your disk: "destroy.pas", "size.pas", and "se-re.pas". All use the material just covered. We shall shortly look at a related program; its purpose is to delete all empty files in a directory tree. If you are not thoroughly confused at this point, then note that disk procedures that execute standard DOS commands: "Rename", "Erase", MkDir", "RmDir", and "ChDir" are in the "System" unit, not the "DOS" unit. Also in the System unit is a "GetDir" procedure that returns the current directory in a specified drive.

Let's get on to the example. As usual with a longer program, we'll take it piece by piece, starting with **uses** and **var** declarations, and a procedure that either accepts the parameter (if the program is called with one) as the name of the directory to be searched, or prompts for one:

Program 4.1.1

```
{$I-}
program  del_empty_files;   { 08/12/94 }
{ Deletes all empty files in a directory tree }

   uses   DOS, CRT;

   var   old_dir,
         new_dir: DirStr;
           level: Integer;

   procedure   get_name;

     procedure   normalize;

       begin
       { Flush blanks }
       while   new_dir[Length(new_dir)] = ' '   do
          Delete(new_dir, Length(new_dir), 1);
       while   new_dir[1] = ' '   do
          Delete(new_dir, 1, 1);
       { Change   'X:'   to   'X:\" }
       if   (Length(new_dir) = 2) and
            (new_dir[2] = ':')   then
               new_dir := new_dir + '\';
```

```
        end;    { normalize }

    begin    { get_name }
    new_dir := ParamStr(1);
    normalize;
    if  new_dir = ''  then
        begin
        ClrScr;
        WriteLn('Program  DelEmpty.exe');
        WriteLn;
        WriteLn('Usage of DelEmpty:');
        WriteLn;
        WriteLn(' 1)  DelEmpty  Path  <Enter>');
        WriteLn(
        '      Example:  DelEmpty  D:\WORD <Enter>');
        WriteLn;
        WriteLn(' 2)  DelEmpty  <Enter>');
        WriteLn('      You will be prompted for the');
        WriteLn('      directory to be searched.');
        WriteLn;
        WriteLn;
        Write(
 'Enter FULL path name of directory to be searched: ');
        ReadLn(new_dir);
        normalize;
        WriteLn;
        end;
    WriteLn;
    end;    { get_name }
```

This should seem pretty straightforward. We'll take things out of order now, and look at the main action of the program:

```
    begin    { del_empty_files }
    ClrScr;
    WriteLn('Program  delempty.exe');
    { Save the current directory name }
    GetDir(0, old_dir);
    old_dir := FExpand(old_dir);
    get_name;    { Returns new_dir }
    WriteLn;
    level := 0;
    process_directory(new_dir);
    ChDir(old_dir);
    Write('Done: Press <Enter>: ');
```

```
ReadLn;
end.   { del_empty_files }
```

Notice that the current directory is saved by a call to *GetDir*. Its first parameter 0 means the current drive. (1 means drive A:, 2 drive B:, etc.) The main action in the program is in the procedure "process_directory" which we'll come to in a moment. After it is executed, *ChDir* is called to change back to the directory from which the program was called.

"Process_directory" is called with parameter "new_dir", which it expands into a full path name and changes to that directory. Note that the System unit's "ChDir" is slightly stronger than DOS's "CHDIR" (abbreviated "CD") in that it allows a change of drive to be included. Let's examine the first 20 or so lines of "process_directory":

```
procedure  process_directory(dir: DirStr);

    var   serch_rec: SearchRec;
          io_result: Word;
               path: PathStr;
                 gg: file;
               attr: Word;

    begin
    WriteLn('Directory level = ', level);
    if   level < 0   then
       begin
       Write('Programming error; press <Enter>: ');
       Halt;
       end;
    dir := FExpand(dir);
    ChDir(dir);
    if   IOResult <> 0   then   { A safeguard! }
       begin
       WriteLn('Sorry, but ', dir, ' is not an');
       WriteLn(
    'existing directory.  Program will be halted.');
       WriteLn(
  'Check the directory name carefully and try again.');
       Write('Press <Enter>: ');
       ReadLn;
       Exit;
       end;
```

Most of "process_directory" is error messages. Keeping track via "level" of the level of subdirectory is a little insurance against incorrect programming. It could be dropped after we are sure that the program is flawless.

Next we call "FindFirst" to find the first entry in the directory. As you know, when DOS lists a directory, unless it is the root directory of a drive, the first two entries are "." and "..", and we must account for these. Let's look at that much of the procedure:

```
FindFirst('*.*', AnyFile, serch_rec);
while  DosError = 0  do
   begin
   if  serch_rec.name = '.'  then    { root }
      begin
      FindNext(serch_rec);    { '..' }
      FindNext(serch_rec);    { First real file }
      end
```

Now we come to the main parts of the procedure; first how we handle an empty file:

```
else
   begin
   { Is it an empty file ? }
   if  (level >= 0) and
       (serch_rec.attr <> $10) and
       (serch_rec.size=0)   then
      begin
      path := FExpand(serch_rec.name);
      Assign(gg, path);
      Erase(gg);
      Write(path,
             ' deleted; press <Enter>: ');
      ReadLn;
      end;
```

Note that *Erase* takes a file variable as its argument, and we code accordingly.

It remains to handle a subdirectory. This we do by a *recursive* call of "process_directory":

```
attr := serch_rec.attr;
if  attr and $10 = $10   then
   begin    { Directory }
   Inc(level);
   process_directory(serch_rec.name);
```

```
            { Recursive call! }
            Dec(level);
            end;
        FindNext(serch_rec);
          end;
      end;    { while }
  if  dir <> new_dir  then  ChDir('..');
  end;    { process_directory }
```

I hope this, with the three programs referenced in Appendix 1, gives you enough material in case you are interested in this type of program.

An executing program can "shell to DOS". This is a little complicated to understand, but easy to program, so long as you follow the rules. It's not really shelling to DOS as we normally understand it; you don't get a DOS prompt. But it is possible to interrupt a program to execute external program(s), and then return to point of interruption. Let's examine an example, which also gives us a chance to use the *GetEnv* function in the first part of the program:

Program 4.1.2

```
{ $M $4000,0,0} { Free some memory (real mode) }
program  test_exec;    { 08/13/94 }

  uses  CRT, DOS;

  var  ss, tt: string;
          k: Word;

  begin
  ClrScr;
  ss := GetEnv('PATH');
  WriteLn('DOS path:');
  k := Length(ss);
  if  k <= 70  then  WriteLn(ss)
  else
     begin
     k := 70;
     while  ss[k] <> ';'  do  Dec(k);
     tt := Copy(ss, 1, k);
     WriteLn(tt);
     tt := Copy(ss, k + 1, Length(ss) -  k);
     if  tt <> ''  then  WriteLn(tt);
     end;
  WriteLn;
  WriteLn('Command Processor: ', GetEnv('COMSPEC'));
  ReadLn;
```

The "DOS" unit's **function** "GetEnv" returns a string with the corresponding DOS environment variable's value. Here it has been used to display the DOS "path", neatly formatted, and to display the full path of the command processor, usually 'C:\COMMAND.COM'.

The compiler directive at the beginning has been downgraded to a comment because I tested the program in protected mode. But note that the forthcoming calls of "Exec" may require some memory, which a real mode program takes all for itself unless you have such a $M compiler directive. If a real mode program calls "Exec" and it lacks such a $M directive, it will very likely hang the computer hopelessly—this is a warning!

Let's get on with the rest of the program:

```
WriteLn(
    'Now execute commands outside of the program');
SwapVectors;
Exec('c:\command.com', '/c dir c:\dos /2 ');
Exec('c:\command.com', '/c copy zxywvuts stuvwxyz');
Exec('c:\command.com', '/c rd ch5');
Exec('d:\uts\ldir.com', ' /d');
Exec('d:\uts\numlock.exe', '');
SwapVectors;
ReadLn;
end.   { test_exec }
```

The *SwapVectors* procedure *must* be invoked immediately before and immediately after a sequence of *Exec* calls. Don't sweat over the details of why; the idea is that the calling program can resume with its interrupt handlers in precisely the state they were in before the call of "Exec".

The DOS unit "Exec" procedure has the declaration

function Exec(path: PathStr; parameters: **string**);

The first three calls of "Exec" show precisely how it is used to call internal command.com procedures; note the "/c " heading the list of parameters. The other two calls show how other (non DOS) programs are called; in particular, the suffix ".com", ".exe", or ".bat" of an executable program *must* be included.

The disk file "doserror.pas" contains a procedure for displaying an error message when a DosError is detected. You may find some useful information in it.

Section 2. DOS Functions

Once in a while it seems necessary to access computer memory directly. DOS has a number of special addresses, and we next look at a little program that accesses the *keyboard status byte*. Its absolute memory address, in the usual segment:offset hex notation is 0040:0017. Remember that the offset of this byte from 0 is computed in hex by

```
  00400
 +00017
 ------
  00417
```

Programs compiled in protected mode wreck havoc with absolute segment addresses, mapping them all over the place. Borland Pascal provides several segment variables that can be relied on to give the correct address, whether in real or protected mode. One of these is "Seg0040".

The following program simply turns off the NumLock key, in case it is on. Some computers always boot with the NumLock key on; in such a case a call to program "numlock" in the autoexec.bat file will guarantee that NumLock is off when the computer is booted. The program is brief, but contains some ideas you may find useful, plus a list of all the bits of the keyboard status byte.

Program 4.2.1

```
program  numlock;  { 07/09/94 }

{ Turns off NumLock

  The keyboard status byte is at absolute address
  $0040:$0017.  Seg0040 is used as a protected mode
  segment, which may be different from $0040.

  Status byte bits if set:
    7: Insert on         3: Alt Key Down
    6: Caps Lock on      2: Ctrl key down
    5: Num Lock on       1: Left Shift key down
    4: Scroll Lock on    0: Right shift key down }

var  pp: ^Byte;
```

```
begin
pp := Ptr(Seg0040, $0017);
pp^ := (pp^ and $DF);
                  { $DF = 11011111; turns bit 5 off }
end.
```

Numlock uses a pointer variable and the procedure "Ptr" to make it point to the status byte in question. This procedure is useful for accessing any memory address. Its declaration in the System unit is

```
procedure  Ptr(segment, offset: Word);
```

The location $0040:$0017 is one of many special memory locations. You will find a list of many of these addresses in the data file "ROM_addr.dat". The main point of this and the next section is to discuss interrupts. These are low level calls to the procedures supplied by MS-DOS and by the BIOS (basic input output services). Most assembler programs are full of these calls; Borland Pascal provides two procedures, "MSDos" and "Intr", for using DOS and BIOS services directly. Interrupts depend on the state of the CPU registers, and return results in those registers. The System unit has a type "Registers" declared as follows:

```
type  Registers = record
                    case Integer of
                      0: (AX, BX, CX, DX,
                          BP, SI, DI, DS,
                          ES, Flags: Word);
                      1: (AL, AH, BL, BH,
                          CL, CH, DL, DH: Byte);
                  end;
```

"Registers" is the kind of record called a free union Section 3.5 on variant parts. For instance the space used for the Word BP can be interpreted as its low and high order bytes BL and BH. The names, of course, mimic the CPU register names. The disk file "registers.dat" contains some basic information on 80x86 registers.

Let's examine the DOS unit declarations of the procedures MSDos and Intr:

```
procedure  MSDos(var  regs: Registers);
procedure  Intr(number: Byte;
                var  regs: Registers);
```

The two calls

```
MSDos(regs);     Intr($21, regs);
```

are equivalent; I shall use the latter only. There are about 200 to 300 DOS functions and subfunctions under interrupt $21. They do everything there is to do with files and disks. Actually, there is hardly any reason to use these services, because Pascal procedures (using all these low level services) provide much more convenient (and safer) ways to do the jobs. Obviously, you will have to consult a DOS book such as Dettmann, T, *DOS Programmer's Reference* 2/e, Que Corp, 1989, for details on all the DOS functions. We shall take one simple example here, function $2C for reading time from the system clock.

Program 4.2.2

```
program gettime;   { 08/13/94 }

   uses  DOS;

   var  regs: Registers;

   begin
   regs.AH := $2C;   { function $2C, get system time }
   Intr($21, regs);
   Write('The time is ');
   with  regs  do
      begin
      Write(CH, ':');   { hours }
      Write(CL, ':');   { minutes }
      Write(DH, '.');   { seconds }
      if  DL < 10  then  Write(0);
      WriteLn(DL);      { hundredths }
      end;
   ReadLn;
   end.   { gettime }
```

When "Intr" is called, the CPU registers may very well have values that are needed after the call. Both "Intr" and "MSDos" make a backup of the CPU registers by pushing them onto the DOS stack, copy regs into those registers, call the interrupt, then restore the CPU registers by popping them from the stack.

Section 3. BIOS Interrupts

There are many other interrupts besides $21, the DOS functions. In this section we are going to look at two BIOS interrupts, $16 and $10. In Chapter 2 we had Program 2.3.3 for reading scan codes of key presses, and it used Pascal's "ReadKey" function. It had its limits, for instance, it could not read key F11, which Borland has not yet recognized. The following program uses interrupt $16, a low level function for reading keyboard characters.

Program 4.3.1

```
program  scan_codes;  { 08/13/94 }
    { Reads the enhanced keyboard via BIOS interrupt
      $16, function $10.
      If not an enhanced keyboard, use function $00. }

uses  DOS;

var  ch: Char;
     reg: Registers;

begin
WriteLn(
    'Program for character and scan (extended) codes.');
WriteLn('Press <Ctrl-Break> to halt program.');
repeat
  WriteLn;  Write('Next key: ');
  reg.ah := $10;   { function $10: get keystroke }
  Intr($16, reg);  { keyboard interrupt }
  WriteLn('Character code returned: ', reg.al,
          '   Scan code: ', reg.ah);
until  false;
end.   { scan_codes }
```

If you will test this program with an enhanced keyboard (which almost all AT's have), you will find its superiority to the previous Program 2.3.3. The disk file "scancode.dat" contains a listing of all scan codes I know at the moment.

BIOS interrupt $10 contains all low level CRT (screen) functions. There are functions special to the PS/2 computer and others that are of no interest to us. Many of the functions apply both to text modes and to graphics modes; we'll deal with graphics in the next chapter. For now, we'll set up a unit with some of the functions of interrupt $10 and concentrate on their applications to text modes. An important point to keep in mind is that variables are zero-based, unlike the one-based procedures for the text screen that Pascal uses. Pascal assumes the text screen goes from upper left (1, 1) to lower right (80, 25). Thus GotoXY(80, 1) moves the cursor to the upper right corner of the text screen. For interrupt 10 text modes, the upper left is (0, 0) and the lower right (79, 24).

The next important point is that there are actually 8 text screens! You can do operations on any of these at any time, no matter which is currently displayed. Pascal uses only page 0, and all of the System and CRT units' text procedures refer to page 0 only.

As usual, we start the listing of the unit with its interface:

Program 4.3.2

```
unit  intr_10;    { 08/16/94 }
   { Text video services of BIOS interrupt 10
     There are 8 text pages.
     Bounds:  page: 0..7              Upper left:
              x: 0..79                 (0, 0)
              y: 0..24                                     }
interface

    procedure  set_cursor_shape(top, bottom: Byte);
    procedure  set_cursor_position(page, x, y: Byte);
    procedure  get_cursor(page: Byte;
                        var  x, y, top, bottom: Byte);
    procedure  display_page(page: Byte);
    procedure  scroll_up(x0, y0, x1, y1, n, attr: Byte);
    procedure  scroll_down
                  (x0, y0, x1, y1, n, attr: Byte);
    procedure  write_chars(page, x, y: Byte;
                          ch0: Char;  attr: Byte;
                          multiplicity: Word);
    procedure  teletype_string(page, x, y: Byte;
                            ss: string;  attr: Byte);
    procedure  write_string(page, x, y: Byte;
                          ss: string;
                          attr: Byte);
```

We start the implementation with the three procedures for changing the cursor shape and position, and reading its status, and a procedure for changing the display page. This material is pretty routine, but read it carefully to get the flavor.

```
implementation

    uses  DOS;

    var  reg: Registers;

    procedure  set_cursor_shape(top, bottom: Byte);
       { Applies to all pages.
         Have $00 <= top <= bottom <= $0D }

       begin
       with  reg  do
```

```
        begin
        AH := $01;
        CH := top;       { Top scan line }
        CL := bottom;    { Bottom scan line }
        end;
     Intr($10, reg);
     end;

procedure  set_cursor_position(page, x, y: Byte);

   begin
   with  reg  do
      begin
      AH := $02;
      BH := page;
      DH := y;    { Row }
      DL := x;    { Col }
      end;
   Intr($10, reg);
   end;

procedure  get_cursor(page: Byte;
                      var  x, y, top, bottom: Byte);

   begin
   reg.AH := $03;
   reg.BH := page;
   Intr($10, reg);
   with  reg  do
      begin   { Output }
      y := DH;    { Row }
      x := DL;    { Col }
      top := CH;
      bottom := CL;
      end;
   end;   { get_cursor }

procedure  display_page(page: Byte);

   begin
   reg.AH := $05;
   reg.AL := page;
   Intr($10, reg);
   end;
```

Scrolling up (down) within a window loses one line at the top (bottom) and adds a blank line at the bottom (top) with the given attributes.

```
procedure  scroll_up(x0, y0, x1, y1, n, attr: Byte);
   { Scrolls window up n lines, or blanks window
      with the given attribute. }

   begin
   with   reg   do
      begin
      ah := $06;
      al := n;    { 0 to blank the window }
      bh := attr;
      cl := x0;   ch := y0;
      dl := x1;   dh := y1;
      end;
   Intr($10, reg);
   end;


procedure  scroll_down(x0, y0, x1, y1, n,
                       attr: Byte);

   begin
   with   reg   do
      begin
      ah := $07;
      al := n;    { 0 to blank the window }
      bh := attr;
      cl := x0;   ch := y0;
      dl := x1;   dh := y1;
      end;
   Intr($10, reg);
   end;
```

The next three procedures write characters and strings. Each of the 2000 (80×25) screen positions contains a character byte and an attribute byte. The attribute byte contains the color of the character, the background color, and whether it is blinking. The first comment below gives details.

The procedures "write_char" and "write_string" have no effect on the cursor position, and they treat control characters as printable characters, not as instructions. Procedure "teletype_string" is different. It moves the cursor to the end of the string and it responds to some control codes, like CR.

```
procedure  write_chars(page, x, y: Byte;
                          ch0: Char;  attr: Byte;
                          multiplicity: Word);
    { Attr byte bits:
          7               6  5  4         3  2  1  0
          Blink        Back color      Fore color

    Cursor does not move }

    begin
    set_cursor_position(page, x, y);
    with  reg  do
       begin
       AH := $09;
       AL := Ord(ch0);
       BH := page;
       BL := attr;
       CX := multiplicity;
       end;
    Intr($10, reg);
    end;    { write_chars }

procedure  teletype_string(page, x, y: Byte;
                              ss: string;  attr: Byte);
    { Control characters operate.
      Cursor updated. }

    begin
    with  reg  do
       begin
       AH := $13;
       AL := $01;     { Update cursor }
       BH := page;
       BL := attr;
       CX := Length(ss);
       DH := y;
       DL := x;
       ES := Seg(ss[1]);
       BP := Ofs(ss[1]);
       end;
    Intr($10, reg);
    end;    { teletype_string }
```

```
procedure  write_string(page, x, y: Byte;
                             ss: string;
                             attr: Byte);
   { Control characters printed.
     Cursor does not move. }

   var  j: Word;

   begin    { WriteString }
   for  j := 1  to  Length(ss)  do
      begin
      write_chars(page, x, y, ss[j], attr, 1);
      Inc(x);
      end;    { for j }
   end;    { write_string }

end.   { unit  intr_10 }
```

It remains to play around with these procedures, and the following two programs use most of them. You may find some useful ideas even here. It is particularly nice that you can prepare a whole page while no one is looking and then spring it.

Program 4.3.3

```
program  test_Intr_10;    { 08/16/94 }
   { Demo of BIOS interrupt 10, Video services. }

   uses  DOS, CRT, intr_10;

   var    top, bottom,
        x, y, attr, k: Byte;
                 ss: string;

   begin    { test_Intr_10 }
   ClrScr;
   WriteLn('Tests of interrupt 10, video services.');
   WriteLn('Testing only text screen services.');
   Write('Press <Enter> to begin test 0; ');
   ReadLn;
   for  k := 0  to  7   do
      begin    { Fill pages }
      { Clear top row }
      write_chars(k, 0, 0, ' ', Black, 80);
      { Fore color k + 8, back color 16*k, odd k blink }
      attr := 128*(k mod 2) + k shl 4 + k + 8;
```

```
       write_chars(k, 0, 1, Chr(177), attr, 80*24);
       write_string(k, 0, 0,
                      'Page ' + Chr(48 + k) + '     ',
                  k + 8);
       display_page(k);
       Delay(1000);    { 2 sec }
       end;
  write_string(7, 40, 4, 'Press <Enter>:  ', Yellow);
  ReadLn;

  write_string(4, 30, 0, 'Test 1, on page 4', Yellow);
  write_string(4, 20, 10, 'Hello world!', Yellow);
  write_string(4, 20, 15, 'Press <Enter>:  ', Yellow);
  display_page(4);
  ReadLn;

  write_string(3, 30, 0, 'Test 2, on page 3', Yellow);
  set_cursor_shape($00, $06);
  { Clear field }
  write_chars(3, 0, 1, #219, White, 80*24);
  ss := '';
  for k := 1 to 31   do  ss := ss + Chr(k);
  write_string(3, 0, 9, ss, White);
  write_string(3, 0, 10, ss, White);
  write_string(3, 0, 11, ss, White);
  write_string(3, 20, 13, 'Hello again world!',
                           Yellow);
  teletype_string(3, 20, 15, 'Press <Enter>: ',
                             Yellow);
  set_cursor_position(3, 48, 0);
  display_page(3);
  ReadLn;

  write_string(1, 30, 0, 'Test 3, on page 1    ',
                          Yellow);
  write_chars(1, 0, 1, #228, Red, 80*24);
  set_cursor_shape($04, $0B);
  set_cursor_position(1, 48, 0);
  get_cursor(1, x, y, top, bottom);
  display_page(1);
  ReadLn;
```

```
write_string(0, 30, 0, 'Test 4, on page 0      ',
                          Yellow);
set_cursor_shape($00, $0D);
attr := Blue shl 3 + White;
write_string(0, 0, 2, 'On page 1:', attr);
Str(top, ss);
ss := 'Top line of cursor:     ' + ss;
write_string(0, 0, 3, ss, attr);
Str(bottom, ss);
ss := 'Bottom line of cursor: ' + ss;
write_string(0, 0, 4, ss, attr);
Str(y, ss);
ss := 'Cursor is in row:      ' + ss;
write_string(0, 0, 5, ss, attr);
Str(x, ss);
ss := 'Cursor is in column:   ' + ss;
write_string(0, 0, 6, ss, attr);
teletype_string(0, 20, 24, 'Press <Enter>:   ',
                          Yellow);
set_cursor_position(0, 48, 0);
display_page(0);
ReadLn;

write_string(1, 30, 0, 'Test 5, on page 1      ',
                          Yellow);
teletype_string(1, 20, 24, 'Press <Enter>: ',
                          Yellow);
set_cursor_shape($00, $02);
set_cursor_position(1, 48, 0);
display_page(1);
ReadLn;

{ Restore normal state }
set_cursor_shape($0C, $0D);    { Two bottom lines }
TextColor(LightGray);
TextBackground(Black);
display_page(0);
ClrScr;
end.    { test_Intr_10 }
```

The other test program shows the use of scrolling:

Program 4.3.4

```
program  test_text_video;    { 08/22/94 }

  uses  DOS, CRT, intr_10;
```

```pascal
   const   ss: array[1..10] of string[40] =
             ('  ooooooooooooooooooooooooooooooooooo  ',
              'oo  ooooooooooooooooooooooooooooooooo  oo',
              'oooo  ooooooooooooooooooooooooooo  oooo',
              'oooooo  ooooooooooooooooooooooo  oooooo',
              'oooooooo  ooooooooooooooooooo  oooooooo',
              'oooooooooo  ooooooooooooooo  oooooooooo',
              'oooooooooooo  ooooooooooo  oooooooooooo',
              'oooooooooooooo  ooooooo  oooooooooooooo',
              'oooooooooooooooo  oooo  oooooooooooooooo',
              'oooooooooooooooooo    oooooooooooooooooo');

   var   x, y, n, attr, col, row: Byte;

begin   { test_text_video }
scroll_up(0, 0, 79, 24, 0, Black);
attr := Yellow + 16*Red;
for   col := 0  to   39  do
   begin
   for   row := 1  to   10  do
      write_string(0, col, row, ss[row], attr);
   for   row := 1  to   10  do
      write_chars(0, 0, row, ' ', Black, col);
   end;
for   row := 0 to 9  do
   begin
   Delay(70);
   scroll_down(0, 0, 79, 24, 1, Black);
   end;
for   col := 39 downto 0  do
   begin
   for   row := 11  to   20  do
      write_string(0, col, row, ss[row - 10], attr);
   for   row := 11  to   20  do
      write_chars(0, 40 + col, row, ' ',
                  Black, 39 - col);
   end;
for   row := 0 to 9  do
   begin
   Delay(70);
   scroll_up(0, 0, 79, 24, 1, Green);
   end;
teletype_string(0, 0, 23, 'Press space: ', Yellow);
ReadLn;
ClrScr;
end.   { test_text_video }
```

Exercises

1. You may have noticed that Borland Pascal does not allow you to write characters to the text screen with a bright background. When the call TextBackground(C) is made with C one of the bright colors (8..15), the result is the corresponding dark color, with the character blinking. Explore function $10 of interrupt $10 to work around this problem.

Section 4.   The Mouse

Actually, the mouse interrupt $33 is a DOS interrupt not a BIOS interrupt. That is, interrupt $33 is in the (software) operating system, not in the BIOS chip. But we must use "Intr" to access interrupt $33 as "MSDos" only refers to the functions of interrupt $21.

As it is somewhat difficult to drag this material out of the literature, I'll include possibly more than I should in a unit of most of the useful mouse functions. We'll test what we can on the text screen now, and on the graphics screen in the next chapter. As usual, we begin with the **interface** section.

To each computer screen mode there corresponds a mouse *virtual screen*. The virtual screen is a grid over the screen whose dimensions may or may not equal the pixel dimensions of the screen. For the screen modes that we use the virtual screens (widths first) are as follows: VGA 16 color, 80×25 text mode (mode $02): virtual 640×200; VGA 16 color, 640×480 graphics mode (mode $12): virtual 640×480 ; VGA 16 color, 640×350 graphics mode (mode $10): virtual 640×350.

Program 4.4.1

```
unit   mouse;    { 09/03/94 }

interface

   type   Str6 = string[6];
          proc = procedure;

   var       old_mask: Word;
          old_address: Pointer;
```

```
procedure   reset_mouse(var   mouse_OK: Boolean;
                        var   button: Byte);
procedure   show_cursor;
procedure   hide_cursor;
procedure   get_mouse_status(var   button: Byte;
                             var   x, y: Word);
procedure   mouse_gotoXY(x, y: Word);
procedure   get_mouse_button_press
               (var   button: Byte;
                var   count, x, y: Word);
function    double_click(timeout: Word): Boolean;
procedure   get_mouse_button_release
               (var   button: Byte;
                var   count, x, y: Word);
procedure   set_cursor_x_lim(min_x, max_x: Word);
procedure   set_cursor_y_lim(min_y, max_y: Word);
procedure   set_graph_cursor_shape(
               hot_x, hot_y: Integer;
               address: Pointer);
procedure   set_text_cursor_shape
               (screen_mask, cursor_mask: Word);
procedure   set_event_handler
               (call_mask: Word;   address: Pointer);
procedure   get_relative_move(var   x, y: Integer);
procedure   set_hide_cursor_window
               (x0, y0, x1, y1: Word);
procedure   set_mouse_event_interrupt
               (call_mask: Word;   address: Pointer);
procedure   get_mouse_state_size(var   size: Word);
procedure   save_mouse_driver_state
               (address: Pointer);
procedure   restore_mouse_driver_state
               (address: Pointer);
procedure   set_subrout_masks_addresses
               (call_mask: Word;   p: proc;
                OK: Boolean);
procedure   set_mouse_sensitivity
               (hor, ver, double: Word);
procedure   get_mouse_sensitivity
               (var   hor, ver, double: Word);
procedure   set_mouse_polling_rate(rate: Word);
procedure   get_mouse_page(var   page: Word);
procedure   set_mouse_page(page: Word);
procedure   reset_mouse_software
               (var   installed: Boolean;
                var   no_buttons: Word);
```

```
procedure  set_mouse_language(language: Word);
procedure  get_mouse_language
              (var  language: Word);
procedure  get_mouse_info(var  ver: Str6;
                              var  port, IRQ: Byte);
procedure  get_max_virtual_coords
              (var  disabled: Boolean;
               var  hor, ver: Word);
```

What all this means will come out in the implementation. Let's start with procedures that reset the mouse to its defaults and center the mouse cursor on the screen, show the cursor, and hide it:

implementation

```
   uses  DOS, CRT;

   var  regs: Registers;

   procedure  reset_mouse(var  mouse_OK: Boolean;
                              var  button: Byte);
     { Sets defaults:
         Cursor: screen center
         Shape: Block (text), arrow (graphics)
         Call mask: $0000
         Hor mickey/pixel ratio: 8 to 8
         Ver mickey/pixel ratio: 16 to 8
         Double speed threshold: 64 mickey/sec
         Page 0
         Full page allowed for cursor }

     begin
     regs.AX := $00;
     Intr($33, regs);
     mouse_OK := Odd(regs.AX);   { AX = 0 or $FFFF }
     button := regs.BX;
     end;

   procedure  show_cursor;

     begin
     regs.AX := $01;
     Intr($33, regs);
     end;
```

```
procedure  hide_cursor;

    begin
    regs.AX := $02;
    Intr($33, regs);
    end;
```

Associated with the mouse cursor is a "hot spot", usually (but not necessarily) in the cursor. We shall always refer to the coordinates of this hot spot by (x, y), origin as usual the upper left corner of the screen. The unit of cursor movement is the "mickey", explained in the first comment below. The next procedure, "get_mouse_status" returns the status of the mouse buttons and the location of the mouse cursor. Procedure "mouse_gotoXY" moves the cursor to wherever you want it, and "get_mouse_button_press" is like "get_mouse_status" except that it includes some history of button presses since last it was called.

```
procedure  get_mouse_status
           (var  button: Byte;
                              var  x, y: Word);
{ button byte bits:
     7  6  5  4  3        2          1            0
        unused         center     right        left
  The "hot point" (x, y) varies in "mickeys" on a
  "virtual screen".
  In most text modes,   0 <= x <= 640,  0 <= y <= 199,
  so the mouse returns 8 times its text position.
  In most graphics modes, (x, y) corresponds to the
  graphics coordinates. }

    begin
    regs.AX := $03;
    Intr($33, regs);
    with  regs  do
       begin
       button := BL;   x := CX;   y := DX;
       end;
    end;

procedure  mouse_gotoXY(x, y: Word);
    { Move to closest point to virtual screen (x, y) }
```

```
begin
with  regs  do
   begin  AX := $04;  CX := x;  DX := y;  end;
Intr($33, regs);
end;

procedure  get_mouse_button_press
              (var  button: Byte;
               var  count, x, y: Word);
  { count returns the number of presses of
    the button in question since its last call }

begin
regs.AX := $05;
regs.BL := button;
Intr($33, regs);
with  regs  do
   begin
   button := AL;  count := BX;
   x := CX;  y := DX;
   end;
end;
```

Interrupt $33 does not seem to have any function to check on double clicks, so we have to program one using the previous procedures:

```
function  double_click(timeout: Word):
                                    Boolean;
  { Returns true if a double left click takes place
    within timeout msec }

   var  k, x, y: Word;
        button: Byte;

begin
double_click := false;
{ First make sure button is up }
repeat
   get_mouse_status(button, x ,y);
until  button = 0;   { Released }
{ Get first press and release }
repeat
   get_mouse_status(button, x ,y);
until  button = 1;
```

```
repeat
   get_mouse_status(button, x ,y);
until  button = 0;
{ Start countdown for second press/release }
k := 0;
repeat
   Delay(1);  Inc(k);
   get_mouse_status(button, x ,y);
until  (button = 1) or (k = timeout);
if  k = timeout  then  Exit;
{ Press early; continue counting to release }
repeat
   Delay(1);  Inc(k);
   get_mouse_status(button, x ,y);
until  (button = 0) or (k = timeout);
double_click := (k < timeout);
end;    { double_click }
```

The next procedure that we implement, "get_mouse_button_release", is like "get_mouse_button_press" with a history of releases rather than presses. The following two procedures are used to limit the mouse cursor to a rectangle on the screen from which it cannot escape until the procedures are called again.

```
procedure  get_mouse_button_release
             (var  button: Byte;
              var  count, x, y: Word);

begin
regs.AX := $06;
regs.BL := button;
Intr($33, regs);
with  regs  do
   begin
   button := AL;  count := BX;
   x := CX;  y := DX;
   end;
end;

procedure  set_cursor_x_lim(min_x, max_x: Word);
```

```
begin
with  regs  do
    begin
    AX := $07;  CX := min_x;  DX := max_x;
    end;
Intr($33, regs);
end;

procedure  set_cursor_y_lim(min_y, max_y: Word);

begin
with  regs  do
    begin
    AX := $08;  CX := min_y;  DX := max_y;
    end;
Intr($33, regs);
end;
```

The next two procedures set the shape of the cursor and the location of its hot spot. The graphics cursor is a 16×16 bit image. There are two parts to the cursor specification, the *screen mask* and the *cursor mask*. When the cursor moves over any area, that area is first **and**'ed with the screen mask, then the result **xor**'ed with the cursor mask. It takes a little playing with this idea to see how to use it. We'll do so later. The text screen mouse cursor is a simple block, and the masks determine how it is displayed, as we shall soon see.

```
procedure  set_graph_cursor_shape
               (hot_x, hot_y: Integer;
                address: Pointer);
{ address points to an array of 32 words:
  16 words screen mask, 16 words cursor mask.
  When the 16x16 mouse moves over an area A,
  result = (A and screen_mask) xor cursor_mask.
  The "hot spot" (hot_x, hot_y) is the point reported
  by a click.  (0, 0) is the upper left hand corner of
  the 16x16 mouse cursor.  The range is [-129, 128],
  usually [0, 15]. For the default arrow mouse cursor,
  (hot_x, hot_y) = (0, -1). }

begin
with  regs  do
    begin
    AX := $09;
```

```
          BX := Word(hot_x);      CX := Word(hot_y);
          ES := Seg(address^);    DX := Ofs(address^);
          end;
     Intr($33, regs);
     end;


 procedure   set_text_cursor_shape
                (screen_mask, cursor_mask: Word);
{ Each mask is two bytes:   (character, attribute)
   If   T   is a screen char, attr, then
   result = (T and screen mask) xor cursor_mask.
   Attribute bytes as usual:
     bit 7: blink,
     bits 4..6 background color
     bits 0..3 foreground color.
   Default: screen mask = $77FF, cursor_mask = $7700 }

     begin
     with  regs  do
        begin
        AX := $0A;
        BX := $00;    { software mouse }
        CX := screen_mask;  DX := cursor_mask;
        end;
     Intr($33, regs);
     end;
```

The action of the next procedure is considerably more complex than anything so far, even though its code is brief enough. The mouse driver polls the state of the mouse frequently, order of 50 times a second. If there has been any change in the mouse, move, click, or release since the previous change, control goes to an "event handler", which takes some action. The purpose of "set_event_handler" is to replace this event handler by one of our own. The possibility of doing so is good news; the bad news is that the new event handler *must* be written in assembler (shudder) language (as far as I can see). More on this later. Note that "address" is a pointer to the entry point of the event handler, which then acts like a TSR (terminate and stay resident program). See the comments below for "set_mouse_event_interrupt" for an explanation of the "call_mask".

```
     procedure   set_event_handler
              (call_mask: Word;   address: Pointer);

     begin
     with  regs   do
```

```
    begin
    AX := $0C;  CX := call_mask;
    ES := Seg(address^);  DX := Ofs(address^);
    end;
  Intr($33, regs);
  end;
```

The next two procedures that we implement explain themselves, and "set_mouse_event_interrupt" is a more advanced version of "set_event_handler" that remembers what the previous call_mask and event handler were, so they can be restored. The global variables "old_mask" and "old_address" are specifically for saving this information.

```
  procedure  get_relative_move
                    (var  x, y: Integer);
    { Number of mickeys moved since last call }

    begin
    regs.AX := $0B;
    Intr($33, regs);
    x := Integer(regs.CX);
    y := Integer(regs.DX);
    end;

  procedure  set_hide_cursor_window
                  (x0, y0, x1, y1: Word);
    { If the cursor moves into the window,
      it is hidden until show_cursor called.
      Coordinates virtual, even in text modes! }

    begin
    with  regs  do
      begin
      AX := $10;
      CX := x0;  DX := y0;  SI := x1;  DI := y1;
      end;
    Intr($33, regs);
    end;
```

```
procedure  set_mouse_event_interrupt
              (call_mask: Word;   address: Pointer);
{ This loads an event handler, much like a TSR, to
  handle mouse events.
  Event triggered by call_mask bits:
   4        3        2         1          0
   R rel    R press  L rel     L press     move
   address points to the user interrupt routine.
     Upon entry to that routine,
         AX = condition mask (like call mask, but
              with only the bit causing the event set.
         BX = button state
         (CX, DX) = cursor position
         (DI, SI) = relative mouse movement (mickey's)
         DS contains the mouse driver data segment;
             it may be changed by the event handler
  Call this procedure after finished with
  old_mask and old_address, to restore them.
  Call reset_mouse at end of the program to restore
  the default call mask. }

   begin    { set_mouse_event_interrupt }
   with  regs  do
      begin
      AX := $14;   CX := call_mask;
      ES := Seg(address^);   DX := Ofs(address^);
      end;
   Intr($33, regs);
   with  regs  do
      begin
      old_mask := CX;   old_address := Ptr(ES, DX);
      end;
   end;
```

The next three procedures have a special history mission explained in the following comment:

```
{ If you interrupt one program using the mouse
  and start another, you should store the mouse
  driver state, and restore it when you resume the
  first program.  The next 3 procedures handle this. }
```

```
procedure  get_mouse_state_size(var  size: Word);

   begin
   regs.AX := $15;
   Intr($33, regs);
   size := regs.BX;
   end;

procedure  save_mouse_driver_state
              (address: Pointer);
   { The buffer is pointed to by address; its size
     determined by a call of the previous function. }

   begin
   with  regs  do
      begin
      AX := $16;
      ES := Seg(address^);  DX := Ofs(address^);
      end;
   Intr($33, regs);
   end;

procedure  restore_mouse_driver_state
                  (address: Pointer);

   begin
   with  regs  do
      begin
      AX := $17;
      ES := Seg(address^);  DX := Ofs(address^);
      end;
   Intr($33, regs);
   end;
```

The next procedure allows you to define an event handler for an event like
Shift–Left Click. Up to three such handlers can be active at a time.

```
procedure  set_subrout_masks_addresses
              (call_mask: Word;  p: proc;
               OK: Boolean);
{ This allows additional bits of call_mask to be set:
   bit   key pressed during event (press/release)
   5     Shift
   6     Ctrl    (Combinations like Alt-Shift ...
   7     Alt      not supported)
```

One of these must be set for the handler to work.

Up to 3 of these event handlers may be set simultaneously with this procedure. An additional one
may be set with "set_mouse_event_interrupt".
To disable any of these, call this procedure again
with call_mask = $0000. On entry to the event
handler, registers AX...DI are set as in
"set_mouse_event_interrupt" except that bits 5..7
of AX are 0, i.e, Shift, etc information is not
passed. }

```
begin
with   regs   do
    begin
    AX := $18;
    ES := Seg(p);   DX := Ofs(p);
    end;
Intr($33, regs);
OK := (regs.AX = $18);
end;
```

The next three procedures deal with the sensitivity of the mouse and its (mouse-
port mouse only) polling rate.

```
procedure   set_mouse_sensitivity
                (hor, ver, double: Word);
  { Sets factors to multiply the mickey/pixel ratios
    and double speed threshold, which have default
    values of horizontal: 8/8, vertical: 16/8,
    double: 64
    hor..ver = # mickeys per 8 virtual screen pixels.
    Ranges:    1..100
    Defaults:  50
    When the speed exceeds double, then double speed
    is activated.  Double = 0  resets the default.
    "Reset_mouse" does not reset these factors.
    Note: 1 mickey = 1/200" for older mice, 1/400"
    for modern mice. }
```

```
  begin
  with  regs  do
     begin
     AX := $1A;   BX := hor;
     CX := ver;   DX := double;
     end;
  Intr($33, regs);
  end;

procedure  get_mouse_sensitivity
              (var  hor, ver, double: Word);

  begin
  regs.AX := $1B;   Intr($33, regs);
  with  regs  do
     begin
     hor := BX;   ver := CX;   double := DX;
     end;
  end;

procedure  set_mouse_polling_rate(rate: Word);
  { This applies only to mouse-port mice.
     rates:  1: none  2: 30/sec  4: 50/sec
             8: 100/sec  16: 200/sec
     Only the low byte or rate is used. }

  begin
  regs.AX := $1C;   regs.BX := rate;
  Intr($33, regs);
  end;
```

When you switch the text page or the graphics page, the mouse does not change pages automatically; you must program a change. That is the purpose of the next procedure. The remaining procedures in the unit deal with mouse information of various sorts.

```
procedure  set_mouse_page(page: Word);
   { Page on which cursor is displayed }

  begin
  regs.AX := $1D;   regs.BX := page;
  Intr($33, regs);
  end;
```

```pascal
procedure  get_mouse_page(var  page: Word);

   begin
   regs.AX := $1E;  Intr($33, regs);
   page := regs.BX;
   end;

procedure  reset_mouse_software
                  (var  installed: Boolean;
                   var  no_buttons: Word);
  { Like "reset_mouse", but doesn't reset hardware }

   begin
   regs.AX := $21;  Intr($33, regs);
   installed := regs.AX = $FFFF;
   no_buttons := regs.BX;    { 2 or 3 }
   end;
```

```
{ The next two procedures apply to international mouse
  drivers.  Language:  0 English  1 French   2 Dutch
            3 German   4 Swedish  5 Finnish  6 Spanish
            7 Portuguese         8 Italian
 (Yiddish and Esperanto expected in future versions }
```

```pascal
procedure  set_mouse_language(language: Word);

   begin
   regs.AX := $22;  regs.BX := language;
   Intr($33, regs);
   end;

procedure  get_mouse_language
             (var  language: Word);

   begin
   regs.AX := $23;  Intr($33, regs);
   language := regs.BX;
   end;

procedure  get_mouse_info(var  ver: Str6;
                          var  port, IRQ: Byte);
 { ver: version number of mouse driver
   port: 1: bus, 2: serial, 3: InPort, 4: PS/2, 5: HP
   IRQ: Interrupt request number assigned to mouse }
```

```
var   ss: Str6;

begin
regs.AX := $24;
Intr($33, regs);
with  regs  do
   begin
   Str(BH, ver);  Str(BL, ss);
   ver := ver + '.' + ss;
   port := CH;  IRQ := CL;
   end;
end;


procedure  get_max_virtual_coords
              (var  disabled: Boolean;
               var  hor, ver: Word);
              { ver 6.26 }
begin
regs.AX := $26;
Intr($33, regs);
with  regs  do
   begin
   disabled := BX <> 0;  hor := CX;  ver := DX;
   end;
end;


procedure  get_cursor_masks_counts
              (var  screen, cursor: Byte;
               var  hor, ver: Word);
              { ver 7.02 }
begin
regs.AX := $27;
Intr($33, regs);
with  regs  do
   begin
   screen := AL; cursor := BL;
   hor := CX;  ver := DX;
   end;
end;

end.   { unit mouse }
```

Some of the interrupt $33 functions that I have omitted deal with the light pen; others strike me as not very useful.

Two places for more information on programming the mouse: Microsoft Mouse Programmer's Reference 2/e, Microsoft Press, 1991; Dettmann, T, DOS Programmer's Reference 2/e, Que Corp, 1989, pp 717–740.

Section 5.   Testing the Text Mouse

Now we'll test a number of the text mode procedures in the unit "mouse". Let us begin with the beginning declarations and the procedure "handle_event". It is the event handler that will be invoked by "set_event_handler". As mentioned earlier, apparently this kind of routine must be written in assembler. I apologize for this, particularly as I am an absolute clod at assembler and I am sure the following code shows this. Borland Pascal's compiler includes a small assembler with rather restricted rules. Note the reserved word **far** is really a compiler directive necessary for a "TSR" like "handle_event". The pair of compiler directives {$F+} and {$F–} before and after the procedure would have done the same thing. The reserved word **assembler** tells the compiler that the procedure is to be assembled rather than compiled, but its linking is automatic. The action part of an **assemble** procedure in Borland Pascal is delimited by **asm....end**.

In my opinion, assembler programs are usually difficult to read for one simple reason: the documentation is very stingy. I have tried in this case, I hope the only one in this text, to be generous.

The purpose of this event handler is to tell you if the left or right button is pressed while the cursor is near the upper left of the screen.

Program 4.5.1

```
program  test_text_mouse;    { 09/03/94 }

   uses  CRT, DOS, intr_10, mouse;

   var           ss: string;
           mouse_OK: Boolean;
        count, x, y: Word;
             button: Byte;
                ver: Str6;
          port, IRQ: Byte;

   procedure  handle_event;  far;  assembler;

      type  Str = string[20];

      const  Str2: Str = 'Left click';
             Str8: Str = 'Right click';
              rowL: Byte = 4;
              rowR: Byte = 4;
```

```
asm
mov   AX, $03   { get mouse status }
Int   $33       { Mouse BIOS interrupt }
cmp   CX, 200   { Left of 200? }
jge   @99       { Goto end if CX >= 200 }
cmp   DX, 100   { Above 100? }
jge   @99       { Goto end if DX >= 100 }
test  BX, 1     { Left pressed? }
jnz   @17       { Goto label @17 if BX and 1 <> 0 }
test  BX, 2     { Right pressed? }
jnz   @18       { Goto label @18 if BX and 2 <> 0 }
jmp   @99       { Otherwise goto the end }

@17:            { Left pressed }
mov   AX, Seg Str2 { Must load segment register ES
                    via a general register }
mov   ES, AX    { ES = segment address of str2 }
lea   BP, Str2[1]
                { BP = offset address of str2[1] }
mov   AH, $13   { Function: write string }
mov   AL, $00   { Character only; attribute in BL }
mov   BH, $00   { Page 0 }
mov   BL, $07   { Attribute white on black }
mov   DH, rowL  { Row }
Inc   rowL
mov   DL, 40    { Column }
mov   CX, 10    { Length of str2 }
Int   $10       { BIOS CRT call }
jmp   @99       { Goto the end }

@18:            { Right pressed }
mov   AX, Seg Str8
mov   ES, AX    { ES = segment address of str8 }
lea   BP, Str8[1]  { BP = offset address of str8 }
mov   AH, $13   { Function: write string }
mov   AL, $00   { Character only; attribute in BL }
mov   BH, $00   { Page 0 }
mov   BL, $07   { Attribute white on black }
mov   DH, rowR  { Row }
Inc   rowR
mov   DL, 60    { Column }
mov   CX, 11    { Length of str8 }
Int   $10       { BIOS CRT call }
@99:
end;    { handle_event }
```

The main action of the program tests the various mouse unit procedures, one after the other. By all means play around with the code and test various possibilities. If you, like me, are a lousy assembler programmer, expect a lot of trouble if you make changes to "handle_event", usually the computer hung, until a soft or hard reboot.

If you test the program from the IDE (Integrated Development Environment) twice in a row, you will get different results each time with "handle event". This is not the case when running the program from the DOS prompt.

```pascal
begin
ClrScr;
WriteLn('Testing text mouse.');
reset_mouse(mouse_OK, button);
if not mouse_OK then  Halt;
show_cursor;
get_mouse_info(ver, port, IRQ);
Write('Mouse version ', ver,'   ');
case  port  of
   1: Write('bus');
   2: Write('serial');
   3: Write('mouse');
   4: Write('PS/2');
   5: Write( 'HP');
   end;
WriteLn(' port   IRQ ', IRQ, '   ',
        button, ' button');
WriteLn('To continue, click left mouse button:');
repeat
   button := 0;
   get_mouse_button_press(button, count, x ,y);
until  button = 1;
ClrScr;
WriteLn('Move cursor around before clicking!');
WriteLn('To continue, click right:');
GotoXY(1, 8);
TextColor(Yellow);
TextBackground(Brown);
for  x := 1  to  80*16  do
   Write(Chr(x mod 224 + 31));
TextColor(LightGray);
TextBackground(Black);
repeat
   button := 0;
   get_mouse_button_press(button, count, x ,y);
until  button = 2;
```

```
show_cursor;
GotoXY(1, 3);
WriteLn('To hide cursor, click left:');
repeat
    button := 0;
    get_mouse_button_press(button, count, x ,y);
until  button = 1;
hide_cursor;
WriteLn('Wild cursor: click right:');
repeat
    button := 1;
    get_mouse_button_press(button, count, x ,y);
until  button = 2;
show_cursor;                        { attr char }
set_text_cursor_shape($FFAB, $A200);
{ (T and screen_mask) xor cursor_mask; see comment
  in "set_text_cursor_shape". }
WriteLn('To continue, click both buttons:');
repeat
    button := 0;
    get_mouse_button_press(button, count, x ,y);
until  button = 3;
Window(1, 1, 80, 6);
ClrScr;
Window(1, 1, 80, 25);
WriteLn('Screen will change to page 4          ');
WriteLn(
'Now move the cursor into the brown field.');
WriteLn('To continue, click left:');
repeat
    button := 0;
    get_mouse_button_press(button, count, x ,y);
until  button = 1;
display_page(4);
set_mouse_page(4);
write_chars(4, 0, 0, ' ', 7, 80*25);
{ Clear page 4 }
write_string(4, 0, 20,
             'Page 4: Move cursor around.', Yellow);
write_string(4, 0, 21,
             'To exit, click right: ', Yellow);
repeat
    button := 1;
    get_mouse_button_press(button, count, x, y);
until  button = 2;
set_mouse_page(0);
```

```
display_page(0);
ClrScr;
reset_mouse(mouse_OK, button);
show_cursor;
GotoXY(1, 1);
WriteLn('Page 0: to continue, click right: ');
repeat
   button := 1;
   get_mouse_button_press(button, count, x, y);
until  button = 2;
WriteLn('To change event handler, press <Enter>: ');
WriteLn('Try L and R clicks here and there');
WriteLn('Then  press <Enter> to end program');
ReadLn;
set_event_handler($0A, @handle_event);  { 00001010 }
ReadLn;
reset_mouse(mouse_OK, button);
show_cursor;
ClrScr;
end.    { test_text_mouse }
```

Chapter 5 Graphics

Section 1. The Graph Unit

The *Graph* unit in Borland Pascal offers a great variety of procedures for dealing with the graphics screen. In this section we'll illustrate some of them, starting with procedures that control how text is written to the graphics screen. The default font is a bit mapped font that does not magnify gracefully. Borland Pascal provides four other "stroked" fonts that do magnify very nicely. The following program demos these, and procedures that govern how text will be placed on the screen: *SetTextStyle* and *SetTextJustify*. The font files "*.chr" must be on line for the program to run properly. If a graphics program cannot find a font, it doesn't crash, but simply uses the default font instead.

Program 5.1.1

```
program  font_demo;     { 08/26/94 }
   { Have the stroked font files from BP\BGI available:
          Name                    Word    File
          TriplexFont             1       trip.chr
          SmallFont               2       litt.chr
          SansSerifFont           3       sans.chr
          GothicFont              4       goth.chr }

   uses  CRT, Graph, graphs;

   procedure  show_font(font: Word;   ss: string);

      var  j: Word;

      begin
      for  j := 1  to  7  do
         begin
         SetTextStyle(font, HorizDir, j);
         OutTextXY(0, 50*j, ss);
         end;
      SetTextStyle(font, VertDir, 3);
      OutTextXY(600, 0, 'Press <Enter>:');
      ReadLn;  ClearDevice;
      end;    { show_font }

   procedure  show_justify;
         { Leftext = 0, CenterText = 1, RightText = 2
           BottomText = 0, CenterText = 1, TopText = 2 }
```

```
    var   j, k, 1: Word;

    begin
    SetTextStyle(SansSerifFont, HorizDir, 1);
    for  j:= 0  to  2  do
       for  k := 0  to  2  do
          begin
          Rectangle(250, 20 + 120*j + 30*k,
                    389, 45 + 120*j + 30*k);
          SetTextJustify(j, k);
          OutTextXY(320, 30 + 120*j + 30*k ,
                    'IJKLMNXYZ');
          end;
    OutTextXY(420, 430, 'Press space:');
    end;

begin
open_graph;
show_font(DefaultFont,    'Default Font');
show_font(TriplexFont,    'TriplexFont');
show_font(SmallFont,      'SmallFont');
show_font(SansSerifFont,  'SansSerifFont');
show_font(GothicFont,     'GothicFont');
show_justify;
ReadLn;
close_graph;
end.
```

There is a method to incorporate one or more font files into your program, and also to incorporate "*.bgi" graphics drivers. The mechanism is explained in detail in the file "bgi_chr.pas" on your disk. I should mention that an extra copy of "graphs.pas" (to open/close graphics) is on the disk in the same subdirectory as the files of this section.

Remark Borland supplies you with six other fonts, but their names do not seem to be recognized by the compiler—you refer to them by number. They are 5 (script font), 6 (simplex font), 7 (triplex script font), 8 (complex font), 9 (European font), 10 (bold font). The names are fanciful, not very informative, and you should try these fonts at various sizes to decide if you ever want to use any of them (I never have). If you want to refer to them by names like "SansSerifFont", then simple add **const** declarations like

```
    const   BoldFont = 10;
```

As pointed out in one of the appendices, sometimes during testing a graphics program you have a crash that leaves the screen in graphics mode. On your disk you will find a little program "clear.pas" that may succeed to restoring text mode in such situations.

The following program, "test_graph_unit", demos several things. First, how to save in memory a portion of a graph (*GetImage*) and later restore it (*PutImage*) possibly in another position (useful for animation). The latter procedure allows an image to be combined logically in various ways with what is on the screen already. Only a straight copy (*CopyPut*) and xor (*XORPut*) are allowed for most graphing procedures; "PutImage" is more generous.

Next, "test_graph_unit" demos various fill styles for filling regions. We'll have some comments on this shortly. Finally, the program demos changing the color palette.

Program 5.1.2

```pascal
program  test_graph_unit;    { 08/26/94 }

  uses   CRT, Graph;

  var  grdriver, grmode: Integer;
                   Ptr: Pointer;
                  size: Word;
                    ch: Char;

  procedure  test_get_put_image;

     var    Ptr: Pointer;
           size: Word;

     begin
     SetFillStyle(SolidFill, LightCyan);
     SetBkColor(Brown);
     FillEllipse(100, 210, 20, 20);
     Circle(140, 210, 20);
     Rectangle(80, 190, 160, 230);
     size := ImageSize(80, 190, 160, 230);
     GetMem(Ptr, size);
     GetImage(80, 190, 160, 230, Ptr^);
     SetColor(White);
     OutTextXY(20, 20, 'Press <Enter>:');
     ReadLn;

     SetFillStyle(SolidFill, Blue);
     Bar(280, 40, 400, 400);
```

```
PutImage(300,  80, Ptr^, CopyPut);
PutImage(300, 140, Ptr^, OrPut);
PutImage(300, 200, Ptr^, XORPut);
PutImage(300, 260, Ptr^, NotPut);
PutImage(300, 320, Ptr^, AndPut);
OutTextXY(450, 100, 'CopyPut');
OutTextXY(450, 160, 'OrPut');
OutTextXY(450, 220, 'XorPut');
OutTextXY(450, 280, 'NotPut');
OutTextXY(450, 340, 'AndPut');
ReadLn;
ClearDevice;
FreeMem(Ptr, size);
end;    { test_get_put_image }
```

Let's break the listing here for some comments. The function *ImageSize* computes how large a buffer is needed to store the desired portion of the screen. You should always use this to compute the buffer size, as the saved chunk includes some header information, so its size is not exactly 16 (colors) times the number of pixels. Procedure "GetImage" does the copying. You decide if any of the restoring modes available to "PutImage" besides "CopyPut" and "XORPut" are useful. We continue:

```
procedure  test_fill_styles;

    procedure  fill(x, y, style: Integer;
                    color: Word;
                    ss: string);

        begin
        OutTextXY(x, y - 15, ss);
        SetFillStyle(style, color);
        Rectangle(x, y, x + 120, y + 120);
        Bar(x + 1, y + 1, x + 119, y + 119);
        end;    { fill }

    begin    { test_fill_styles }
    SetBkColor(Black);
    fill(  0,  20, EmptyFill,      1,  'EmptyFill');
    fill(150,  20, SolidFill,      2,  'SolidFill');
    fill(300,  20, LineFill,      15,  'LineFill');
    fill(450,  20, LtSlashFill,    3,  'LtSlashFill');
    fill(  0, 170, SlashFill,      6,  'SlashFill');
    fill(150, 170, BkSlashFill,   14,  'BkSlashFill');
    fill(300, 170, ltbkslashfill,
```

```
                            14,   'LtBkSlashFill');
fill(450, 170, HatchFill,      9,   'HatchFill');
fill(  0, 320, XHatchFill,    10,   'XHatchFill');
fill(150, 320, InterleaveFill,
                            11,   'InterleaveFill');
fill(300, 320, WideDotFill,   12,   'WideDotFill');
fill(450, 320, CloseDotFill,
                            13,   'CloseDotFill');
OutTextXY(GetMaxX div 2, GetMaxY - 15,
          'Press <Enter>: ');
ReadLn;
end;   { test_fill_styles }
```

When you run "test_graph_unit", pay close attention to the yellow filled regions. Borland has managed to have some errors in these fill styles since Turbo Pascal 3.0, or possibly earlier. *LtBackSlashFill* is particularly chaotic. The good news is that all this can be fixed if you have access to some program for patching files (or are willing to write dedicated ones for this purpose). The Norton Utilities "Disk Editor" is great for this. On your disk is a file "patchbgi.dat" with complete instructions for patching each of the graph drivers "*.bgi".

It is possible to modify the 16 individual colors available at any time. Any modification is effected immediately. The idea is that each color is a mix of the primaries Red, Green, Blue, and the shade is determined by how brightly each component shines. There are 64 possible intensities for each primary color. The intensity is a "Word", but only its 6 least significant bits are used. The following procedure, "test_palette", makes some random changes in intensities; if you are interested in playing with this type of programming, perhaps it will give you a start. Note that "test_palette" begins with a declaration of a *structured constant* of array type, and note the syntax of its initialization. Then program "test_graph_unit" ends with its main action.

```
procedure  test_palette;

   const  col_num: array[0..15] of Word
             = (0, 1, 2, 3 ,4 ,5, 20, 7,
                56, 57, 58, 59, 60, 61, 62, 63);

   var  k: Word;

   begin   { RGB }
   ClearDevice;
   SetBkColor(Black);
```

```pascal
    Randomize;
    SetRGBPalette(col_num[LightRed], 63, 10, 10);
    SetFillStyle(SolidFill, LightRed);
    Bar(0, 0, 159, 479);
    SetRGBPalette(col_num[LightGreen], 9, 63, 9);
    SetFillStyle(SolidFill, LightGreen);
    Bar(160, 0, 319, 479);
    SetRGBPalette(col_num[LightBlue], 0, 0, 63);
    SetFillStyle(SolidFill, LightBlue);
    Bar(320, 0, 479, 479);
    SetRGBPalette(col_num[Brown], 40, 30, 0);
    SetFillStyle(SolidFill, Brown);
    Bar(480, 0, 639, 479);
    SetColor(White);
    OutTextXY(  0, 0, 'LightRed');
    OutTextXY(160, 0, 'LightGreen');
    OutTextXY(320, 0, 'LightBlue');
    OutTextXY(480, 0, 'Brown');
    Delay(2000);
    OutTextXY(320, 460, 'Press <Enter>:');
    repeat
       SetRGBPalette(col_num[LightRed],
                     63, Random(40), Random(40));
       Delay(100);
       SetRGBPalette(col_num[LightGreen],
                     Random(40), 63, Random(40));
       Delay(100);
       SetRGBPalette(col_num[LightBlue],
                     Random(40), Random(40), 63);
       Delay(100);
       SetRGBPalette(col_num[Brown], 30 + Random(20),
                     20 + Random(20), 0);
       Delay(100);
    until  KeyPressed;
    ReadLn;
    end;    { test_palette }

begin    { test_graph_unit }
grdriver := VGA;
grmode := VgaHi;
InitGraph(grdriver, grmode, '');
test_get_put_image;
test_fill_styles;
test_palette;
CloseGraph;
end.    { test_graph_unit }
```

A problem with some graphics monitors is that they change from text mode to graphics mode and vice versa with a noticeable screen jerking. The best way to handle this is to insert a short delay before the new mode is shown, maybe a quarter of a second. The following program helps you make a rough estimate of that delay (and maybe complain to your dealer):

Program 5.1.3

```pascal
program  blackout;    { 07/25/94 }

   uses  CRT, Graph;

   var  grdriver, grmode: Integer;

   procedure introduction;

      begin
      ClrScr;
      WriteLn(
'Text to Graphics to Text blackout tests');
      WriteLn;
      WriteLn('The graphics mode will be started.');
      WriteLn(
'Numbers 1, 2, 3,... will be written to the screen');
      WriteLn('at 1/10 sec intervals.');
      WriteLn('Note which one you see first.');
      WriteLn(
'It, times 1/10, measures the amount of screen');
      WriteLn(
'blackout in seconds, going from text to graphics');
      WriteLn('The 640 x 480 graphics mode is used.');
      WriteLn;
      Write('Press <Enter> to start/stop the test: ');
      ReadLn;
      end;    { introduction }

   procedure  Text2graph_test;

      var  j, k, m, n: Word;
                    ss: string;
```

```pascal
begin
j := 1;
k := 0;
m := 0;
n := 0;
grdriver := VGA;
grmode := VgaHi;
InitGraph(grdriver, grmode, '');
repeat
   SetFillStyle(SolidFill, j);
   Bar(k, n, k + 40, n + 50);
   Str(m, ss);
   OutTextXY(k + 10, n + 60, ss);
   Delay(100);
   Inc(j);
   if  j = 16   then   j := 1;
   Inc(k, 40);
   Inc(m);
   if  m mod 15 = 0   then
      begin
      Inc(n, 75);
      k   := 0;
      end;
until  KeyPressed;

ReadLn;
CloseGraph;
end;    { text2graph_test }

procedure  Graph2text_test;

   var  j, k: Word;

   begin
   ClrScr;
   j := 0;
   repeat
      Write(j:4);
      Delay(100);
      Inc(j);
      if  j mod 15 = 0   then WriteLn;
   until  KeyPressed;
   WriteLn;
   ReadLn;
   WriteLn('This tested the blackout in returning');
   WriteLn('from graph mode to text mode.');
```

```
    WriteLn(
  'The numbers were written at 1/10 sec intervals.');
    end;    { graph2text_test }

  begin    {  blackout }
  introduction;
  Text2graph_test;
  Graph2text_test;
  WriteLn;
  Write('Press <Enter>: ');   ReadLn;
  end.
```

Section 2. Applications: Fractals, Life

We begin with a modest program to draw the Mandelbrot set. The theory can be summarized as follows. The Mandelbrot set is the set of all complex numbers c with the following property: The sequence of complex numbers z defined by

```
z := 0;   z := sqr(z) + c
```

is bounded. The basic idea is to color white those points c for which the z approach infinity, leaving the points of the Mandelbrot set black. More colorful drawings are made by giving different colors to points that lead to divergence according to how soon the iterates z leave a definite region. The following program does this, and you can modify it easily. The euclidean norm is used to measure the divergence rate. If you change to another norm using absolute values instead of squares, you both speed the program (which *is* slow) and make the figure different if not more attractive. It is remarkable that such a short program can produce such a complicated figure.

Program 5.2.1

```
{$N+}
program  fractal_set;    { 08/25/94 }
  { Plots a Mandelbrot set (black).  Ref:
    Falconer, K, Fractal Geometry, Wiley (1990) p. 206 }

  uses  CRT, Graph, graphs;

  type  Real = Extended;

  const  p0 = 0.016;  q0 = 0.016;
```

```pascal
var   zero_x, zero_y: Integer;
                 ch: Char;
   x, y, p, q, u, v: Real;
   color, iteration: Word;
               j, k: Integer;
             finite: Boolean;

begin
open_graph;
OutTextXY(0, 460, 'Press <Esc> to halt');
SetLineStyle(DashedLn, 0, ThickWidth);
Rectangle(120, 40, 520, 440);
zero_x := 320;  zero_y := 240;
for  j := -200  to  200  do
   begin
   if  KeyPressed  then
      begin
      ch := ReadKey;  close_graph;  Break;
      end;
   for  k := -200  to  200  do
      begin   { z = x + yi }
      p := p0*j;  q := q0*k;    { r = p + qi }
      x := 0.0;  y := 0.0;    { z0 := 0 }
      iteration := 0;
      finite := true;
      while  finite  and  (iteration <= 1000)  do
         begin   { z := z*z + r }
         Inc(iteration);
         u := (x + y)*(x - y);  v := 2.0*x*y;
         x := u + p;  y := v + q;
         finite := (Sqrt(x) + Sqrt(y)) < 100.0;
         end;
      if  (not finite) and
            (iteration <= 1000)  then
         begin   { Color points that --> infinity }
         case  iteration mod 4  of
            0: color := Yellow;
            1: color := LightRed;
            2: color := White;
            3: color := Cyan;
            end;
         PutPixel(j + zero_x, zero_y - k, color);
         end;
      end;
   end;
OutTextXY(320, 460, 'Press Space');
```

```
ch := ReadKey;
close_graph;
end.    { fractal_set }
```

We now come to Conway's famous "Game of Life". We'll play on a torus rather than on a rectangle, as this eliminates the problem of how to treat the boundary—the torus has none.

The Game of Life is not a game at all. It is a population of many generations which you watch passively. The torus is divided into cells by two families of rulings. Thus each cell has exactly 8 neighboring cells. (Remember that this is a torus.) The fundamental domain that we draw in the plane is a rectangle divided into many squares. In any generation, each cell is either alive or dead. The first generation is determined by chance. After that the generation succeeding a given generation is determined by rules, which apply simultaneously to all cells: (1) A live cell with 0 or 1 live neighbors dies of loneliness. (2) A live cell with 4 or more live neighbors dies of overcrowding. (3) A dead cell with exactly 3 live neighbors is reborn. (4) Otherwise there is no change of state.

To do the change of state simultaneously on all cells, we keep two copies of the fundamental domain. To make a snappy change from one generation to the next, we are going to use two graphics pages. This is possible in medium resolution VGA mode, 640×350, which provides quite adequate resolution. The drawing can be made on the hidden page while the page in view is static; then the pages can be switched, giving an almost instantaneous change of generation.

We need to halt operations if a generation becomes stable, i.e, the next generation is unchanged. This can happen, and occasionally generations show a periodicity of period two. To check for this, we keep a *third* copy, the grandparents of the current generation. I do not know if higher periods are possible; you might try some experiments, maybe with smaller planes than the 40×28 plane of the program.

Let us start with the **const** and **var** declarations and the computational procedures:

Program 5.2.2

```
program game_of_life;    { 07/24/94 }
   { Conway's game of life, on a torus }

   uses   CRT, DOS, Graph;
```

```pascal
const           hor = 40;   { No of cells: hor X ver }
                ver = 28;
                { Abs limits:  hor = 40, ver = 28 }
               hor1 = hor + 1;   { No of rulings }
               ver1 = ver + 1;   { .. }
         cell_width = 15;
        cell_height = 11;     { Center the field }
              min_x = (640 - hor*cell_width) div 2;
              min_y = (350 - ver*cell_height) div 2
                    - 6;
        field_width = hor1*cell_width;
       field_height = ver1*cell_height;
              max_x = min_x + field_width;
              max_y = min_y + field_height;
           min_prob = 0.1;
        { Cells initialized alive with }
        prob_factor = 0.5;
        { prob = min_prob + prob_factor*ramdom }

var  old2_gen,
   { 3 generations, to check for period 2 }
      old_gen,
      new_gen: array[0..ver, 0..hor] of 0..1;
    gen_count: Word;
         prob: Real;
           ch: Char;

{ Graphing variables.
  A circle drawn for each live cell }
      x_center: array[0..hor] of Word;
      y_center: array[0..ver] of Word;
        radius,
          page: Word;   { Two VGA pages used }
            ss: string[10];

{ Computing procedures }

procedure  init_cells;

   var  j, k: Word;
```

```
begin
gen_count := 0;
{ Make all cells dead }
for  j := 0  to   ver   do
   for  k := 0  to   hor   do
       begin
       old_gen[j, k] := 0;
       old2_gen[j, k] := 0;
       if  Random <= prob  then
          new_gen[j, k] := 1
       else  new_gen[j, k] := 0;
       end;
   end;   { init_cells }

procedure  next_generation;

   var        j, k, m,
       prev_j, next_j,
       prev_k, next_k: Word;

   begin
   old2_gen := old_gen;  { grandparents }
   old_gen := new_gen;   { parents }
   for  j := 0  to   ver   do
      begin
      if  j = 0  then   prev_j := ver
      else  prev_j := j - 1;
      if  j = ver  then   next_j := 0
      else  next_j := j + 1;
      for  k := 0  to   hor   do
          begin
          if  k = 0  then   prev_k := ver
          else  prev_k := k - 1;
          if  k = hor  then   next_k := 0
          else  next_k := k + 1;
          { Determine number of live neighbors }
          m :=    old_gen[prev_j, prev_k]
                + old_gen[prev_j, k    ]
                + old_gen[prev_j, next_k]
                + old_gen[j,      prev_k]
                + old_gen[j,      next_k]
                + old_gen[next_j, prev_k]
                + old_gen[next_j, k    ]
                + old_gen[next_j, next_k];
          { Implement the Rules of Life }
          if  (old_gen[j, k] = 1) and
```

```
                     ((m <= 1) or (m >= 4))   then
                   new_gen[j, k]  := 0
            else  if   (old_gen[j, k] = 0) and
                         (m = 3)   then
                   new_gen[j, k]  := 1
            else  new_gen[j, k]  := old_gen[j, k];
          end;    { for  k }
      end;    { for  j }
  end;    { next_generation }

{ Three tests }
function  dead: Boolean;

  var  j, k: Word;

  begin
  dead := true;
  for  j := 0  to  ver  do
     for  k := 0  to  hor  do
         if  new_gen[j, k] = 1  then
            begin  dead := false;  Exit;  end;
  end;    { dead }

function  stable: Boolean;

  var  j, k: Word;

  begin
  stable := true;
  for  j := 0  to  ver  do
     for  k := 0  to  hor  do
         if  old_gen[j, k] <> new_gen[j, k]  then
            begin  stable := false;  Exit;  end;
  end;    { stable }

function  period2: Boolean;

  var  j, k: Word;

  begin
  period2 := true;
```

```
for  j := 0  to  ver  do
    for  k := 0  to  hor  do
        if  old2_gen[j, k] <> new_gen[j, k]  then
            begin  period2 := false;  Exit;  end;
end;    { period2 }
```

This brings us to the graphics part of the program. Please pay attention first to the procedure *SetViewport*. Its fifth (Boolean) parameter to turn clipping on or off is of no importance here. Its first four parameters give the upper left and lower right corners of a graphical window (viewport) and all output procedures after that are relative to the new viewport. Its upper left corner has coordinates (0, 0). Thus "SetViewport" saves for you a lot of figuring offsets, and is a most useful procedure.

Next note the procedures *SetActivePage* and *SetVisualPage*, called with parameter 0 or 1. The *active page*, 0 or 1, is the one you are drawing to; the *visual page* is the one shown on the screen. They may or may not be the same.

```
procedure  init_screen;

    var  GraphDriver,
             GraphMode: Integer;
                 j, k: Word;

    begin
    GraphDriver := VGA;
    GraphMode := VGAMed;
    { Note that VGAHi mode cannot be used;
      it has one page. }
    page := 0;
    InitGraph(GraphDriver, GraphMode, '');
    if  GraphResult <> grOK  then  Halt;
    { Initialize disks }
    for  k := 0  to  hor  do
        x_center[k]
            := k*cell_width + cell_width div 2;
    for  j := 0  to  ver  do
        y_center[j]
            := j*cell_height + cell_height div 2;
    radius := 4;
    end;    { init_screen }

procedure  display_field;

    var  j, k: Word;
```

```pascal
procedure   rule_plane;

   var   j, k: Word;

   begin
   SetViewport(0, 0, GetMaxX, GetMaxY, ClipOff);
   SetColor(White);
   OutText('Life on a Torus.    Generation: ');
   OutTextXY(0, GetMaxY - 8,
   'H: Halt    Any other key: Renew Life');
   Str(gen_count, ss);   OutText(ss);
   SetColor(Brown);
   SetViewport
      (min_x, min_y, max_x, max_y, ClipOff);
   for   k := 0   to   ver1   do
      begin
      j := k*cell_height;    { horizontal rules }
      Line(0, j, field_width, j);
      end;
   for   j := 0   to   hor1   do
      begin
      k := j*cell_width;   { vertical rules }
      Line(k, 0, k, field_height);
      end;
   end;    { rule_plane }

begin    { display_field }
if   gen_count <> 0   then   next_generation;
Inc(gen_count);
page := 1 - page;
{ Work in the background }
SetActivePage(page);
ClearDevice;
rule_plane;
SetColor(LightCyan);
for   j := 0   to   ver   do
   for   k := 0   to   hor   do
      if   new_gen[j, k] = 1   then
         Circle(x_center[k], y_center[j],
            radius);
{ Now show result }
SetVisualPage(page);
end;    { display_field }
```

```
begin    { game_of_life }
init_screen;
repeat    { new game }
   Randomize;
   prob := min_prob + prob_factor*Random;
   {Initial message }
   OutTextXY(0, 0,
               'Conway''s Game of Life (on a torus)');
   WriteLn;
   OutTextXY(0, 15,
     'Live cells inserted at random,');
   Str(prob:3:3, ss);
   OutTextXY(0, 30, 'with probability '+ ss);
   OutTextXY(0, 60, 'Press any key to start: ');
   ch := ReadKey;
   ClearDevice;
   init_cells;
   repeat    { next generation }
      display_field;
      if  dead  then
         begin
         SetViewport(400, 1, GetMaxX, GetMaxY,
                     ClipOff);
         SetColor(White);
         OutText('Colony is dead.');
         repeat  until  KeyPressed;    { wait }
         end
      else  if  stable  then
         begin
         SetViewport(400, 1, GetMaxX, GetMaxY,
                     ClipOff);
         SetColor(White);
         OutText('Colony is stable.');
         repeat  until  KeyPressed;    { wait }
         end
      else  if  period2  then
         begin
         SetViewport(400, 1, GetMaxX, GetMaxY,
                     ClipOff);
         SetColor(White);
         OutText('Colony has period 2.');
         repeat  until  KeyPressed;    { wait }
         end;
      if  KeyPressed  then
          begin  ch := ReadKey;  Break;  end;
      Delay(0);    { Try 50, 100, 200 }
```

```
until   false;    { next generation }
SetViewport(0, 0, GetMaxX, GetMaxY, ClipOff);
ClearDevice;
SetColor(White);
   if   Upcase(ch) = 'H'   then   Break;
until   false;    { new game }
CloseGraph;
Delay(400);    { To prevent flicker }
end.    { game_of_life }
```

I hope this program gives you, your friends, and your family some pleasure, including some interesting graphics techniques.

Section 3. Basic Graphics

To really understand computer graphics, you must come to grips with its fundamentals. In this and the following two sections we are going to cover some very basic graphics, using the Graph unit only to open/close the graphics screen when we test our results on the screen. But more important for us is writing graphics images directly to the printer, in very high resolution. In later sections, we'll deal with the VGA at the BIOS and hardware levels.

The first thing to know about computer graphics is that it is deceptive sleight-of-hand. Accuracy to about 3 significant digits is more than the eye can see. For instance, suppose a screen has about 1000 pixels to 10 inches. Then the pixels are about 1/100 of an inch apart. An occasional error of one pixel simply will not make any visible difference. In addition, the discrete grid of pixels approximates a continuous field, and even simple figures like circles are very jerky if inspected with a magnifying glass. Yet in spite of this, computer graphics is wonderful for visualizing geometry.

In this section we are going to do our own graphics from scratch. By doing so, we can learn much about computer graphics. Consider the following project: The high resolution VGA screen is 640×480. Even the SVGA is only 1024×768. Sometimes higher resolution is desirable. We are going to write a unit for 2000×2000 computer graphics. The result of graphing a figure will be in memory, but the unit will include routines for dumping the image to a printer or a disk file, and even for testing (at lower resolutions) on the screen. This is an important facility actually. Dumping to a file is fast, but you cannot inspect what you have created until you print the file. But printing a graphics file is painfully slow, so the screen is the best testing ground for debugging and testing.

Our first problem is how to allocate memory for a figure. We represent each point of our 2000×2000 plane by one bit, so we require 500,000 bytes, much more than the 64K limit for any variable size. We cut our plane by horizontal slices into 8 rectangles, each 250×2000 bits, or 62,500 bytes, safely under the 64K limit. We'll start the **interface** section of our unit. Note the commented out optional plane size. The VGA in high resolution mode is 640×480 pixels. If the (* comment *) is uncommented, then the 2000×2000 declarations should be commented out in its place. Also, the purpose of x_max...y_min is to center the origin.

Program 5.3.1

```
{$R+,S+,I-}
unit  grafunit;    { 08/24/94 }
    { For testing on the VGA, change the first
      two const declarations as indicated }

interface

  { For testing on the VGA }                           (*
    const   no_rows = 480;      { multiple of 8 }
            no_cols = 640;      { multiple of 8 } *)
    const   no_rows = 2000;     { multiple of 8 }
            no_cols = 2000;     { multiple of 8 }
              x_max = no_cols div 2;
              x_min = - x_max + 1;
              y_max = no_rows div 2;
              y_min = - y_max + 1;
    { x runs horizontally, left to right
      y runs vertically, bottom to top }

    var   xor_plot: Boolean;
          file_name: string[79];

    { Initialization procedures }
    procedure   init_plane;
    procedure   recycle_plane;
    procedure   zero_plane;
    { Printing procedures }
    procedure   dump_to_screen;
    procedure   dump_to_file;
    procedure   dump_to_printer;
    { Drawing procedures }
    procedure   plot(x, y: Integer);
        { Plots a single point in plane }
        { x_min <= x <= x_max,   y_min <= y <= y_max }
    procedure   horizontal_seg(x0, x1, y: Integer);
    procedure   vertical_seg(x, y0, y1: Integer);
    procedure   border;
```

```
function    clip(var  x0, y0, x1, y1: Integer):
                           Boolean;
{ Returns false if segment (x0, y0)(x1, y1)
  is entirely outside the plane. Otherwise
  returns true and replaces (x0, y0) & (x1, y1) by
  points in the plane, defining the same segment. }
procedure  segment(x0, y0, x1, y1: Integer);
  { Draw segment (x0, y0) (x1, y1) }
procedure  solid_rectangle(x0, y0, x1, y1: Integer);
procedure  Circle(x0, y0, r: Integer);
```

As set up, our graphing domain is [-999, 1000]×[-999, 1000]. The **implementation** contains all details of how the image is stored; this is of no concern to the user, who is supposed to know only the **interface**. It also contains declarations of several strings that will be used later for printer control. These are aimed at printers that use the HP Printer Control Language (PCL)—many desk jet and laser printers. You will have to make changes for other printers of course.

implementation

```
uses  CRT, DOS, Graph;

const          max_row = no_rows - 1;
               max_col = no_cols - 1;
         bytes_per_row = no_cols div 8;
        rows_per_strip = no_rows div 8;
                  mask : array[0..7] of Byte
                       = (128, 64, 32, 16,
                            8,  4,  2,  1);
     { HP-PCL printer control codes }
const  reset_printer = #27'E';
        graf_resol300 = #27'*t300R';    { 300 dpi }
           graf_start = #27'*r1A'; { Start graphics }
        graf_send250  = #27'*b250W';
                        { Send 250 bytes }
            graf_stop = #27'*rB';   { End graphics }
        clear_margins = #27'9';
     { Set left & top margins; center image on page.
       Allow for standard 1/2" top, 1/4" L margins.
       Image side = 2000/300 = 6.667",   Paper: 11x8.5
        720*[(11 - 6.667)/2 - .5] = 1200
        720*[(8.5 - 6.667)/2 - .25] = 480 }
        hor_posit = #27'&a480H';    { hor 480/720" }
        ver_posit = #27'&a1200V';   { ver 1200/720" }
        { 1" = 72 points; the unit is deci-point. }
```

```
{ The plane has "no_rows rows" of "no_cols" bits. Each
  row has "bytes_per_row" bytes.  The plane itself is
  divided into 8 strips, each with "rows_per_strip"
  rows.   The number of bytes per strip must not exceed
  64K.   "Mask" accesses individual bits of a byte. }

type    row = array[0..bytes_per_row - 1]
                   of Byte;
        strip = array[0..rows_per_strip - 1] of row;
        Str25 = string[25];

var     plane: array[0..7] of ^strip;
           { top to bottom }
      { The origin is (almost) centered, so to the user,
        the plane is defined by
          x_min <= x <= x_max   y_min <= y <= y_max
```

The first comment explains how "plane" is organized. By using "^strip" instead of "strip", we open up EMS memory for "plane". Conventional 640K memory is probably inadequate.

Given any point (x, y) we must locate quickly its byte and bit in "plane". To avoid a great deal of computation, we initialize lookup tables once and for all. To continue where we broke off, middle of a comment:

```
      Lookup tables avoid division calculations and
      provide the mapping  (x, y) --> plane:

        (x, y) -->   row       y_coord[y].row
                 of strip      y_coord[y].strip
                    bit        x_coord[x].bit
                 of bite       x_coord[x].bite   }

      x_coord: array[x_min..x_max] of
                   record
                      bite: Word;
                      { 0..(bytes_per_row - 1) }
                       bit: Byte;    { 0..7 }
                   end;
      y_coord: array[y_min..y_max] of
                   record
                        strip: Byte;    { 0..7 }
                          row: Word;
                        { 0..(rows_per_strip - 1) )
                   end;
```

The following procedure initializes these tables and allocates memory by 8 calls of New:

```
procedure  init_plane;

    var  strip_no: Byte;
         x, y, z: Integer;

    begin
    for  strip_no := 0  to  7  do
       New(plane[strip_no]);
    { Initialize  x_coord and y_coord }
    for  x := x_min  to  x_max  do
      with  x_coord[x]  do
         begin
         z := x - x_min;
         Byte := z div 8;  bit  := z mod 8;
         end;
    for  y := y_min  to  y_max  do
      with  y_coord[y]  do
         begin
         z := y_max - y;
         strip := z div rows_per_strip;
         row   := z mod rows_per_strip;
         end;
    zero_plane;
    end;    { init_plane }

procedure  recycle_plane;

    var  strip_no: Byte;

    begin
    for  strip_no := 0  to  7  do
       Dispose(plane[strip_no]);
    end;    { recycle_plane }

procedure  zero_plane;

    var  strip_no: Byte;  k: Word;

    begin
    k := SizeOf(plane[0]^);
```

```
     for   strip_no := 0   to   7   do
        FillChar(plane[strip_no]^, k, 0);
     end;
```

We must "Dispose" of this memory when finished using it; that is the purpose of "recycle_plane". The procedure "zero_plane" blanks the whole plane very fast, by using the standard procedure *FillChar* which, like *Move*, is a low level procedure. Its syntax should be clear from the example. (*FillChar* is dangerous because it does no checking, so misuse can overwrite memory.)

We have our plane in memory; our next goal is to dump it where we can see it, or save it for later viewing. The easiest is to dump our small model plane to the screen. We use the Graph unit's *InitGraph* to open the graphics screen, its procedure *PutPixel* to copy point-by-point "plane" onto the screen, and finally *CloseGraph* to return to text mode.

```
     procedure   dump_to_screen;

        var   gd, gm: Integer;
                 x, y: Word;

     begin
     gd := Detect;
     InitGraph(gd, gm, '');
     for   x := 0   to   no_cols - 1   do
        begin
        for   y := 0   to   no_rows - 1   do
           if   (plane[y div rows_per_strip]^
                        [y mod rows_per_strip, x div 8]
                  and mask[x mod 8]) <> 0   then
                     PutPixel(x, y, White);
        end;
     ReadLn;
     CloseGraph;
     end;     { dump_to_screen }
```

The next procedure dumps the plane to a disk file, for later printing. This can be done by means of DOS's COPY, but be warned to use its /B (binary) option before the name of the (input) disk file. For high speed, we declare the output file as type **file**, an untyped file, which we write to with "BlockWrite". Note that there is a running progress report on the screen, to keep you informed that your computer is working, not hung. Also the loop has an escape clause, really unnecessary as this has been well tested, but...

```
procedure  dump_to_file;
{ Set up for HP and other printers using HP-PCL }

    var  strip_no, row_no: Word;
                        hh: file;
                        ss: Str25;

    begin    { dump_to_file }
    WriteLn('Dumping plot to file ', file_name);
    Assign(hh, file_name);
    Rewrite(hh, 1);
    { Initializations }
    if  IOResult <> 0  then  Halt;
    ss := reset_printer;
    BlockWrite(hh, ss[1], Length(ss));
    ss := clear_margins;
    BlockWrite(hh, ss[1], Length(ss));
    ss := graf_resol300;
    BlockWrite(hh, ss[1], Length(ss));
    ss := hor_posit;
    BlockWrite(hh, ss[1], Length(ss));
    ss := ver_posit;
    BlockWrite(hh, ss[1], Length(ss));
    ss := graf_start;
    BlockWrite(hh, ss[1], Length(ss));
    { Dump image }
    ss := graf_send250;
    for  strip_no := 0  to  7  do
       begin
       if  KeyPressed  then  Halt;
       for  row_no := 0  to  rows_per_strip - 1  do
          begin
          BlockWrite(hh, ss[1], Length(ss));
          BlockWrite(hh, plane[strip_no]^[row_no],
                        bytes_per_row);
          if  row_no mod 50 = 0  then
             WriteLn('Strip ', strip_no,
                    ' Row ', row_no);
          end;    { for row_no }
       WriteLn;
       end;    { for strip_no }
    ss := graf_stop;
```

```
BlockWrite(hh, ss[1], Length(ss));
ss := #12 + reset_printer;   { #12: FormFeed }
BlockWrite(hh, ss[1], Length(ss));
Close(hh);
WriteLn('Be sure to use');
WriteLn('  COPY /B ', file_name, '.prn');
WriteLn('with the /B option!');
end;   { dump_to_file }
```

Finally, we dump "plane" directly to the printer. There is a problem here with using "Write" to send individual bytes to the printer: Some DOS functions interpret the control character Ctrl–Z = #26 as "end of file". This is probably an anachronism by now, but the fact is that DOS does not like to send Ctrl–Z to a "character device" like a printer, and Borland Pascal's "Write" apparently uses DOS services. (Recall that some uses of COPY require the /B option to tell DOS that the file is binary, not text, and Ctrl–Z should be passed.) We get around this by using the even lower level BIOS service for sending a character to the printer.

```
procedure  dump_to_printer;
{ Set up for HP-PCL printers
  Use BIOS to get around DOS's interpretation of
  #26 = SUB = Ctrl-Z as "end of file". }

  var  strip_no, row_no, k: Word;
                       hh: file of Char;
                     regs: Registers;

  procedure  write_string(ss: Str25);

    var  k: Word;

    begin
    for  k := 1  to  Length(ss)  do
      Write(hh, ss[k]);
    end;   { write_string }

  begin   { dump_to_printer }
  WriteLn('Dumping plot to printer');
  Assign(hh, 'prn');
  Rewrite(hh);
  if  IOResult <> 0  then  Halt;
  write_string(reset_printer);
  write_string(clear_margins);
  write_string(graf_resol300);
  write_string(hor_posit);
```

```
write_string(ver_posit);
write_string(graf_start);
{ print rows }
regs.DX := $00;    { Select printer 0 = prn }
for  strip_no := 0  to  7  do
   begin
   if  KeyPressed  then  Halt;
   for  row_no := 0  to  rows_per_strip - 1  do
      begin
      write_string(graf_send250);
      for  k := 0  to  bytes_per_row - 1  do
         begin
         { Write byte }
         regs.AH := $00;    { print byte in AL }
         regs.AL := plane[strip_no]^[row_no, k];
         Intr($17, regs);
         { BIOS printer services }
         end;
      if  row_no mod 50 = 0  then
         WriteLn('Strip ', strip_no,
                ' Row ', row_no);
      end;    { for  row_no }
   WriteLn;
   end;    { for strip_no }
write_string(graf_stop);
write_string(#12 + reset_printer);  { form feed }
Close(hh);
end;    { dump_to_printer }
```

Each pass through the inner loop sends a control string to the printer announcing how many graphics bytes will follow, and then the actual graphics bytes. You may think that one character at a time transmission is slow, but the bottleneck is the printer, not the computer, which is always close to infinitely faster.

Note that procedures dump_to_screen and dump_to_printer use the constant graf_send250, which is defined for our large model plane, not the small model screen version. So don't try dumps of the small model plane to screen or printer without some modifications!

We have come to the heart of the matter, actually drawing something. The most basic operation is plotting a single point, and we do this in two possible ways according to the Boolean variable "xor_plot": either setting the corresponding bit, or *xor*'ing with what is already in that bit.

```
{ Drawing procedures }
   procedure  plot(x, y: Integer);
      { Plots a single point in plane }
      { x_min <= x <= x_max,   y_min <= y <= y_max }

      var  b, strip_no: Byte;
                  j, k: Word;

      begin
      with  x_coord[x]  do
         begin  k := bite;  b := bit;  end;
      with  y_coord[y]  do
         begin  strip_no := strip;  j := row;  end;
      if  xor_plot  then
         plane[strip_no]^[j, k]
               := plane[strip_no]^[j, k] xor mask[b]
      else  plane[strip_no]^[j, k]
                  := plane[strip_no]^[j, k] or mask[b];
      end;   { plot }
```

Everything we draw will require many calls of "plot", so this is a critical time sink. The lookup tables "x_coord" and "y_coord" limit arithmetic operations to **xor** only; in particular, no divisions. Drawing horizontal or vertical segments is routine, as are their applications: drawing a border for the plane, drawing a solid rectangle:

```
procedure  horizontal_seg(x0, x1, y: Integer);
   { No clipping done; use with care! }

   var  x: Integer;

   begin
   if  x0 <= x1  then
      for  x := x0  to  x1  do  plot(x, y)
      else  for  x := x1  to  x0  do  plot(x, y);
      end;   { horizontal_seg }

procedure  vertical_seg(x, y0, y1: Integer);
   { No clipping done; use with care! }

   var  y: Integer;
```

```
begin
if  y0 <= y1  then
    for  y := y0  to  y1  do  plot(x, y)
  else  for  y := y1  to  y0  do  plot(x, y);
  end;   { vertical_seg }
```

```
procedure  border;

  begin
  horizontal_seg(x_min, x_max, y_min);
  horizontal_seg(x_min, x_max, y_max);
  vertical_seg(x_min, y_min, y_max);
  vertical_seg(x_max, y_min, y_max);
  end;   { border }
```

```
procedure  solid_rectangle(x0, y0, x1, y1: Integer);
  { lower left (x0, y0); upper right (x1, y1) }

  var  y: Integer;

  begin
  if  y0 <= y1  then
    for  y := y0  to  y1  do
        horizontal_seg(x0, x1, y)
  else  for  y := y1  to  y0  do
      horizontal_seg(x0, x1, y);
  end;   { solid_rectangle }
```

Exercises

1. "Solid_rectangle" uses "horizontal_seg" heavily. Perhaps you can write a much faster version of "horizontal_seg".

Section 4.   Drawing a Segment

The first problem with drawing the segment from (x0, y0) to (x1, y1) is that one or both of its endpoints may lie out of bounds, in fact the whole segment may not show at all. This happens all the time in programs that graph functions: curves or surfaces. If the segment does cross into the legal plane space, then we need a procedure that replaces segment endpoints outside of the plane by points on the boundary of the plane that yield the visible part of the segment. This is the purpose of the Boolean-valued function "clip", a subtle algorithm worthy of careful study. Note that it uses a scalar type with identifiers "left", . . . Each

endpoint is assigned a set of these identifiers. If the set is empty, the endpoint
is in the plane.

```
function  clip(var  x0, y0, x1, y1: Integer):
                    Boolean;
  { Cohen Sutherland algorithm }

  type  region
          = set of (left, right, below, above);

  var  reg0, reg1: region;
              t: Integer;
              b: Boolean;

  procedure  get_region(x, y: Integer;
                         var  reg: region);

     begin
     reg := [];
     if  x < x_min  then  reg := [left]
     else  if  x > x_max  then  reg := [right];
     if  y < y_min  then  reg := reg + [below]
     else  if  y > y_max  then
        reg := reg + [above];
     end;

  begin    { clip }
  repeat
     get_region(x0, y0, reg0);
     get_region(x1, y1, reg1);
     if  (reg0 = []) and (reg1 = [])   then
        { both points inside or on boundary }
        begin  clip := true;  Exit;  end
     else  if  reg0*reg1 <> []   then
        { both points left or both above, etc }
        begin  clip := false;  Exit;  end;
     { Either reg0 <> [] or reg1 <> [] }
     if . reg1 <> []   then
        begin
        b := left in reg1;
        if  b or (right in reg1)   then
           begin    { interpolate }
           if  b  then  t := x_min
           else  t := x_max;
           y1 := Round(y1 + (y0 - y1)
```

```
                                     * ((t - x1) / (x0 - x1)));
              x1 := t;        { Group to avoid overflow }
              end
         else
              { (above in reg1) or (below in reg1) }
              begin     { interpolate }
              if  below in reg1  then   t := y_min
              else   t := y_max;
              x1 := Round(x1 + (x0 - x1)
                             * ((t - y1) / (y0 - y1)));
              y1 := t;
              end;
         end
    else    { reg1 = [], reg0 <> [] }
         begin
         b := left in reg0;
         if  b  or  (right in reg0)    then
              begin    { interpolate }
              if  b   then   t := x_min
              else   t := x_max;
              y0 := Round(y0 + (y1 - y0)
                             * ((t - x0) / (x1 - x0)));
              x0 := t;
              end
         else
              { (above in reg0) or (below in reg0) }
              begin    { interpolate }
              if  below in reg0   then   t := y_min
              else   t := y_max;
              x0 := Round(x0 + (x1 - x0)
                             * ((t - y0) / (y1 - y0)));
              y0 := t;
              end;
         end;
   until  false;
   end;    { clip }
```

Next we have the critically important procedure, "segment" for drawing a line segment. This brings us to a basic principle of computer graphics: do only integer arithmetic in loops, never floating point arithmetic, which is much slower. The problem is to find all points of the plane *close* to the geometrical line segment, and to do it *fast*.

What might be your first guess on how to approximate a segment: solve a linear equation (including multiplications and a division) and then round the result? This is not satisfactory.

To fix ideas, suppose that x0 < x1, y0 < y1, and the (nonnegative) slope is at most 1. Then we want exactly one point on each vertical line of the plane from x0 to x1. Usually there is just one, but there might be a tie with two points, one on each side of the line—it doesn't matter which of the two we choose.

We use two strategies. If the slope is between 1/2 and 1, we use the basic algorithm of J. Bresenham, which (in our case) moves from left to right, determining at each move if the y-level stays the same or increases by one. If the slope is between 0 and 1/2 we use a variation (probably also due to Bresenham) called a *run-length slice* algorithm. The idea is that the "segment" consists of runs of horizontal pixels, so we move *up* from y0 to y1, and determine at each y the length (of only two possible) of the next run of horizontal pixels. For instance, if the slope is 3/11, then each horizontal run has 3 or 4 pixels. If we draw three, then there is an error of 2/3. When the accumulated error exceeds one, the next run is 4 pixels, and the error is decremented.

It is these tests that take the most time in "segment". To minimize tests, the algorithm works symmetrically, from both ends towards the middle. Organizing all cases takes some thought, and the resulting algorithm, one of the most basic in computer graphics, is worth much study. For clarity, there are four subprocedures, each called once. It is advisable to put these *in line* (incorporate them into the body of "segment") to save the overhead of procedure calls. While the subprocedures are not in loops, any graphing program will usually call "segment" many times.

```
procedure  segment(x0, y0, x1, y1: Integer);
  { Draw segment (x0, y0) (x1, y1)
    Bresenham's algorithm: J E Bresenham, Algorithm
    for computer control of a digital plotter,
    IBM Systems Journal 4(1), 1965, pp 25-30;
    J D Foley, et al, Computer Graphics 2/e,
    Addison-Wesley 1990, pp. 73, ff; B Wyvill,
    Symmetric double-step line algorithm, Graphics
    Gems, Academic Press, 1990, pp 101-104; M Abrash,
    Graphics Programming, (Run-length slice)
    Dr Dobb's J #194, Nov. 1992, pp 171 ff.
    Note that each case is treated symmetrically, to
    halve the number of tests. }

  var    dx, two_dx,
         dy, two_dy,
           abs_slope,
       Error, dError,
          eps, sum, t: Integer;
```

```
procedure   run_length_big_y;
   { |slope| > 2 }

   var   y00, y01, y10, y11: Integer;

   begin
   { Run-length slice algorithm.
     Main idea: The number of points in
     each vertical segment is either [dy/dx]
     or [dy/dx] + 1.  So each test results in
     several points plotted. }
   if  y1 < y0   then
       begin    { switch points }
       t := x0;   x0 := x1;   x1 := t;
       t := y0;   y0 := y1;   y1 := t;
       end;
   if  x1 > x0   then   eps := 1   else   eps := -1;
   { Initialization }
   abs_slope := dy div dx;    { [dy/dx] }
   dError := 2*(dy mod dx);
   sum := y0 + y1;
   Error := dy mod two_dx;
   y00 := y0;
   y10 := y1;
   y01 := y0 + dy div two_dx;
   while   Abs(x1 - x0) >= 2   do
       begin
       if  Error >= dx   then
           begin
           Inc(y01);
           Error := Error - two_dx;
           end;
       vertical_seg(x0, y00, y01);
       y11 := sum - y01;
       vertical_seg(x1, y11, y10);
       { Prepare for next segments }
       x0 :=   x0   + eps;   x1 :=   x1 - eps;
       y00 := y01 + 1;   y10 := y11 - 1;
       y01 := y01 + abs_slope;
       Error := Error + dError;
       end;    { while }
   { Final segment }
   if  x0 = x1   then   vertical_seg(x0, y00, y10)
   else
       begin
       sum := sum div 2;
```

```
      vertical_seg(x0, y00, sum);
      vertical_seg(x1, sum + 1, y10);
      end;
   end;    { run_length_big_y }

procedure  run_length_big_x;
   { |slope| < 1/2 }

   var  x00, x01, x10, x11: Integer;

   begin
   { Run-length slice algorithm }
   if  x1 < x0  then
      begin    { switch points }
      t := x0;   x0 := x1;   x1 := t;
      t := y0;   y0 := y1;   y1 := t;
      end;
   if  y1 > y0  then   eps := 1  else   eps := -1;
   { Initialization }
   abs_slope := dx div dy;    { [dx/dy] }
   dError := 2*(dx mod dy);
   sum := x0 + x1;
   Error := dx mod two_dy;
   x00 := x0;   x10 := x1;
   x01 := x0 + dx div two_dy;
   while  Abs(y1 - y0) >= 2  do
      begin
      if  Error >= dy  then
         begin
         Inc(x01);   Error := Error - two_dy;
         end;
      horizontal_seg(x00, x01, y0);
      x11 := sum - x01;
      horizontal_seg(x11, x10, y1);
      { Prepare for next segments }
      y0 := y0  + eps;   y1 := y1 - eps;
      x00 := x01 + 1;
      x10 := x11 - 1;
      x01 := x01 + abs_slope;
      Error := Error + dError;
      end;    { while }
   { Final segment }
   if  y0 = y1  then
      horizontal_seg(x00, x10, y0)
   else
```

```
     begin
     sum := sum div 2;
     horizontal_seg(x00, sum, y0);
     horizontal_seg(sum + 1, x10, y1);
     end;
   end;   { run_length_big_x }

procedure  bresenham_big_y;
   { 2 >= |slope| >= 1 }

   begin    { Basic Bresenham algorithm }
   if  x1 < x0   then
      begin   { switch points }
      t := x0;   x0 := x1;   x1 := t;
      t := y0;   y0 := y1;   y1 := t;
      end;
   if  y0 < y1   then   eps := 1   else   eps := -1;
   Error := -dy;
   while   Abs(y1 - y0) >= 2   do
      begin
      plot(x0, y0);   plot(x1, y1);
      Error := Error + two_dx;
      if  Error >= 0   then
         begin
         Error := Error - two_dy;
         Inc(x0);   Dec(x1);
         end;
      y0 := y0 + eps;   y1 := y1 - eps;
      end;   { while }
   if  y0 = y1   then   plot(x0, y0)
   else
      begin
      plot(x0, y0);   plot(x1, y1);
      end;
   end;   { bresenham_big_y }

procedure  bresenham_big_x;
   {  1/2 <= |slope| < 1 }

   begin    { Basic Bresenham algorithm }
   if  y1 < y0   then
      begin    { switch points }
      t := x0;   x0 := x1;   x1 := t;
      t := y0;   y0 := y1;   y1 := t;
      end;
   if  x0 < x1   then   eps := 1   else   eps := -1;
```

```
    Error := -dx;
    while  Abs(x1 - x0) >= 2  do
        begin
        plot(x0, y0);   plot(x1, y1);
        Error := Error + two_dy;
        if  Error >= 0  then
            begin
            Error := Error - two_dx;
            Inc(y0);   Dec(y1);
            end;
        x0 := x0 + eps;   x1 := x1 - eps;
        end;    { while }
    if  x0 = x1   then   plot(x0, y0)
    else
        begin
        plot(x0, y0);   plot(x1, y1);
        end;
    end;    { bresenham_big_x }

begin    { segment }
if   not clip(x0, y0, x1, y1)   then   Exit;
if   x0 = x1   then   vertical_seg(x0, y0, y1)
else   if   y0 = y1   then
    horizontal_seg(x0, x1, y0)
else
    begin    { x0 <> x1   and   y0 <> y1 }
    dx := Abs(x1 - x0);
    dy := Abs(y1 - y0);
    two_dx := 2*dx;   two_dy := 2*dy;
    if   dy > 2*dx   then   run_length_big_y
    else   if   2*dy < dx   then   run_length_big_x
    { Remaining:   1/2 <= |slope| <= 2 }
    else   if   dx <= dy   then   bresenham_big_y
    else   bresenham_big_x;
    end;    { x0 <> x1   and   y0 <> y1 }
end;    { segment }
```

Section 5. Plotting a Circle

Drawing a circle with center $(x0, y0)$ and radius r is another fundamental algorithm of computer graphics. There are several ways to do so; probably the most efficient is Bresenham's algorithm. You should do the hand calculation to verify that the change in the variable s is as given in the two cases. Once again, to reduce the number of tests, we only compute one-eighth of the circle, and use symmetries for the other 7/8-ths. The code is a little fussy at the ends of the arc; that is because if xor_plot = True, we must be careful not to draw any point twice; the second draw will turn it off. (Note that `circle` is only for drawing a complete circle that is entirely within the "plane"; there is no clipping here.)

```
procedure   circle(x0, y0, r: Integer);
{ J. Bresemham's algorithm: JE Bresenham, A linear
  algorithm for incremental digital display of
  circular arcs, CACM 20, 1977, pp 100-106.
  Take (x0, y0) = (0, 0).   Assume (x, y) in the
  first octant of the circle.   The next point
  is either (x, y + 1) or   (x - 1, y + 1),
  the one closer to the circle.   The quantity
     s = [x^2 + (y+1)^2 - r^2]
       + [(x-1)^2 + (y+1)^2 - r^2]
  is negative in the first case, else positive. }

  var  x, y, s: Integer;
           xx, yy: array[1..8] of Integer;
                k: Word;

  begin
  if  (x0 + r > x_max) or (x0 - r < x_min) or
           (y0 + r > y_max) or (y0 - r < y_min)    then
       Exit;
  plot(x0 + r, y0);   plot(x0 - r, y0);
  plot(x0, y0 + r);   plot(x0, y0 - r);
  x := r;  y := 0;   s := 3 - 2*r;
  for  k := 1  to   4   do
       begin
       yy[k]  := y0;   xx[4 + k]  := x0;
       end;
  xx[1]  := x0 + r;   xx[2]  := xx[1];
  xx[3]  := x0 - r;   xx[4]  := xx[3];
  yy[5]  := y0 + r;   yy[6]  := y0 - r;
  yy[7]  := yy[5];    yy[8]  := yy[6];
  while  y < x - 1   do
       begin
       Inc(yy[1]);   yy[3]  := yy[1];
       Dec(yy[2]);   yy[4]  := yy[2];
       Inc(xx[5]);   xx[6]  := xx[5];
       Dec(xx[7]);   xx[8]  := xx[7];
       if  s <= 0   then   s := s + 4*y + 6
       else
            begin
            s := s + 4*(y - x) + 10;   Dec(x);
```

```
        Dec(xx[1]);   xx[2] := xx[1];
        Inc(xx[3]);   xx[4] := xx[3];
        Dec(yy[5]);   yy[7] := yy[5];
        Inc(yy[6]);   yy[8] := yy[6];
      end;
    Inc(y);
    if  x = y  then
        for  k := 1  to  4  do  plot(xx[k], yy[k])
      else  for  k := 1  to  8  do
        plot(xx[k], yy[k]);
    end;    { while }
  end;    { circle }

begin
xor_plot := false;
end.    { unit  grafunit }
```

To develop a unit like "grafunit" obviously requires a great deal of testing. It would be unthinkable to send a plane to the printer each time any little correction is made in the code. I believe I averaged 60 tests per hour over many hours while writing the unit above; this might have taken 5 – 10 minutes per test printing graphics on an actual printer. That is why I found "dump_to" screen invaluable.

Now let's run some tests of "grafunit". The following program will get you started and a few exercises will carry you further.

Program 5.5.1

```
{$R+,S+}
program  testgrfu;    { 08/24/94 }

  uses  grafunit;

  var  k: Word;

  procedure  black_hole;

      var  x0, y0, x1, y1: Integer;
                    i: Word;

      begin
      x0 := 0;   y0 := 0;   i := 1;
      Write('Plotting:  ');
      repeat
          x1 := Round(0.5 * i * Cos(i));
          y1 := Round(0.5 * i * Sin(i));
```

```
        if   (Abs(x1) >= x_max)
             or (Abs(y1) >= y_max)   then   Break;
        segment(x0, y0, x1, y1);
        x0 := x1;   y0 := y1;
        Inc(i);   Write('.');
    until  false;
    WriteLn;
    end;   { black_hole }

begin   { testgrfu }
WriteLn('Initializing');
init_plane;
border;
black_hole;
file_name := 'c:dump.prn';
dump_to_file;
dump_to_printer;
recycle_plane;
Write('Press <Enter> to quit: ');   ReadLn;
end.   { testgrfu }
```

Exercises

1. Add a grid to the plane, just dots at the intersections of the grid lines, not the lines themselves, which would clutter the plane.
2. Add a procedure to draw a series of concentric circles.
3. Add a procedure to draw a spiral.

Section 6. The Graphics Mouse

We now look at examples of programming the mouse in graphics mode. The first program, *grafmous*, uses the units graphs from Section 1 and the unit *mouse* from Chapter 4, Section 4. Let us examine the declarations section of the unit. Note particularly the typed two-dimensional array constant "masks", how it is initialized, and the accompanying comment showing visually what the screen and cursor masks mean:

Program 5.6.1

```pascal
program  test_graph_mouse;     { 08/22/94 }

  uses  CRT, Graph, graphs, mouse;

  const  x_min = 220;   x_max = 420;
         y_min = 290;   y_max = 390;

         masks: array[0..1, 0..15] of Word
              = ( ($E1FF, $EDFF, $EDFF, $EDFF,
                   $EDFF, $EC00, $EDB6, $EDB6,
                   $0DB6, $6FFE, $6FFE, $6FFE,
                   $7FFE, $7FFE, $7FFE, $0000) ,

                  ($1E00, $1200, $1200, $1200,
                   $1200, $13FF, $1249, $1249,
                   $F249, $9001, $9001, $9001,
                   $8001, $8001, $8001, $FFFF) );
```

```
{ Screen mask:          Cursor mask:        Cursor mask:
FEDCBA9876543210    FEDCBA9876543210    FEDCBA9876543210
1110000111111111    0001111000000000       XXXX
1110110111111111    0001001000000000      X   X
1110110111111111    0001001000000000      X   X
1110110111111111    0001001000000000      X   X
1110110111111111    0001001000000000      X   X
1110110000000000    0001001111111111      X   XXXXXXXXXX
1110110110110110    0001001001001001      X   X   X   X   X
1110110110110110    0001001001001001      X   X   X   X   X
0000110110110110    1111001001001001    XXXX  X   X   X   X
0110111111111110    1001000000000001    X   X               X
0110111111111110    1001000000000001    X   X               X
0110111111111110    1001000000000001    X   X               X
0111111111111110    1000000000000001    X                   X
0111111111111110    1000000000000001    X                   X
0111111111111110    1000000000000001    X                   X
0000000000000000    1111111111111111    XXXXXXXXXXXXXXXXXX
  }

  var      ss, tt: string;
        not_mouse: Boolean;
            x, y: Word;
          button: Byte;
        k, count: Word;
```

Remember the order in which the mouse cursor is written to the screen: first what is on the screen in the 16×16 cursor area is **and**'ed with the screen mask, then the result is **xor**'ed with the cursor mask. Now look at the comment showing the bits of the screen mask. Clearly *and*'ing with the screen will change the boundary of the hand to black and leave everything else unchanged. The cursor mask is the opposite of the screen mask, bits 1 and 0 transposed. So **xor**'ing with the cursor mask will change the boundary of the hand to white and "not" every other bit.

The default graphics cursor is the arrow you see in many graphics programs, like *MS Windows*. Now examine the rest of "test_graph_mouse" and run it:

```
begin
GotoXY(1, 6);
Write('To start graphics, click both buttons');
repeat
    button := 2;
    get_mouse_button_press(button, count, x, y);
until   button = 3;

open_graph;
SetColor(Yellow);
SetBkColor(DarkGray);
SetFillStyle(SolidFill, Red);
OutTextXY(0, 0,
          'Testing graphics mouse.   Move around!');
reset_mouse(not_mouse, button);
Bar(x_min, y_min, x_max, y_max);

show_cursor;
OutTextXY(0, 20, 'Click left to hide cursor');
repeat
    button := 0;
    get_mouse_button_press(button, count, x ,y);
until   button = 1;
hide_cursor;
OutTextXY(0, 40,
 'Click right for the finger; move it into the red!');
repeat
    button := 1;
    get_mouse_button_press(button, count, x ,y);
until   button = 2;
set_graph_cursor_shape(4, 0, @masks);
show_cursor;
repeat
    get_mouse_status(button, x, y);
until   (x >= x_min) and (x <= x_max) and
```

```
        (y >= y_min) and (y <= y_max);
   set_hide_cursor_window(x_min, y_min, x_max, y_max);
   OutTextXY(0, 60,
      'Click left: restrict and show cursor');
   repeat
      button := 0;
      get_mouse_button_press(button, count, x ,y);
   until  button = 1;   { Left pressed }
   set_cursor_x_lim(x_min, x_max);
   set_cursor_y_lim(y_min, y_max);
   OutTextXY(0, 80,
      'Try to move the cursor out of jail:');
   show_cursor;
   OutTextXY(0, 100, 'Double click left to quit:');
   repeat  until  double_click(500);
   close_graph;

   reset_mouse(not_mouse, button);
   ClrScr;
   end.   { test_mouse }
```

Our next program is one of the war horses of mouse programming, the *piano*. It does give us some practice in defining screen regions that respond differently to a mouse click, and it also gives us reason to examine how the computer speaker can be controlled.

This speaker, I regret to report, is of the lowest possible fidelity. It has no volume control; it resonates very differently to different frequencies. Don't expect much quality sound!

We could use the procedures "Sound" and "NoSound" from the CRT unit to generate notes in the program "piano", but "Sound" has rather poor intonation because it only allows whole number frequencies. Instead, we'll develop our own procedures in a unit "speaker".

The speaker cone has two positions. We send it a square wave of given frequency. During each cycle the cone is moved to the on position for half of a cycle, then moved to the off position for the other half of the cycle. This moves air, which tends to smooth the square wave somewhat, producing its unpleasant note.

Your computer contains a chip called the 8255, or Programmable Peripheral Interface (PPI) which controls I/O. One of its ports (#61) controls how the timer chip 8254 drives the speaker. The comments in "speaker" give details. One point to note is that once the sound is started, it continues until turned off. This is useful: we program a sound to start when a mouse key is pressed, and stops when it is released.

Program 5.6.2

```
unit  speaker;    { 08/22/94 }
{ Theory: The 8255 chip (PPI: Programmable Peripheral
  Interface) controls I/O, and has various ports,
  including I/O port $61, a byte size port with bits
  b7..b0.

  If b0 = 1, then channel 2 of the 8254 timer chip
  drives the speaker.   (There are two other channels
  used by the system.)

  The speaker is on if b1 = 1, off if b1 = 0.   The idea
  is to turn on b0 and b1 without changing any other
  bits.   The 8254 timer chip port $43 must be sent $B6
  to notify the timer that the next two bytes (count)
  are a word inversely proportional to the frequency,
  and that the output should be a square wave.   The
  timer turns the speaker on/off every count*freq0 sec.
  There is no volume control! }

interface

   procedure  sound(freq: Real);
   procedure  no_sound;

implementation

   procedure  sound(freq: Real);

      const  freq0 = 1.19318e6;    { Hz }
      { From the Borland Pascal RTL "sound" source code,
        CRT.ASM: $1234DD = 1193181.
        From Norton, P, Programmer's Guide to the IBM
        PC, MicroSoft Press, 1985. p. 148: 1,193,180.
        Willen, D C and Krantz, J I, 8088 Assembly
        Language Programming: The IBM PC, Sams & Co.,
        1983, p.105: 1.19318 MHz. }

      var  count: Word;
              b: Byte;
```

```
  begin    { sound }
  if  freq <= 19.0  then  Exit;
  { freq0/ (2^16 - 1) = 18.2 }
  count := Round(freq0/freq);
{ The speaker is on only if bits 1 and 0 of
  port $61 are set.  The other bits have other uses.
  Changing them could disable the keyboard, change
  the RAM configuration, or cause worse disasters. }
   b := port[$61];   { speaker port }
   if  b and $03 = 0  then    { speaker off }
      begin    { turn it on }
      b := b or $03;   { Only bits 0, 1 changed }
      port[$61] := b;
      port[$43] := $B6;   { 10 11 011 0 }
      { Program channel 2, 2 bytes of count number,
        square wave, binary counting }
      end;
   { Send the timer the 2 bytes of the note. }
   b := count and $FF;
   port[$42] := b;   { low byte }
   b := count shr 8;
   port[$42] := b;   { high byte }
   end;   { sound }

procedure  no_sound;

   var  x: Byte;

   begin
   x := port[$61];
   x := x and $FC;   { 11111100 }
   port[$61] := x;
   end;

end.    { unit  speaker }
```

As it is somewhat difficult to drag out of the literature what the various ports are, I have put all that I know about them in the data file "ports.dat".

We come to the program "piano". Its declaration section includes the shape of the mouse cursor that will bang keys. The keyboard extends almost 5 octaves, and the 12 frequencies of each octave are stored in a real two-index array. There are 7 white keys and 5 black keys in each piano octave, and it is convenient to keep separate arrays for these frequencies by color of the keys. Don't mind that a couple of dummy keys have been added to the black key array; that just shortens the program a little. Here are the preliminary declarations:

Program 5.6.3

```
program  piano;    { 08/22/94 }

   uses  CRT, speaker, DOS, mouse, Graph, graphs;

   const  masks: array[0..1, 0..15] of Word
                       = ( ($FE7F, $FE7F, $FC3F, $FC3F,
                            $F81F, $F81F, $F00F, $F00F,
                            $E007, $E007, $C003, $C003,
                            $FE7F, $FE7F, $FE7F, $FE7F),
                           ($0180, $0180, $03C0, $03C0,
                            $07E0, $07E0, $0FF0, $0FF0,
                            $1FF8, $1FF8, $3FFC, $3FFC,
                            $0180, $0180, $0180, $0180) );
{ Screen mask              Cursor mask
  1111111001111111         0000000110000000
  1111111001111111         0000000110000000
  1111110000111111         0000001111000000
  1111110000111111         0000001111000000
  1111100000011111         0000011111100000
  1111100000011111         0000011111100000
  1111000000001111         0000111111110000
  1111000000001111         0000111111110000
  1110000000000111         0001111111111000
  1110000000000111         0001111111111000
  1100000000000011         0011111111111100
  1100000000000011         0011111111111100
  1111111001111111         0000000110000000
  1111111001111111         0000000110000000
  1111111001111111         0000000110000000
  1111111001111111         0000000110000000 }
   const  width = 18;    { width of a white key }
          no_white_keys = 35;
          x1 = no_white_keys * width;   y1 = 160;
          { keyboard from (0, 0) to (x1, y1) }
          black_key_low = 100;
          half_black_width = width div 3;
          quit_x = 500;   quit_y = 400;

   var   frequency: array[0..4, 0..11] of Real;
        white_freq: array[0..4, 0..6] of Real;
        black_freq: array[0..4, 0..6] of Real;
          mouse_OK: Boolean;
              x, y: Word;
            button: Byte;
```

Temperament basics: the well-tempered scale has all frequency ratios of two successive notes (half tones) equal.   As there are 12 notes to an octave, and frequency doubles in an octave, each frequency is the 12-th root of 2 times the previous frequency. We initialize frequencies accordingly:

```
procedure  init_frequency;

    const
            white_map: array[0..6] of Word
                     = (0, 2, 4, 5, 7, 9, 11);
            black_map: array[0..6] of Word
                     = (0, 1, 3, 0, 6, 8, 10);

    var  semitone_ratio: Real;
                    i, j: Word;
                    x: Real;

    begin
    semitone_ratio := Exp(Ln(2.0)/12.0);
                    { 2^(1/12) }
    frequency[1, 9] := 440.0;    { A }
    x := 440.0;
    for  j := 8  downto  0  do
        begin
        x := x/semitone_ratio;
        frequency[1, j] := x;
        end;
    x := 440.0;
    for  j := 10  to  11  do
        begin
        x := x*semitone_ratio;
        frequency[1, j] := x;
        end;
    for  j := 0  to  11  do
        begin
        frequency[0, j] := 0.5*frequency[1, j];
        frequency[2, j] := 2.0*frequency[1, j];
        frequency[3, j] := 4.0*frequency[1, j];
        frequency[4, j] := 8.0*frequency[1, j];
        end;
    for  i := 0  to  4  do
        begin
        for  j := 0  to  6  do
```

```
        begin
        white_freq[i, j]
            := frequency[i, white_map[j]];
        black_freq[i, j]
            := frequency[i, black_map[j]];
        end;
    end;
end;    { init_frequency }
```

Procedure "draw_keyboard" draws the keyboard and initializes the rest of the screen, in particular it marks a region to click on for quiting:

```
procedure  draw_keyboard;

    var  i, k: Word;

    begin
    i := 0;
    SetFillStyle(SolidFill, LightBlue);
    repeat
        Bar(i, 0, i + width, y1);
        Line(i, 0, i, y1);
        Inc(i, width);
    until  i >= x1;
    k := 6;
    i := 0;
    SetFillStyle(SolidFill, LightRed);
    repeat
        Inc(k);
        if  k = 7  then  k := 0;
        if  k in [1..2, 4..6]    then
            begin
            Bar(i - half_black_width, 0,
                i + half_black_width, black_key_low);
            Rectangle(i - half_black_width, 0,
                        i + half_black_width,
                        black_key_low);
            end;
        Inc(i, width);
    until  i > x1;
    Rectangle(0, 0, x1, y1);
    Rectangle(quit_x, quit_y, 639, 479);
    MoveTo(quit_x + 40, quit_y + 35);
    OutText('Q U I T');
```

```
OutTextXY(20, 300,
            'Hold down left mouse button');
OutTextXY(20, 320, 'on any key you choose.');
end;
```

Procedure "poll" checks for a left click; procedure "play" starts a note and continues it until the left button is released.

```
procedure  poll;

   begin
   repeat
      get_mouse_status(button,  x, y);
      if  button and $01 <> 0   then   Exit
      else  NoSound;
   until  false;
   end;

procedure  play;

   var  key, octave: Word;
            z, w: Integer;

   begin   { play }
   key := x div width;     { 0..34 }
   octave := key div 7;    { 0..4 }
 { Compute the note, adjusted to the lowest octave. }
   z := x mod (7*width);
   w := Round(z/width);
   { nearest line between white notes }
   if  (y <= black_key_low) and
       (Abs(width*w - z) <= half_black_width) and
       (w in [1, 2, 4..6])   then
          sound(black_freq[octave, w])
   else  sound(white_freq[octave, key mod 7]);
   repeat
      get_mouse_status(button, x, y);
   until  button and $01 = $00;
   NoSound;
   end;   { play }
```

Finally the main action pulls this all together:

```
begin   { piano }
init_frequency;
```

```
   reset_mouse(mouse_OK, button);
   if  not mouse_OK   then   Halt;
   open_graph;
   reset_mouse(mouse_OK, button);
   set_graph_cursor_shape(7, 0, @masks);
   show_cursor;
   set_mouse_sensitivity(32, 64, 64);
   mouse_gotoXY(320, 300);
   draw_keyboard;
   repeat
      poll;
      if  (x = 0) or (y = 0)   then
         begin  NoSound;  poll;  end;
      if  (x <= x1) and (y <= y1)   then   play
      else  if  (x >= quit_x) and
                (y >= quit_y)   then
         Break;
      NoSound;
   until  false;
   close_graph;
   end.   { piano }
```

Section 7. VGA Graphics

We have two missions here: test some of the interrupt $10 procedures in graphics mode; explore the VGA hardware.

   The following two programs require the unit "intr_10" from Chapter 4, Section 3. Be sure it is available to compile these programs. The first, "test_VGA", writes some text to the graphics screen, including the symbols for ASCII 1 – 31.

Program 5.7.1

```
program  test_VGA;    { 09/04/94 }

uses  CRT, Graph, intr_10;

var  GraphDriver, GraphMode: Integer;
                          k: Word;
               x, y, attr,
                   x1, y1,
              top, bottom,
          mode, cols, page: Byte;
                   ss, tt: string[80];

   begin   { test_VGA }
   GraphDriver := Detect;
```

```
InitGraph(GraphDriver, GraphMode, '');
SetBkColor(DarkGray);
SetColor(LightBlue);
ss := 'Graphics mode';
FillEllipse(320, 240, 20, 20);
x := 0;   y := 0;
write_string(0, x, y, ss, White);
set_cursor_position(0, 0, 5);
y := 5;
attr := 16*Brown + LightCyan;
for  k := 0  to  15  do
   begin
   Str(k:3, ss);
   x := 3*k;
   write_string(0, x, y, ss, attr);
   end;
for  k := 0  to  15  do
  write_chars(0, 3*k + 2, 6, Chr(k), attr, 1);
set_cursor_position(0, 0, 8);
y := 8;
for  k := 0  to  15  do
   begin
   Str((16 + k):3, ss);
   x := 3*k;
   write_string(0, x, y, ss, attr);
   end;
for  k := 0  to  15  do
   write_chars(0, 3*k + 2, 9, Chr(16 + k), attr, 1);
ss := '0123456';
x := 0;   y := 20;
write_string(0, x, y, ss, 11);
get_cursor(0, x, y, top, bottom);
Str(x, ss);   Str(y, tt);
Inc(y);
ss := 'At end of "0123456", (x, y) = (' + ss
      + ', ' + tt + ')';
write_string(0, x, y, ss, White);
ss := 'Press <Enter>: ';
x := 0;   y := 25;
attr := 16*DarkGray + Yellow;
write_string(0, x, y, ss, attr);
get_cursor(0, x, y, top, bottom);
repeat
   write_chars(0, x, y, #220, Yellow, 1);
   Delay(100);
   write_chars(0, x, y, #223, LightBlue, 1);
```

```
    Delay(100);
until   KeyPressed;
ReadLn;
CloseGraph;
ClrScr;
end.    { test_VGA }
```

The VGA medium resolution mode (640×350) has two pages (high resolution mode has only one page). Program "test_graph_pages" checks this out, and perhaps introduces some graphics possibilities you have not used before.

Program 5.7.2

```
program   test_graph_pages;    { 09/04/94 }

   uses   Crt, Dos, intr_10, Graph;

   var   GraphDriver,
            GraphMode: Integer;
               ss, tt: string;
                    k: Integer;

   begin    { test_graph_pages }
   GraphDriver := VGA;
   graphmode := VGAMed;    { 640x350, 2 display pages }
   InitGraph(GraphDriver, GraphMode, 'H:\');
   { 'H:\' is the directory path for *.BGI files;
      Use '' for current directory }
   write_string(0, 20, 0, 'Page 0', Yellow);
   write_string(0, 20, 1, 'Press <Enter>: ', Yellow);
   write_string(1, 30, 0, 'Page 1', Magenta);
   write_string(1, 30, 1, 'Press <Enter>: ', Magenta);
   SetVisualPage(0);    { Show page 0 }
   SetActivePage(1);    { Work on page 1 }
   ss := 'XX';
   for  k := 0  to  25  do
      begin    { Write some stuff on page 1 }
      Str(k, tt);
      tt := tt + '  ' + ss;
      if  k <= 9  then   tt := ' ' + tt;
      write_string(1, 0, k, tt, LightGreen);
      Line(0, 13 + 14*k, 8*Length(tt), 13 + 14*k);
      ss := ss + 'XX';
      end;
   SetColor(White);
   SetFillStyle(ltbkslashfill, LightRed);
```

```
Bar(500, 0, 600, 349);
ReadLn;
SetVisualPage(1);    { Show page 1 }
SetActivePage(0);    { Work on page 0 }
SetColor(LightMagenta);
SetFillStyle(XHatchFill, LightCyan);
FillEllipse(500, 175, 137, 100);
ReadLn;
SetVisualPage(0);    { Show page 0 }
ReadLn;
SetVisualPage(1);    { Show page 1 }
ReadLn;
RestoreCRTMode;    { Back to text mode }
Delay(400);    { Avoid filcker }
end.    { test_graph_pages }
```

Finally we come face to face with the VGA hardware. Borland's "Graph" unit has procedures "GetImage" and "PutImage", discussed in Section 1, for saving a portion of the graphics screen into a buffer, and later restoring it, possibly somewhere else. These are fine for a small region. But if you try to save and restore the whole screen, you will find these procedures painfully slow. By going to the hardware, we can do these operations very much faster.

The following program illustrates the method. Of course saving to memory rather than storing the screen to a disk file is much faster, however, the VGA screen uses 153,600 bytes, so the memory option may not be possible for a large program, especially in real mode. The program assumes disk storage, but the memory option is included in (* *) comments. The first part of the program, declarations and a procedure for filling the screen with 16 colors, is routine stuff:

Program 5.7.3

```
program  save_VGA;    { 09/04/93 }

uses  DOS, CRT, Graph;

const      pixels = 38400;    { 640*480 div 8 }
        file_name = 'picture';

type       plane = array[1..pixels] of Byte;
(*    plane_ptr = ^plane;    *)

var        storage: file;
            result: Word;
        graph_driver,
          graph_mode: Integer;
```

```
              (* Alternative to use memory instead of a file:
                     hold: array[0..3] of plane_ptr;
                          j: integer;    *)

procedure   fill_screen;

   const   dx = 640 div 4;   dy = 480 div 4;

   var   i, j, x, y: Integer;
                color: Word;

   begin
   color := 0;   y := 0;
   for   i := 0  to  3  do
      begin
      x := 0;
      for   j := 0  to  3  do
         begin
         SetFillStyle(SolidFill, color);
         Bar(x, y, x + dx, y + dy);
         Inc(x, dx);   Inc(color);
         end;
      Inc(y, dy);
      end;
   end;    { fill_screen }
```

Procedure "copy_screen" copies the screen into memory. The VGA is organized into 4 planes, red, green, blue, and gray. Each pixel corresponds to a bit in each plane. If all 4 bits are on, the pixel is white. Only the red bit on corresponds to a red pixel. This and the gray bit on makes a light red pixel, etc.

I/O port $03CE contains the index to the *VGA graphics control registers*. Set this index to 4 to read a plane. Then I/O port $03CF must also be set—to the plane to be read: 0...3. Then the actual data starts at absolute address $A000:$0000. In protected mode $A000 must be replaced by the selector SegA000. To read the plane, you simply copy 38,400 bytes by "BlockWrite" for a file, or "Move" for memory.

```
procedure   copy_screen;

   var   i: Byte;

   begin
   Assign(storage, file_name);
   Rewrite(storage, 1);
```

```
    { Port $3CE is the index to the graphics control
      registers.   Index 4: Read Plane Select. }
    port[$3CE] := 4;
    for  i := 3  downto  0  do
      { Port $3CF selects a plane (0...3). }
      begin
      port[$3CF] := i;
      BlockWrite(storage, mem[SegA000:$0000],
                  pixels, Result);
      if  result <> pixels  then
          begin
          WriteLn('Error: i = ', i);
          ReadLn;  Halt;
          end;
(*        Move(Mem[SegA000:$0000], hold[i]^, pixels);  *)
      end;
    Close(storage);
    end;   { copy_screen }
```

The reverse process requires accessing the *VGA sequencer*, which controls the timing of VGA functions. The index to its registers is in I/O port $03C4. If set to 2, this enables writing to the color planes. The I/O port $03C4 then determines which plane you write to. Actually you can write to more than one plane at a time. The low 4 bits of port $03C4 each control one plane, and 0 to 4 of these bits can be set. For instance value 6 = 0110 forces writing to planes 1 and 2 simultaneously.

```
procedure  restore_screen;

    var  i, j, k: Byte;

    begin
    Assign(storage, file_name);
    Reset(storage, 1);
    { Port $3C4 is the index to the Sequencer
      registers.   Index 2: Color Plane Write Enable. }
    port[$3C4] := 2;
    j := 8;    { 1000 }
    { Port $3C5 selects which plane, according
      to the bits }
(*    k := 3;   *)
    repeat
    { set: 0001, 0010, 0100, 1000 and combinations. }
        port[$3C5] := j;
        { SegA000: segment (selector) for the VGA/EGA. }
```

```
        BlockRead(storage, mem[segA000:$0000],
                pixels, Result);
        if   result <> pixels   then   Break;
(*       move(hold[k]^, mem[segA000:$0000], pixels);   *)
        j := j shr 1;    { 0100, 0010, 0001, 0000 }
(*      dec(k);    *)
    until  j = 0;
    Close(storage);
    { Must now restore all 4 planes for writing.
      Load the Sequencer, color plane write enable. }
    port[$3C5] := $0F;
    end;     { restore_screen }
```

Now we finish the program:

```
begin    { save_VGA }
ClrScr;
(* for  j := 0  to  3  do  new(hold[j]);
write('Memory available = ', memavail);   readln;   *)
graph_driver := Detect;
InitGraph(graph_driver, graph_mode, '');
if  GraphResult <> 0   then   Halt;
fill_screen;
copy_screen;
ReadLn;
CloseGraph;
WriteLn('Text screen, press <Enter>: ');
ReadLn;
InitGraph(graph_driver, graph_mode, '');
restore_screen;
OutText('Graph screen');
ReadLn;
CloseGraph;
end.    { save_VGA }
```

You can do some fun experiments with this program. For instance, read the planes back in a different order.

Section 8. Graphing 3-Space

We shall expand the unit "graphs", to include various projections of 3-space into the graphics screen. We begin with a preliminary unit "bgi_chr":

Program 5.8.1

```
unit  bgi_chr;     { 12/27/94 }

interface

    procedure EgaVgaDriverProc;
    procedure SmallFontProc;
    procedure SansSerifFontProc;

implementation

    procedure EgaVgaDriverProc; external;
    {$l egavga.obj }
    procedure SmallFontProc; external;
    {$l litt.obj }
    procedure SansSerifFontProc; external;
    {$l sans.obj }

end.
```

The purpose of this small unit is to link the VGA graphics driver and two fonts into executable programs *.exe, so that the driver file "egavga.bgi" and the font files "litt.chr" and "sans.chr" do not have to be present to run the programs. Note that Borland's utility program "bin_obj" must be run first to create object files. The details are in the comments in "bgi_chr.pas" (on disk).

Let's begin the listing of the "graphs" unit with its interface section. Much of this unit develops procedures for graphics based on the origin in the center of the screen, the x-axis horizontal right, the y-axis vertical up, as is usual in mathematics. It really helps a lot to make these normalizations; accordingly we make "norm_" versions of many standard graphics procedures.

The **type** "projection" is the basis for projecting a point **x** of 3-space into the plane. Besides a possible translation that fixes a local origin, the projection due to a projection matrix P is

$$\begin{bmatrix} x \\ y \\ z \end{bmatrix} \longrightarrow P \begin{bmatrix} x \\ y \\ z \end{bmatrix} = \begin{bmatrix} p_{11} & p_{12} & p_{13} \\ p_{21} & p_{22} & p_{23} \end{bmatrix} \begin{bmatrix} x \\ y \\ z \end{bmatrix}$$

Program 5.8.2

unit graphs;   { 01/02/95 }

interface

   uses   DOS, CRT, Graph, bgi_chr;

   type   vector = **array**[1..3] **of** Real;
   type projection = **array**[1..2] **of** vector;
   var abs_x_center,
        abs_y_center: Integer;

   procedure   open_graph;
   procedure   close_graph;
   procedure   pause;
   procedure   norm_out_text_XY(ss: **string**;
                                   x, y: Integer;
                                   color: Word);
   procedure   norm_put_pixel(x, y: Integer;
                                 color: Word);
   procedure   norm_line(x0, y0, x1, y1: Integer;
                      color: Word);
   procedure   norm_arrow(**const** x0, y0, x1, y1,
                         width: Integer; color: Word);
   procedure   norm_parallelogram
                  (x0, y0, x1, y1, x2, y2: Integer;
                    color: Word);
              { The 4th vertex is opposite (x0, y0) }

   procedure   norm_circle(x0, y0: Integer;
                          radius, color: Word);

   procedure   norm_disk(**const** x, y: Integer;
                        const radius: Word;
                        const color: Word);

   procedure   norm_arc(x0, y0: Integer;
                       theta0, theta1,
                       radius, color: Word);

   procedure   compute_isometric_matrix
                     (**var** P: projection);
   procedure   compute_dimetric_matrix
                     (alpha: Real; **var** P: projection);

```
procedure   compute_oblique_matrix
                   (alpha: Real; var   P: projection);
procedure   project(const   P: projection;
                   const   x, y, z: Real;
                   const   u0, v0: Integer;
                   var   u, v: Integer);
```

The implementation begins with "open_graph" and "close_graph", somewhat changed from earlier versions. Note in particular that the font is set to the SansSerif font, and that the center of the screen is set to the origin of the normalized coordinate system. Procedure "pause" is convenient for pausing and prompting to continue. Procedure "norm_out_text_XY" is useful for writing legends on the screen in the normalized coordinates. Note that it, as do most of these procedures, includes a color parameter, so it is not necessary to change color before and after calling the procedure.

implementation

```
   var   hold_color: Word;

   procedure   open_graph;

      var   GraphDriver,
            GraphMode: Integer;

      begin
      GraphDriver := VGA;
      GraphMode := VGAHi;    { 640 x 480 }
      InitGraph(GraphDriver, GraphMode, '');
      SetBkColor(Black);
      SetColor(White);
      SetTextStyle(SansSerifFont, HorizDir, 3);
      SetTextJustify(LeftText, BottomText);
      abs_x_center := (GetMaxX + 1) div 2;    { 310 }
      abs_y_center := (GetMaxY + 1) div 2;    { 240 }
      end;

   procedure   close_graph;

      var   gr: Integer;

      begin
      CloseGraph;
      gr := GraphResult;
```

```
if  gr <> 0   then
    begin
    WriteLn('GraphResult = ', gr);
    ReadLn;
    Halt;
    end;
end;

procedure  pause;

    var   ch: Char;

    begin
    OutTextXY(400, 475, 'Press <Space>:');
    ch := ReadKey;
    end;

procedure  norm_out_text_XY(ss: string;
                            x, y: Integer;
                            color: Word);

    begin
    hold_color := GetColor;
    SetColor(color);
    OutTextXY(x + abs_x_center,
            abs_y_center - y, ss);
    SetColor(hold_color);
    end;
```

The next three procedures plot points, line segments, and arrows. You should read the subroutine "arrow" carefully, as it is not obvious in advance how to draw an arrowhead in any direction. Note that a rotation of $\pi/2$ is involved.

```
procedure  norm_put_pixel(x, y: Integer;
                          color: Word);

    begin
    PutPixel(x + abs_x_center,
            abs_y_center - y, color);
    end;

procedure  norm_line(x0, y0, x1, y1: Integer;
                     color: Word);
```

```
  begin
  hold_color := GetColor;
  SetColor(color);
  Line(x0 + abs_x_center, abs_y_center - y0,
       x1 + abs_x_center, abs_y_center - y1);
  SetColor(hold_color);
  end;

procedure  norm_arrow(const  x0, y0, x1, y1,
                      width: Integer; color: Word);

  procedure  arrow(const  x0, y0, x1, y1,
                   width: Integer);

    var                  u, v, h,
          base_x, base_y, x, y: Real;

    begin
    Line(x0, y0, x1, y1);     { The shaft }
    u := x1 - x0;   v := y1 - y0;
    h := Sqrt(Sqr(u) + Sqr(v));
    u := width*u/h;   v := width*v/h;
    { (u, v) is a vector of length width
      in the direction of the shaft. }
    base_x := x1 - 3.0*u;   base_y := y1 - 3.0*v;
    { (base_x, base_y) is the base of the head.
      (-v, u) is orthogonal to the shaft and
      of length w. }
    x := base_x - v;   y := base_y + u;
    { (x, y) is one vertex of the head. }
    Line(Round(x), Round(y), x1, y1);
    x := base_x + v;   y := base_y - u;
    { Now (x, y) is the opposite vertex. }
    Line(Round(x), Round(y), x1, y1);
    end;    { arrow }


  begin    { norm_arrow }
  hold_color := GetColor;
  SetColor(color);
  arrow(x0 + abs_x_center, abs_y_center - y0,
        x1 + abs_x_center, abs_y_center - y1,
        width);
  SetColor(hold_color);
  end;    { norm_arrow }
```

Drawing a parallelogram involves some routine vector algebra, while drawing circles, disks, and circular arcs is absolutely routine.

```
procedure   norm_parallelogram
                      (x0, y0, x1, y1, x2, y2: Integer;
                      color: Word);
             { The 4th vertex is opposite (x0, y0) }


    begin
    norm_line(x0, y0, x1, y1, color);
    norm_line(x0, y0, x2, y2, color);
    x0 := x1 + x2 - x0;
    y0 := y1 + y2 - y0;
    norm_line(x0, y0, x1, y1, color);
    norm_line(x0, y0, x2, y2, color);
    end;    { norm_parallelogram }

procedure   norm_circle(x0, y0: Integer;
                            radius: Word;   color: Word);

    begin
    hold_color := GetColor;
    SetColor(color);
    Circle(x0 + abs_x_center, abs_y_center - y0,
            radius);
    SetColor(hold_color);
    end;

procedure   norm_disk(const  x, y: Integer;
                       const  radius: Word;
                       const  color: Word);

    begin
    SetFillStyle(SolidFill, color);
    FillEllipse(x + abs_x_center, abs_y_center - y,
                radius, radius);
    end;

procedure   norm_arc(x0, y0: Integer;
                     theta0, theta1,
                     radius, color: Word);

    begin
    hold_color := GetColor;
```

```
SetColor(color);
Arc(x0 + abs_x_center, abs_y_center - y0,
    theta0, theta1, radius);
SetColor(hold_color);
end;
```

Now we come to projections. We consider only projections from a *point at infinity*. Think of a light source very far away, so all the rays of light coming from the source are parallel to each other. Each point of the space figure we are drawing casts a shadow on the plane of the screen, and that shadow makes up our graph, or projection. These are the drawings of mathematics and much of engineering. The drawings are *not in perspective*. Perspective drawing is projection from a *finite* point in space; that is the difference.

There are two main classes of (parallel) projections, *orthogonal projections* and *oblique projections* . In orthogonal projections, the rays from the projection point (at infinity) are orthogonal (perpendicular) to the plane of the figure (screen). In oblique projections, the rays are not orthogonal to the plane, and usually the space y- and z-axes are taken in the plane of the figure.

It simplifies matters to think of (x, y, z) as a point of space and (u, v) as a point of the drawing plane. *Isometric projection* is an orthogonal projection in which the x, y, z-axes are at equal angles to the u, v-plane. It is called "isometric" because distances on the three coordinate axes are shrunk the same amount. The projection is often taken so that the z-axis projects onto the v-axis. The details of the math are in the comments to "graphs".

Dimetric projection is another orthogonal projection. The z-axis projects onto the v-axis and the x- and y-axes make equal but opposite angles with the u-axis. (Dimetric projection includes isometric projection as a special case.)

```
procedure  compute_isometric_matrix
                  (var  P: projection);

    begin   { compute_isometric_matrix }
    P[1, 1] := -1.0/Sqrt(2.0);
    P[1, 2] := -P[1, 1];
    P[1, 3] :=  0.0;
    P[2, 1] :=  -1.0/Sqrt(6.0);
    P[2, 2] :=  P[2, 1];
    P[2, 3] := -2.0* P[2, 1];
    end;    { compute_isometric_matrix }
```

```
procedure   compute_dimetric_matrix
                    (alpha: Real; var   P: projection);

    var   t: Real;

    begin     { compute_dimetric_matrix }
    alpha := Pi*alpha/180.0;      { radians }
    P[1, 1] := -1.0/Sqrt(2.0);
    P[1, 2] := -P[1, 1];
    P[1, 3] := 0.0;
    t := Sin(alpha)/Cos(alpha);
    P[2, 1] := t*P[1, 1];
    P[2, 2] := P[2, 1];
    P[2, 3] := Sqrt(1.0 - Sqr(t));
    end;     { compute_dimetric_matrix }
```

The oblique projection we code has the y- and z-axes on top of the u- and v-axes, while the positive x-axis projects onto a ray at angle α with the negative v-axis.

```
procedure   compute_oblique_matrix
                    (alpha: Real; var   P: projection);

    begin     { compute_oblique_matrix }
    alpha := Pi*alpha/180.0;      { radians }
    P[1, 1] := - Sin(alpha);
    P[1, 2] := 1.0;
    P[1, 3] := 0.0;
    P[2, 1] := - Cos(alpha);
    P[2, 2] := 0.0;
    P[2, 3] := 1.0;
    end;     { compute_oblique_matrix }
```

Finally, the procedure "project" actually carries out the projection of (x, y, z) onto (u, v). It includes a localization of the origin.

```
procedure   project(const   P: projection;
                    const   x, y, z: Real;
                    const   u0, v0: Integer;
                    var   u, v: Integer);

    begin
    u := u0 + Round(P[1, 1]*x + P[1, 2]*y
                    + P[1, 3]*z);
```

```
    v := v0 + Round(P[2, 1]*x + P[2, 2]*y
                      + P[2, 3]*z);
  end;    { project }
```

The initialization section, executed once—when the main program is started, registers the graphics driver and the fonts, i.e, loads them onto the heap.

```
  begin
  if  RegisterBGIDriver(@EGAVGADriverProc) < 0   then
     begin
     ClrScr;
     WriteLn(GraphErrorMsg(GraphResult));
     Write('Press <Enter>; ');   ReadLn;
     Halt;
     end;

  if  RegisterBGIFont(@SmallFontProc) < 0   then
     begin
     ClrScr;
     WriteLn(GraphErrorMsg(GraphResult));
     Write('Press <Enter>; ');   ReadLn;
     Halt;
     end;

  if  RegisterBGIFont(@SansSerifFontProc) < 0   then
     begin
     ClrScr;
     WriteLn(GraphErrorMsg(GraphResult));
     Write('Press <Enter>; ');   ReadLn;
     Halt;
     end;

  end.    { unit graphs }
```

We are going to use a little linear algebra for solutions of 3×3 linear systems. "Lin_alg" is a brief unit with this material. The implementation is quite primitive compared with what we shall do in a later chapter.

Program 5.8.3

```pascal
unit  lin_alg;    { 01/02/95 }

interface

   uses  graphs;

   type  matrix_3 = array[1..3, 1..3] of Real;

   procedure  solve_system(M: matrix_3;   rhs: vector;
                                  var  u: vector);
      { Solves the 3x3 system  M*u = rhs }

implementation

   function  determinant_3(B: matrix_3): Real;

      begin
      determinant_3 :=
             B[1, 1]*B[2, 2]*B[3, 3]
           + B[1, 2]*B[2, 3]*B[3, 1]
           + B[1, 3]*B[2, 1]*B[3, 2]
           - B[1, 1]*B[2, 3]*B[3, 2]
           - B[1, 3]*B[2, 2]*B[3, 1]
           - B[1, 2]*B[2, 1]*B[3, 3];
      end;    { determinant_3 }

   procedure  solve_system(M: matrix_3;   rhs: vector;
                                  var  u: vector);
      { Quick and dirty Cramer's rule solution }

      var      det0: Real;
           det_vec: vector;
              j, k: Word;
                 B: matrix_3;

      begin    { solve_system }
      det0 :=  determinant_3(M);
      if  det0 = 0.0  then  Exit;
      for  j := 1  to  3  do
         begin
         B := M;
         for  k := 1  to  3  do  B[k, j] := rhs[k];
         det_vec[j] := determinant_3(B);
         u[j] := det_vec[j] / det0;
```

```
        end;
    end;    { solve_system }

end.   { unit  lin_alg }
```

The following figures illustrate isometric and dimetric projections of a sphere. The programs creating the figures are in the file "f5_8_1–2.pas", (which uses a unit on ellipses, etc, in the next section).

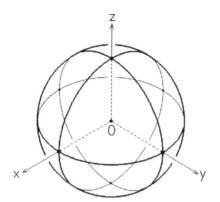

Figure 8.1.   An isometric projection

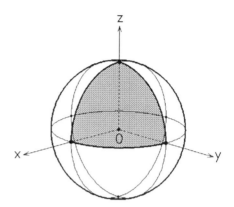

Figure 8.2.   A dimetric projection

Exercises

1. Test the unit "lin_alg.pas".

Section 9. Ellipses and Ellipsoids

Many space figures have edges that are arcs of circles or arcs of ellipses. When projected, these edges project into arcs of ellipses, and the accurate drawing of these arcs is an important part of technical drawing. I regret to report that many technical artists seem unable to get this fundamental skill right. Let us look at two figures that illustrate important principles. (See the disk file "F5_9_1–2.pas".)

In Figure 9.1, a plane ellipse is drawn in standard position relative to the coordinate axes. The four tangents to the ellipse at the points where it intersects the axes are drawn. Each tangent is parallel to a coordinate axis.

Here is the single critical fact, obvious, but apparently ignored by many technical artists. *If this figure is carried into space, tilted, and then projected onto a plane, parallel lines are preserved.* Figure 9.2 shows an oblique projection of the ellipse, with the four tangents drawn quite correctly, each parallel to a coordinate axis. If you simply consider this as a *plane* figure, it is an ellipse with some lines added. *The ellipse is tilted to the coordinate axes.* This tilt is what many technical artists ignore; they cannot seem to resist placing each ellipse drawing template straight on the paper, major axis horizontal!

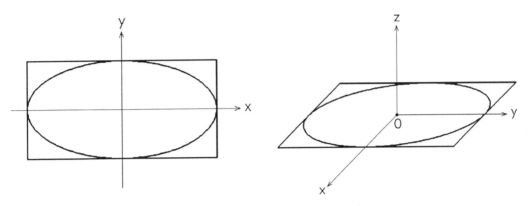

Figure 9.1.   Plane ellipse

Figure 9.2.   Projection of the ellipse

Figure 9.3 is a typical example of what I am talking about. It is taken from a standard text, Repka, J, Calculus, Wm. C. Brown, 1994, Figure 15.4.1(ii), page 901. This is at the beginning of a section on quadric surfaces, and almost all of the figures in that and related sections are wrong in the same way. (Other examples abound; for instance Anton, H, Calculus 5/e, J Wiley, 1995, pp. 699–709; Salas, S L and Hille, E, Calculus 7/e, J Wiley, 1995, pp. 904–908; Stewart, J, Calculus, Brooks/Cole, 1991, pp. 673–679.) Figure 9.3 is almost a copy of the source cited, which is supposed to be an illustration of the ellipsoid

$$x^2 + \frac{y^2}{9} + \frac{z^2}{4} = 1$$

with the cross-sections by the three coordinate planes shown. Look at the section E_{xy}, a long thin ellipse. Where it intersects the positive y-axis, its tangent must

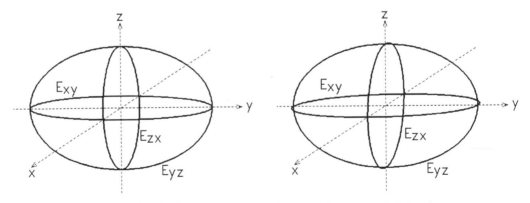

Figure 9.3.   Incorrect drawing!

Figure 9.4.   Corrected drawing

be parallel to the x-axis. But this tangent is *vertical*, not near the direction of the y-axis! The three cross-sections are drawn correctly in Figure 9.4. Examine it carefully; the differences are subtle, but important. (See disk file "F5_9_3–6.pas".)

Figure 9.5 shows just the cross-section E_{xy} with the tangent drawn at its intersection with the positive x-axis. Finally, Figure 9.6 shows the first octant portions of the three cross-sections. I hope the point is clear; it gets us started on a unit aimed at doing such drawings correctly. I suggest another look at Figure 9.2; ask yourself how you would sketch the ellipse inscribed centrally in a parallelogram. Well, draw a parallelogram; draw the two axes through its center, parallel to its edges (conjugate axes of the ellipse). The ellipse is tangent to the sides of the parallelogram at each of the 4 points where an axis meets a side. Isn't this enough to get a fairly correct sketch started?

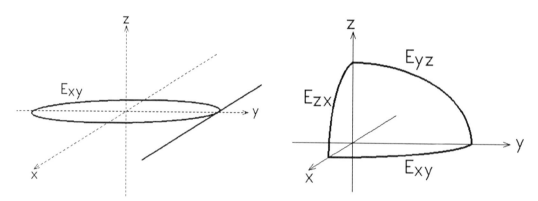

Figure 9.5.   One critical tangent   Figure 9.6.   First octant portion

The unit "ellipses" contains procedures for drawing ellipses and elliptical arcs defined in several ways, all that is needed for the kind of figures we are talking about. (The disk listing contains three further procedures: "ellipse1", "ellipse2", "ellipse3", which are useful computations for ellipses defined by various geometric conditions. They make extensive use of the unit "lin_alg".) Let's start with its interface:

Program 5.9.1

```
unit  ellipses;    { 01/02/95 }

interface

    uses   CRT, DOS, Graph, graphs, lin_alg;

    type       Real = Extended;
           matrix_2 = array[1..2, 1..2] of Real;
```

```
procedure    rotate(aa, bb, cc: Real;
                var   alpha, a, c: Real);

procedure    norm_ellipse_arc(x0, y0: Integer;
                              alpha: Real;
                              a, b: Real;
                              theta0, theta1: Real;
                              color: Word);

procedure    norm_tangent_ellipse
                (const   x0, y0, x1, y1, x2, y2,
                 theta0, theta1: Real;
                 color: Word);

procedure    project_ellipsoid(const   A:    matrix_3;
                               const   P: projection;
                               var    Q: matrix_2);
```

Procedure "rotate" codes the standard formulas from plane analytic geometry for rotating coordinates through angle α so that a (tilted) central ellipse $ax^2 + 2bxy + cy^2 = 1$ becomes standard form $a'u^2 + c'v^2 = 1$. (The notation is slightly different in the code.)

The procedure "norm_ellipse_arc" is modelled on the special cases "norm_circle" and "norm_arc" in the unit "graphs". Parameters a and b are the major and minor radii. The major axis is α forward of the x-axis, translated to the center (x_0, y_0) of the ellipse. Before this rotation, the ellipse arc is

$$(x, y) = (a \cos \theta, b \sin \theta) \qquad \theta_0 \leq \theta \leq \theta_1$$

The procedure "norm_tangent_ellipse" draws an arc of an ellipse, given the center and a pair of conjugate axes. Precisely, (x_0, y_0) is the center of the ellipse. The full ellipse passes through (x_1, y_1) and (x_2, y_2). Its tangent at (x_1, y_1) is parallel to $\overline{(x_0, y_0)(x_2, y_2)}$ and its tangent at (x_2, y_2) is parallel to $\overline{(x_0, y_0)(x_1, y_1)}$. It is assumed that the terminal angles of the elliptic arc satisfy $\theta_0 < \theta_1$. The implementations of these procedures are systemic applications of analytic geometry, routine, not exciting in any way. We'll postpone a discussion of "project_ellipsoid" for a bit.

implementation

```
procedure  rotate(aa, bb, cc: Real;
                var  alpha, a, c: Real);

begin
if  aa = cc  then  alpha := 0.25*Pi
else
    begin
    alpha := 0.5*Arctan(2.0*bb/(aa - cc));
    while  alpha < 0.0  do
       alpha := alpha + 0.5*Pi;
    while  alpha > 0.5*Pi  do
       alpha := alpha - 0.5*Pi;
    end;
a := aa * Sqr( Cos(alpha) )
        + bb * Sin(2.0*alpha)
        + cc * Sqr( Sin(alpha) );
c := aa * Sqr( Sin(alpha) )
        - bb * Sin(2.0*alpha)
        + cc * Sqr( Cos(alpha) );
end;    { rotate }

procedure  norm_ellipse_arc(x0, y0: Integer;
                            alpha: Real;
                            a, b: Real;
                            theta0, theta1: Real;
                            color: Word);

var     c, s, theta: Real;
                  k: Integer;
        dtheta, t0, t1: Real;
        u0, v0, u1, v1: Integer;
            hold_color: Word;

begin
hold_color := GetColor;
SetColor(color);
c := Cos(alpha);   s := Sin(alpha);
if  theta0 > theta1  then
    begin
    theta := theta0;
    theta0 := theta1;
    theta1 := theta;
    end;
theta := theta0;
```

```
dtheta := 0.005*(theta1 - theta0);
t0 := a*Cos(theta);   t1 := b*Sin(theta);
u1 := x0 + Round(c*t0 - s*t1);
v1 := y0 + Round(s*t0 + c*t1);
for  k := 1  to  200  do
   begin
   u0 := u1;   v0 := v1;
   theta := theta + dtheta;
   t0 := a*Cos(theta);   t1 := b*Sin(theta);
   u1 := x0 + Round(c*t0 - s*t1);
   v1 := y0 + Round(s*t0 + c*t1);
   norm_line(u0, v0, u1, v1, color);
   end;
SetColor(hold_color);
end;    { norm_ellipse_arc }

procedure  norm_tangent_ellipse
              (const  x0, y0, x1, y1, x2, y2,
               theta0, theta1: Real;
               color: Word);

var  u1, u2, v1, v2: Real;
        c, s, theta: Real;
                  k: Word;
     dtheta, t0, t1: Real;
     z1, w1, z2, w2: Integer;
           x22, y22: Real;

procedure  affine_transformation
              (x, y: Real;  var  z, w: Integer);

   begin
   z := Round(x0 + x*u1 + y*u2);
   w := Round(y0 + x*v1 + y*v2);
   end;    { affine_transformation }

begin    { norm_tangent_ellipse }
if  theta1 < theta0  then  Exit;
u1 := x1 - x0;   v1 := y1 - y0;
u2 := x2 - x0;   v2 := y2 - y0;
dtheta := 0.005*(theta1 - theta0);
theta := theta0;
x22 := Cos(theta);   y22 := Sin(theta);
affine_transformation(x22, y22, z2, w2);
repeat
   z1 := z2;   w1 := w2;
```

```
     theta := theta + dtheta;
     if   theta > theta1   then   theta := theta1;
     x22 := Cos(theta);   y22 := Sin(theta);
     affine_transformation(x22, y22, z2, w2);
     norm_line(z1, w1, z2, w2, color);
  until   theta = theta1;
  end;   { norm_tangent_ellipse }
```

Let us once again look at a central ellipsoid in space, axes parallel to the coordinate axes. Again we start with its three cross-sections by the coordinate planes (Figure 9.7). This is not completely satisfying because the projection of the outer boundary of the whole ellipsoid is obviously lacking. The purpose of the procedure "project_ellipsoid" is to find this projection, which we might think of as the *shadow* of the ellipsoid. Figure 9.8 shows this shadow added. (See disk file "F5_9_7–8.pas" for these figures.)

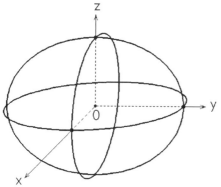

Figure 9.7.   Sections of an ellipsoid

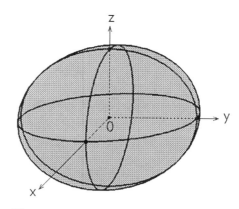

Figure 9.8.   The same, with shadow of whole ellipsoid

The input to "project_ellipsoid" is a positive definite 3×3 (symmetric) matrix A and a 2×3 projection matrix P of rank 2. The output is a positive definite 2×2 matrix Q. The matrix A defines a central solid ellipsoid $\mathbf{x}'A\mathbf{x} \le 1$. The matrix Q determines a solid ellipse $\mathbf{u}'Q\mathbf{u} \le 1$ in the u, v-plane.

The projection P projects in the direction of a (column) vector $\mathbf{v}$, which is determined up to a constant multiple. The simplest such $\mathbf{v}$ is the vector cross product of the rows of P, so "cross_product" is part of the computation. The shadow of the solid ellipsoid in the direction of $\mathbf{v}$ is a solid elliptic cylinder. We locate this shadow by finding all lines with direction $\mathbf{v}$ that are tangent to the ellipsoid. If $\mathbf{x}$ is a point on such a tangent line, then its point of tangency is $\mathbf{x} + t\mathbf{v}$ for a t such that the equation

$$(\mathbf{x} + t\mathbf{v})'A(\mathbf{x} + t\mathbf{v}) = 1$$

has a double zero (where $'$ denotes transpose). Now this quadratic equation for t expands to

$$(\mathbf{v}'A\mathbf{v})t^2 + 2(\mathbf{v}'A\mathbf{x})t + (\mathbf{x}'A\mathbf{x} - 1) = 0$$

The condition on $\mathbf{x}$ for the equation to have a double zero is that its discriminant equals 0:

$$(\mathbf{v}'A\mathbf{x})^2 - (\mathbf{x}'A\mathbf{x} - 1)(\mathbf{v}'A\mathbf{v}) = 0$$

which can be rewritten

$$\mathbf{x}'A\mathbf{x} - \frac{(\mathbf{v}'A\mathbf{x})^2}{\mathbf{v}'A\mathbf{v}} = 1 \qquad \text{that is} \quad \mathbf{x}'A\mathbf{x} - \frac{\mathbf{x}'A\mathbf{v}\mathbf{v}'A\mathbf{x}}{\mathbf{v}'A\mathbf{v}} = 1$$

In other words, the elliptic cylinder in the direction $\mathbf{v}$ generated by the ellipsoid is described by

$$\mathbf{x}'B\mathbf{x} = 1 \qquad \text{where} \quad B = A - \frac{A\mathbf{v}\mathbf{v}'A}{\mathbf{v}'A\mathbf{v}}$$

Note that $B\mathbf{v} = 0$ so that $\text{rank}(B) \leq 2$. Actually $\text{rank}(B) = 2$ because if $\text{rank}(B) \leq 1$, then we can find a nonzero vector $\mathbf{w}$ satisfying both $B\mathbf{w} = 0$ and $\mathbf{v}'A\mathbf{w} = 0$ (at most two constraints on $\mathbf{w}$). But then $0 = B\mathbf{w} = A\mathbf{w} - \mathbf{0}$. This is impossible as A is positive definite, hence nonsingular.

It will be convenient to introduce the *Grammian* matrix

$$G = PP' = \begin{bmatrix} \mathbf{p}'_1 \\ \mathbf{p}'_2 \end{bmatrix} [\mathbf{p}_1, \ \mathbf{p}_2] = \begin{bmatrix} \mathbf{p}'_1\mathbf{p}_1 & \mathbf{p}'_1\mathbf{p}_2 \\ \mathbf{p}'_2\mathbf{p}_1 & \mathbf{p}'_2\mathbf{p}_2 \end{bmatrix}$$

The vectors $\mathbf{p}'_1$ and $\mathbf{p}'_2$ are the rows of P.

If $\mathbf{x} = x_1\mathbf{p}_1 + x_2\mathbf{p}_2 + x_3\mathbf{v}$ is any point of space and $\mathbf{u}$ is its image under P, then (because $P\mathbf{v} = \mathbf{0}$)

$$\mathbf{u} = P\mathbf{x} = P(x_1\mathbf{p}_1 + x_2\mathbf{p}_2) = PP'\begin{bmatrix} x_1 \\ x_2 \end{bmatrix} = G\begin{bmatrix} x_1 \\ x_2 \end{bmatrix}$$

$$\begin{bmatrix} x_1 \\ x_2 \end{bmatrix} = G^{-1}\begin{bmatrix} u_1 \\ u_2 \end{bmatrix} \qquad [x_1, \ x_2] = [u_1, \ u_2]G^{-1}$$

(The third relation follows by transposing the second, and the symmetry of G.) From this we find the shadow of the ellipsoid as follows. If $\mathbf{x}$ on the elliptic cylinder projects to $\mathbf{u}$, then

$$1 = \mathbf{x}'B\mathbf{x} = (x_1\mathbf{p}'_1 + x_2\mathbf{p}'_2)B(x_1\mathbf{p}_1 + x_2\mathbf{p}_2)$$

$$= [x_1, \ x_2]PBP'\begin{bmatrix} x_1 \\ x_2 \end{bmatrix} = [u_1, \ u_2]G^{-1}PBP'G^{-1}\begin{bmatrix} u_1 \\ u_2 \end{bmatrix}$$

Therefore the shadow of the solid ellipsoid is the solid ellipse

$$[u_1,\ u_2]Q\begin{bmatrix} u_1 \\ u_2 \end{bmatrix} \le 1 \qquad \text{where} \quad Q = G^{-1}PBP'G^{-1}$$

The implementation of "project_ellipsoid" follows the above derivation to construct the $2{\times}2$ matrix Q, given the $2{\times}3$ projection matrix P and the $3{\times}3$ positive definite matrix A.

```
procedure  project_ellipsoid(const  A:  matrix_3;
                              const  P: projection;
                              var    Q: matrix_2);

var           v: vector;
              G: matrix_2;
              B: matrix_3;
        h, i, j,
        k, m, n: Word;
           c, t: Real;

procedure  cross_product(const  a, b: vector;
                         var    c: vector);

    begin
    c[1]  := a[2]*b[3]  -  a[3]*b[2];
    c[2]  := a[3]*b[1]  -  a[1]*b[3];
    c[3]  := a[1]*b[2]  -  a[2]*b[1];
    end;   { cross_product }

function  determinant_2(B: matrix_2): Real;

    begin
    determinant_2 := B[1, 1]*B[2, 2]
                     - B[1, 2]*B[2, 1];
    end;   { determinant_2 }

procedure  elliptic_cylinder(const  A: matrix_3;
                             const  v: vector;
                             var    B: matrix_3);
    { A is positive definite and v is a nonzero
      column vector.
          B := A - (Avv'A)/(v'Av)
      The relation x'Ax <= 1 defines a solid
      central ellipsoid.  The relation x'Bx <= 1
      defines the elliptic cylinder through the
      ellipsoid, generator parallel to v. }
```

```
var         s, t: Real;
        h, i, j, k: Word;

  begin
  { Compute 1/v'Av }
  t := 0.0;
  for  i := 1  to  3  do
      for  j := 1  to  3  do
          t := t + v[i]*A[i, j]*v[j];
  t := 1.0/t;
  { B := A - (Avv'A)/(v'Av) }
  for  h := 1  to  3  do
      for  k := 1  to  3  do
          begin
          s := 0.0;
          for  i := 1  to  3  do
              for  j := 1  to  3  do
                  s := s + A[h, i]*v[i]*v[j]
                                      *A[j, k];
          B[h, k] := A[h, k] - t*s;
          end;
  end;    { elliptic_cylinder }

begin    { project_ellipsoid }
cross_product(P[1], P[2], v);
elliptic_cylinder(A, v, B);
for  h := 1  to  2  do
    for  j := 1  to  2  do
        begin    { G := PP' }
        c := 0.0;
        for  i := 1  to  3  do
            c := c + P[h, i]*P[j, i];
        G[h, j] := c;
        end;
{ Q := inv(G)PBP'inv(G)
  First replace G by its inverse }
t := 1.0/determinant_2(G);
c := G[1, 1];   G[1, 1] := t*G[2, 2];
                G[2, 2] := t*c;
G[1, 2] := -t*G[1, 2];   G[2, 1] := -t*G[2, 1];
for  h := 1  to  2  do
    for  n := 1  to  2  do
        begin
        c := 0.0;
        for  i := 1  to  2  do
```

```
for  j := 1  to  3  do
    for  k := 1  to  3  do
        for  m := 1  to  2  do
            c := c + G[h, i]*P[i, j]
                      *B[j, k]*P[m, k]
                      *G[m, n];
    Q[h, n] := c;
    end;
end;    { project_ellipsoid }

end.    { unit  ellipses }
```

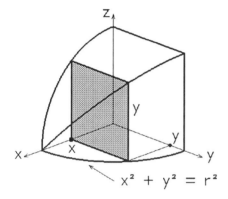

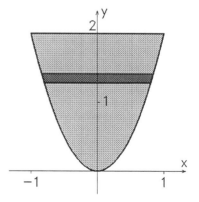

Figure 9.9.   Figure for Exercise 1   Figure 9.10.   Figure for Exercise 2

Exercises

1. Figure 9.9 shows a solid resting on the x, y-plane with a circular base. Each cross-section by a plane orthogonal to the x-axis is a square. Program this drawing.
2. Figure 9.10 shows a spherical shell. Program this drawing.
3. Do unto parabolas what the unit "ellipses" does unto ellipses.
4. Figure 9.11 is a cylindrical tank with parabolic cross-section. Program this drawing.
5. (Cont.) Figure 9.12 is one of the cross-sections. Program this drawing.

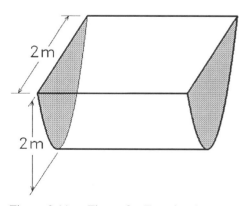

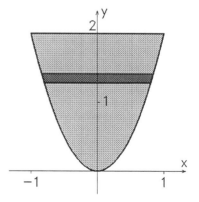

Figure 9.11.   Figure for Exercise 4   Figure 9.12.   Figure for Exercise 5

Chapter 6 Recursive Algorithms

Section 1. Recursion

Recursion is the heart and soul of Pascal. Up to now we have had one or two explicit uses of recursion in our programs. Now we go at it seriously.

A subroutine is *recursive* if it calls itself; that is the simplest definition. Also, *recursive* is a situation in which several procedures call each other. There are situations where it is really hard to write a program that is not recursive, and where writing a recursive program is almost automatic. In fact, writing in recursive style is so easy sometimes that we get carried away and write recursive programs where iterative programs run much faster and use less memory. Remember that when a procedure calls itself (or another procedure) the return state must be saved on the "stack", and this means not only the address of where the calling procedure left off, but all the local variables of the procedure called, and this includes its value parameters.

The following rather easy program has two functions for computing the n-th Fibonacci number. These (long) integers have the defining difference equation that each is the sum of the previous two, and the 0-th and 1-th are 1. The iterative function G implements this with the help of three auxiliary LongInt's. These plus a counting Word variable are all the memory needed.

Look carefully at the recursive function F. The statement returning its value calls the function itself—twice! I strongly recommend that you "play computer" and work this out in detail for a few small n, say, 6 and 10. You will get the idea of how really labor intensive this code is. Of course you should run the program to "see for yourself"!

Program 6.1.1

```
program  fibonacci_numbers;    { 09/09/94 }

    uses    CRT;

    var   n: Word;

    function  F(n: Word): LongInt;    { Recursive }

        begin
        if  KeyPressed  then  Halt;
        if  (n = 0) or (n = 1)  then  F := 1
        else  F := F(n - 2) + F(n - 1);
        end;
```

```
function   G(n: Word): LongInt;    { Iterative }

    var   x, y, t: LongInt;
              k: Word;

    begin
    x := 1;  y := 1;
    for  k := 2  to  n  do
        begin
        t := y;    { t holds old value of y }
        y := x + y;
        x := t;
        end;
    G := y;
    end;    { G }

begin    { fibonacci_numbers }
ClrScr;
n := 20;
TextColor(LightGray + 128);    { blink on }
Write   ('Computing...');
TextColor(LightGray);    { blink off }
WriteLn(' --Be very patient--');
WriteLn;
WriteLn('Recursive: F', n, ' = ', F(n));
WriteLn;
WriteLn('Iterative: F', n, ' = ', G(n));
GotoXY(1, 1);
ClrEol;
GotoXY(1, 7);
Write('Press <Enter> to quit: ');    ReadLn;
end.    { fibonacci_numbers }
```

Warning 1 It is extremely easy to make horrible errors when programming recursively, the kind of errors that hang the computer hopelessly. I strongly recommend the following safeguards be installed until you are *sure* that your code is OK.

1. Compile with the {$S+} option. For good measure use the {$R+} option too; it won't hurt you.
2. Add a **uses**   CRT; clause and insert the line

```
   if  KeyPressed  then  Halt;
```

at the beginning of each procedure (function) that is called recursively, and somewhere in each innermost loop. Then if your program hangs, you just press a key instead of rebooting.

Warning 2 Recursive programs can be very much harder to read and understand than to write. With no effort at all you can write a recursive procedure of a few lines that almost no one can decode. So please be especially diligent with documentation that explains precisely the idea of each recursive algorithm. I confess that I have had to throw out some programs I wrote some time ago for this text because even *I* cannot figure out how they work!

The next example is a simple zero finding program, using the standard bisection technique, which is guaranteed to work on any continuous function with opposite signs at the endpoints of its interval—by the Intermediate Value Theorem. This program follows the wisdom of the warnings above:

Program 6.1.2

```
{$S+}
program  bisection;    { 09/09/94 }

   uses   CRT;

   type   Real = Extended;

   const   eps = 1.0e-17;    { Try -19 for a crash }

   var   x0, x1,
         midpt: Real;
         depth: Word;

   function   F(x: Real): Real;

      begin
      F :=   x*x*x - 2.0;
      end;

   procedure   bisect(var   x0, x1: Real);
      { Assumes F(x) defined and continuous,
        F(x0) < 0.0,    F(x1) > 0.0   }

      begin
      if   KeyPressed   then   Halt;
      Inc(depth);
      midpt := 0.5*(x0 + x1);
      if   F(x1) - F(x0) < eps   then   Exit;
      if   F(midpt) >= 0.0   then   x1 := midpt
      else   x0 := midpt;
```

```
      bisect(x0, x1);
      end;

   begin    { bisection }
   depth := 0;
   x0 := 0.0;
   x1 := 5.0;
   bisect(x0, x1);
   ClrScr;
   WriteLn('Solution of  F(x) = 0:    x '#247,
           midpt:22:18);
   WriteLn('Depth of recursion: ', depth);
   ReadLn;
   end.
```

The following example has two implementations of a very simple form of Euclid's greatest common divisor algorithm, using subtraction only—no division. Again, one form is recursive, the other iterative. It also includes a counter to report the depth of recursive calls. This is often a good idea, particularly when debugging and analyzing performance. Sometimes you get a surprise.

Program 6.1.3

```
{S+}
program   gcd;   { 09/09/94 }

   var   m, n: LongInt;
         depth: Word;

   { Simple greatest common divisor algorithm

     Assume  m, n >= 0,  not both zero

                  / gcd(n, m)          if  m < n
     gcd(m, n) =  | m                  if  n = 0
                  \ gcd(m - n, n)      otherwise }

   function  gcd1(m, n: LongInt): LongInt;
      { recursive }

      begin
      Inc(depth);
      if  n = 0  then   gcd1 := m
      else  if  m < n   then  gcd1 := gcd1(n, m)
      else  gcd1 := gcd1(m - n, n);
      end;
```

```
function  gcd2(m, n: LongInt): LongInt;
   { iterative }

   var  t: LongInt;

   begin
   while  n > 0  do
      begin
      if  m < n  then
         begin  t := m;  m := n;  n := t;  end;
      t := m - n;
      m := n;  n := t;
      end;
   gcd2 := m;
   end;

begin   { gcd }
depth := 0;
WriteLn('Recursive gcd: ', gcd1(200, 62140));
WriteLn('Depth of recursion: ', depth);
WriteLn('Interative gcd: ', gcd2(200, 62140));
ReadLn;
end.
```

If two or more procedures are on the same level, and call each other, it is necessary to declare as **forward** all except the first one. Consider the following example:

Program 6.1.4

```
program  restricted_partitions;   { 09/27/94 }

   function  p1(n, k: Word): LongInt;
      { # partitions of n, highest part = k }

      begin
      if  k > n  then  p1 := 0
      else  if  (k = n) or (k = 1)  then  p1 := 1
      else  p1 := p2(n - k, k);
      end;

   function  p2(n, k: Word): LongInt;
      { # partitions of n, highest part <= k }
```

```
begin
if  k > n  then  k := n;
if  k = 1  then  p2 := 1
else  p2 := p1(n, k) + p2(n, k - 1);
end;

begin
WriteLn(p1(20, 10));
WriteLn(p2(20, 10));
ReadLn;
end.
```

As written, this program will not compile; the compiler reports that "p2" in the body of p1 is an unknown identifier. To fix this, we add the line

```
function  p2(n, k: Word): LongInt;  forward;
```

before the declaration of p1. This line identifies the identifier "p2" to the compiler and notes that p2 will be implemented later. (This is very similar to procedures declared in the interface part of a unit.)

The meaning of "partition" in the program is best explained by a numerical example:

```
p1(5, 3) = 2:   5 = 3+2=3+1+1
p2(5, 3) = 5:   5 = 3+2=3+1+1=2+2+1=2+1+1+1=1+1+1+1+1
```

There are 2 partitions of 5 with highest part 3 and 5 partitions of 5 with highest part at most 3.

The next example is our first which is easy to program recursively, and rather perplexing to do any other way. The problem is to evaluate terms of the sequence defined by (I use linear notation to avoid sub-subscripts)

$$a(1) = 1 \qquad a(n) = n - a(a(n-1)) \quad \text{for} \quad n > 1$$

You should compute the first dozen or so terms by hand to get the idea.

Program 6.1.5

```
program  seq_1;
  { Tabulates values of the sequence  a(1) = 1,
    a(n) = n - a( a(n - 1) ) in a rectangular table,
    15 entries per row. }

  const  n0 = 300;   { Use multiple of 150 }

  var  i, j, k, n: Word;
```

```
function  a(n: Word): Word;

   begin
   if  n = 1  then  a := 1
   else  a := n - a( a(n - 1) );
   end;

begin    { seq_1 }
n := 1;
for  i := 0  to  (n0 div 150) - 1  do
   begin
   Write(' ': 5);
   for  k := 1  to  15  do  Write(k:4);
   WriteLn;
   for  j := 0  to  9  do
      begin
      Write(150*i + 15*j : 3, '   ');
      for  k := 1  to  15  do
         begin  Write(a(n):4);  Inc(n);  end;
      WriteLn;
      end;    { for j }
   WriteLn;
   end;    { for i }
end.    { seq_1 }
```

How easy this is! The code for **function** a(n) is almost literally what the displayed definition says.

I do not know how to program this function iteratively without using some memory for previous results. It requires a little thought to realize that you hardly need to store *all* previous values of the function, just a very limited set of values that are referred to in evaluating a(n) for a *particular* n. The following program does this.

Program 6.1.6

```
program  seq_2;
   { Tabulates values of the sequence  a(1) = 1,
     a(n) = n - a( a(n - 1) )  in a rectangular table,
     15 entries per row. }

   const  n0 = 300;              { Use multiple of 150 }

   var  i, j, k, n: Word;
                    a: array[1..n0] of Word;

   procedure  fill_a;
```

```
    begin
    a[1] := 1;
    for  n := 2 to  n0  do  a[n] := n - a[a[n - 1]];
    end;   { fill_a }

begin
fill_a;
n := 1;
for  i := 0  to   (n0 div 150) - 1  do
    begin
    Write(' ': 5);
    for  k := 1  to  15  do  Write(k : 4);  WriteLn;
    for  j := 0  to  9  do
       begin
       Write(150*i + 15*j : 3, '  ');
       for  k := 1  to  15  do
          begin  Write(a[n]:4);  Inc(n);  end;
       WriteLn;
       end;   { for j }
    WriteLn;
    end;   { for i }
ReadLn;
end.   { seq2
```

Exercises

1. Rewrite *bisection* iteratively.
2. Explain the following brief program:

```
    program  mystery;   { 05/20/94 }

       procedure  explain_me(n: Word);

          begin
          Write(n:4);
          if  n < 150   then   explain_me(2*n);
          Write(n:4);
          end;

       begin
       explain_me(1);
       WriteLn;
       Write  ('Press <Enter> to quit: ');   ReadLn;
       end.
```

3. The program "unknown" on your disk contains a complete program for the following two procedures. The input to each is a keyboard string, terminated with <Enter>.

Predict how each acts.

```
program  figure_this_out;    { 05/14/94 }

    procedure  mystery_1;
    { Note: Eoln means end of line: <Enter> }

        var  ch: Char;

        begin
        Read(ch);
        if  not Eoln  then  mystery_1;
        Write(ch);
        end;    { mystery_1 }

    var  ch: Char;

    procedure  mystery_2;

        begin
        Read(ch);
        if  not Eoln  then  mystery_2;
        Write(ch);
        end;    { mystery_2 }
```

4. Define recursively the even, periodic, real function of period 2 that equals x on [0, 1].
5. Define recursively the periodic, real function of period 2 that equals $-x$ on [−1, 0] and x^2 on [0, 1].
6. Define recursively the function x^n. For given n, how deeply is the recursive call nested?
7. Write a recursive procedure for locating a fixed point of a real function by iterating the function. Test it on cos x and Sqrt(x + 1).
8. Define recursively the value of n-th Legendre polynomial at x:

$$P_0(x) = 1 \qquad P_1(x) = x$$
$$P_n(x) = ((2n - 1)P_{n-1}(x) - (n - 1)P_{n-2}(x))/n$$

9. Declare

```
type  vector = array[1..max] of Word;
```

and let L and U be vectors. The problem is to program a nested **for** loop to execute many times a procedure P whose argument is a vector. The control variables are components of a vector "control". At the very inside of the loop, P(control) is executed. The innermost loop has L[1] ≤ control[1] ≤ U[1] and the outermost loop has L[max] ≤ control[max] ≤ U[max].

Section 2. Applications

We begin with a program, unworthy of much comment, for translating Arabic numerals to Roman numerals:

Program 6.2.1

```pascal
{$S+,R+}
program  arab2roman;    { 05/22/94 }

   uses  CRT;

   var   n: Word;
      ch: Char;

   procedure  convert(n: Word);

      const  no_breaks = 12;
             index: array[0..no_breaks] of Word
                = (1, 4, 5, 9, 10, 40, 50,
                   90, 100, 400, 500, 900, 1000);
             Strings: array[0..no_breaks] of string[2]
                = ( 'I', 'IV', 'V', 'IX', 'X', 'XL', 'L',
                    'XC', 'C', 'CD', 'D', 'CM', 'M' );

      var  Low, High, mean: Word;

      begin
      if  n = 0  then  Exit;
      if  KeyPressed  then  Halt;
      if  n >= index[no_breaks]    then
         begin
         Write(Strings[no_breaks]);
         n := n - index[no_breaks];
         end
      else    { binary search for highest index <= n }
         begin
         Low := 0;  High := no_breaks;
         repeat
            if  KeyPressed  then  Halt;
            mean := (Low + High) shr 1;
            if  n < index[mean]  then  High := mean
            else  if  n >= index[mean]  then
               Low := mean;
         until  High <= Low + 1;
         Write(Strings[Low]);
         n := n - index[Low];
         end;
      convert(n);
      end;    { convert }
```

```
begin     { arab2roman }
ClrScr;
repeat
   Write   ('Enter a positive integer: ');
   ReadLn(n);   Write(n, ' = ');
   convert(n);
   WriteLn;
   Write
      ('Q: Quit;   Any other key: Another number: ');
   ch := UpCase(ReadKey);
   if ch = 'Q'   then   Break;
   WriteLn;   WriteLn;
until   false;
end.
```

Algebraic and RPN Expressions

Now we get down to real business. The following program is an example, in my opinion, of a program that is very difficult to write without recursion. Suppose we have an algebraic structure closed under a single binary operation, which we think of as multiplication, and about which we assume nothing. We have several options for writing the product of two elements x and y. First, algebraic notation (juxtaposition) as usual: xy. Another possibility is *Postfix notation* also called *Reverse Polish Notation* or *RPN*: $xy*$. (Hewlett Packard calculators taught the world RPN.) Because we are not assuming the associative law, or anything like that, we must use parentheses for grouping if using algebraic notation. Note the following:

$$(xy)z = xy * z * \qquad x(yz) = xyz * *$$

indicating that RPN never requires parentheses.

Our task is to write a program that translates algebraic to RPN and vice versa. It turns out that the main problem is to *define* RPN and algebraic expressions precisely. Once done, it turns out that the definitions are naturally recursive, and translate readily into program code. Let us look at the beginning of the program, with a comment giving proper recursive definitions.

Program 6.2.2

```
program  alg_rpn;     { 11/13/94 }

   { Definitions:

      element  = 'a'..'z'
      rpn_expr = element | rpn_expr rpn_expr "*"
```

```
Examples:    ab*
             ab*cd**
             abc**
             ab*cde***f*g*h*
```

```
factor   = element | "(" alg_expr ")"
alg_expr =  factor | factor factor
```

```
Examples:    ab
             c(ab)
             (ab)(cd)
             ((ab)c)d
             ((ab)(cd))(((ef)g)h)    }
```

Read this carefully. The primitives are *elements*, lower case letters. The vertical bar in these definitions means "or". An *RPN expression* is either an *element* or the juxtaposition of two *RPN expressions* followed by *. An *algebraic expression* is either a *factor* or the juxtaposition of two *factors*. A *factor* in turn is either an *element* or an *algebraic expression* inside parens ().

These are truly recursive definitions! Input to the program will be a string, letters and *'s in the RPN case, letters and parens in the algebraic case. After a routine input procedure, there are two procedures for checking consistency expressions. They are pretty direct, but should be understood before continuing.

```
const  low_case = ['a'..'z'];

var  alg, rpn: string;

procedure  get_ss(var  ss: string);

   var  k: Word;

   begin
   ReadLn(ss);
   for  k := Length(ss)  downto  1  do
      if  ss[k] = ' '  then  Delete(ss, k, 1);
   end;

function  check_rpn(var  ss:  string): Boolean;

   var  k, excess: Integer;
```

```
begin
check_rpn := false;
excess := 0;
for  k := 1  to  Length(ss)  do
   begin
   if  ss[k] in low_case  then  Inc(excess)
   else  if  ss[k] = '*'  then  Dec(excess)
   else  Exit;
   if  excess = 0  then  Exit;
   end;
check_rpn := (excess = 1);
end;    { check_rpn }

function  check_alg(var  ss: string): Boolean;

   var  k, open_parens: Integer;

   begin    { check_alg }
   check_alg := false;
   { First check matching parens; all other
     characters low_case only }
   open_parens := 0;
   for  k := 1  to  Length(ss)  do
      begin
      if  ss[k] = '('  then  Inc(open_parens)
      else  if  ss[k] = ')'  then  Dec(open_parens)
      else  if  not (ss[k] in low_case)  then  Exit;
      if  open_parens < 0  then  Exit;
      end;
   if  open_parens <> 0  then  Exit;
   { Now check for three consecutive letters }
   for  k := 3  to  Length(ss)  do
      if  (ss[k] in low_case) and
          (ss[k - 1] in low_case) and
          (ss[k - 2] in low_case)  then    Exit;
   check_alg := true;
   end;    { check_alg }
```

Now we come to the heart of the matter; first going from algebraic to RPN. The innermost procedure "factor" mimics the definition of *factor* to read the next factor out of the input string "alg" at a position (or cursor). If the character at the cursor is '(', then the following *algebraic expression* is processed and the closing) is flushed out. If the character is not '(', but a letter, that is the value added to the output string "rpn".

The enclosing procedure "alg_expr" calls (recursively) its subprocedure "factor" twice and adds a '*' to the output string; that is the definition of *algebraic expression*. Recursive enough for you? Well, anyhow, note how the program is almost written automatically. This is a hint of the way parsers and compilers are written.

```
procedure  alg2rpn(var  alg, rpn: string);

   var  j: Word;

   procedure  alg_expr;

      procedure  factor;

         var  ch:  Char;

         begin
         ch := alg[j];
         if  ch = '('  then
             begin
             Inc(j);  { to enter the expression }
             alg_expr;
             Inc(j);  { to pass the ')' }
             end
         else
             begin    { alg[j] an element }
             rpn := rpn + ch;
             Inc(j);
             end;
         end;    { factor }

      begin    { alg_expr }
      factor;
      factor;
      rpn := rpn + '*';
      if  j = Length(alg)  then  Exit;
      end;    { alg_expr }

   begin    { alg2rpn }
   j := 1;
   rpn := '';
   alg_expr;
   end;    { alg2rpn }
```

Going the other direction is somewhat similar, and I hope you can understand the code.

```
procedure  rpn2alg(var  rpn, alg: string);

   var  j: Word;

   procedure  rpn_expr;

      var  ch: Char;

      begin
      ch := rpn[j];
      if  ch = '*'   then
         begin
         alg := ')' + alg;
         Dec(j);
         rpn_expr;
         rpn_expr;
         alg := '(' + alg;
         end
      else
         begin
         alg := ch + alg;
         Dec(j);
         end;
      if  j = 1  then  Exit;
      end;    { rpn_expr }

   begin    { rpn2alg }
   j := Length(rpn);
   alg := '';
   rpn_expr;
   j := Length(alg);
   if  (alg[1] = '(') and (alg[j] = ')')   then
      begin
      Delete(alg, j, 1);
      Delete(alg, 1, 1);
      end;
   end;    { rpn2alg }
```

Finally we put it together with a main program that does two tests. You should try others of course.

```
begin    { alg_rpn }
WriteLn;
WriteLn;
alg := '(((ab)(cd))((e(fg))h))((ij)k)';
WriteLn('alg = ', alg);
if  not check_alg(alg)   then   Halt;
WriteLn('Translate to RPN:');
alg2rpn(alg, rpn);
WriteLn('rpn = ', rpn);
WriteLn('Check; reverse the translation:');
if  not check_rpn(rpn)   then   Halt;
rpn2alg(rpn, alg);
WriteLn('alg = ', alg);
WriteLn;
WriteLn;
rpn := 'ab*cde***f*g*h*ij*klmn*****';
WriteLn('rpn = ', rpn);
if  not check_rpn(rpn)   then   Halt;
WriteLn('Translate to algbraic:');
rpn2alg(rpn, alg);
WriteLn('alg = ', alg);
WriteLn('Check; reverse the translation:');
if  not check_alg(alg)   then   Halt;
alg2rpn(alg, rpn);
WriteLn('rpn = ', rpn);
WriteLn;
Write ('Press <Enter>: ');   ReadLn;
end.    { alg_rpn }
```

Difference Equations

The idea for this program came from the very origin of Pascal: Jensen, K and
Wirth, N, Pascal User Manual and Report 2/e, Springer-Verlag, 1974, 73–77.

A difference equation whose current term refers to terms far back in the
sequence is a natural candidate for recursive programming. Consider for instance
the initial value problem

$$A(0) = 1, \quad A(n) = A(n \ \textbf{div} \ 2) + A(n \ \textbf{div} \ 3)$$

There is no obvious way to program A(n) iteratively, although it can be done
(Exercise 4). The following program provides an almost automatic recursive
definition of the sequence:

Program 6.2.3

program recursive_solution_diff_eq;   { 05/21/94 }

```
uses   CRT, timer;

var             n: LongInt;
       row, col,
           depth,
       max_depth: Word;

function  A(n: LongInt): LongInt;

    begin
    if   KeyPressed  then   Halt;
    Inc(depth);
    if   depth > max_depth   then   Inc(max_depth);
    if   n = 0   then   A := 1
    else   A := A(n div 2) + A(n div 3);
    Dec(depth);
    end;

begin    { recursive_solution_diff_eq }
repeat
    max_depth := 0;
    col := 0;
    WriteLn;
    Write('Enter a nonnegative  n (0 to quit): ');
    ReadLn(n);
    if   n = 0   then   Halt;
    get_time0;
    depth := 0;
    WriteLn('A(', n, ') = ', A(n));
    get_time1;
    WriteLn;
    WriteLn('Max nesting depth = ', max_depth);
    put_time;
    WriteLn;
until   false;
end.    { recursive_solution_diff_eq }
```

A system of difference equations can also be handled as follows; simply use a vector variable. For instance consider the system

```
A(n)  = A(n div 2)  + B(n div 3)
B(n)  = A(n div 3)  * B(n div 2)  + 1.0
A(0)  = A0,   B(0)  = B0
```

Again, the program almost writes itself.

Program 6.2.4

```
program  system_difference_equations;   { 05/21/94 }

  type  Real = Extended;
        point = array[1..2] of Real;

  var   n, j: Word;
        x0, x: point;

  procedure  P(n: Word;  var  x: point);

    var  y, z: point;

    begin
    if  n = 0   then
       begin
       x[1] := x0[1];  x[2] := x0[2];
       end
    else
       begin
       P(n div 2, y);
       P(n div 3, z);
       x[1] := y[1] + z[2];
       x[2] := z[1]*y[2] + 1.0;
       end;
    end;    { P }

  begin    { system_difference_equations }
  repeat
     WriteLn('Initial point (A0, B0):');
     Write('  A0 = ');   ReadLn(x0[1]);
     Write('  B0 = ');   ReadLn(x0[2]);
     Write  ('Number of terms to compute:  n = ');
     ReadLn(n);
     WriteLn;
     WriteLn('n':4, 'A(n)':20, 'B(n)':20);
     for  j := 0  to  44  do  Write('-');
     WriteLn;
     for  j := 0  to  n  do
        begin
        P(j, x);
        WriteLn(j:4, '   ', x[1]:18, '   ', x[2]:18);
        end;
     WriteLn;
     Write(
```

```
      'Continue: <Enter>;   Halt: <Ctrl-Break>: ');
      ReadLn;
   until   false;
   end.
```

Binary Trees

The *binary tree* is a useful data structure in many applications, including parsing
strings that represent functions. A binary tree is a directed tree in the sense of
graph theory. It contains one special node, its *root* which has no branches coming
into it. Each node has two branches coming out from it, each pointing to another
binary tree, thought of as its children (recursive definition), or **nil**. Clearly then,
we represent nodes by records and branches by pointers. Each node contains as a
minimum two pointers, *left* and *right*, and usually contains useful data in addition.

   We'll show part of a unit for handling binary trees and suggest more of it in
the exercises. This particular unit is biased towards sorting data, so each node
contains a key to its data. The key in each node must be $\geq$ the key of each
of its left descendents and $\leq$ the key of each of its right descendents. (Distinct
nodes may have the same key, e.g, the first letter of a word in the node's data
string.) The data itself may be large, so instead of slinging it around whenever
we move nodes, we put a pointer to the data in each node rather than the data
itself. It is also convenient for some tree operations to be able to find the parent
of a node easily, so we add to each child node a pointer to its parent. Let's write
down the definition:

Program 6.2.5

```
{$S+}
unit   bin_tree;     { 05/16/94 }

interface

    type   binary_tree = ^node;
                  node = record
                           key: Integer;
                           data: Pointer;
                           left,
                           right,
                           parent: binary_tree;
                         end;
```

A list of useful procedures and functions completes the **interface**:

```
procedure   create_tree(var   b: binary_tree);
procedure   destroy_tree(var   b: binary_tree);
function    search(k: Integer;   b: binary_tree):
                                    binary_tree;
   { Searches b for the key k }
function    min_node(b: binary_tree): binary_tree;
function    max_node(b: binary_tree): binary_tree;
function    successor(b: binary_tree): binary_tree;
function    predecessor(b: binary_tree): binary_tree;
function    height(b: binary_tree): Word;
procedure   insert(n: node;   var   b: binary_tree);
     { Assumed: n.left, n.right, n.parent all nil }
procedure   remove(nptr: binary_tree;
                      var   b: binary_tree);
     { nptr points to a node of b. }
```

Now let's implement some of these. Creating and destroying trees is pretty
routine:

implementation

```
procedure   create_tree(var   b: binary_tree);

    begin
    New(b);
    with   b^   do
       begin
       parent := nil;
       left := nil;   right := nil;
       data := nil;
       end;
    end;   { create_tree }

procedure   destroy_tree(var   b: binary_tree);

    begin
    if   KeyPressed   then   Halt;
    if   b^.left <> nil   then   destroy_tree(b^.left);
    if   b^.right <> nil   then
       destroy_tree(b^.right);
    Dispose(b^.data);
    Dispose(b);   b := nil;
    end;   { destroy_tree }
```

It is useful to search a tree for a pointer to a node with a particular key, also to
search for the pointer to the left-most node of the tree:

```
function   search(k: Integer;   b: binary_tree):
                                          binary_tree;
   begin
   while   (b <> nil) and (k <> b^.key)   do
      begin
      if  k < b^.key  then   b := b^.left
      else   b := b^.right;
      end;
   end;    { search }

function   min_node(b: binary_tree): binary_tree;

   begin
   while   b^.left <> nil   do   b := b^.left;
   min_node := b;
   end;    { min_node }
```

The *height* of a binary tree is the number of generations in the tree:

```
function   height(b: binary_tree): Word;

var   j, k: Word;

   begin
   if  KeyPressed   then   Halt;
   if  b = nil   then   height := 0
   else
      begin
      j := height(b^.left);
      k := height(b^.right);
      if  j > k  then   height := 1 + j
      else   height := 1 + k;
      end;
   end;    { height }
```

If we want to insert a new node into a binary tree, we must do it so the condition
on keys is preserved. This is more complicated than the implementations above:

```
procedure   insert(n: node;   var   b: binary_tree);

   var   run, prev: binary_tree;
```

```
begin
if  b = nil   then
   begin
   create_tree(b);
   b^ := n;
   end
else
   begin    { b <> nil }
   prev := nil;
   run := b;
   while  run <> nil   do
      begin
      prev := run;
      if  n.key < run^.key   then
         run :=   run^.left
      else   run := run^.right;
      end;
   n.parent := prev;
   if  n.key < prev^.key   then
      begin
      create_tree(prev^.left);   prev^.left^ := n;
      end
   else
      begin
      create_tree(prev^.right);
      prev^.right^ := n;
      end;
   end;    { b <> nil }
end;    { insert }
```

Removing a node is even trickier because we have to paste together with care what remains:

```
procedure   remove(nptr: binary_tree;
                var   b: binary_tree);

var   x, y: binary_tree;

begin
if  nptr^.left = nil   then
   begin
   if  nptr^.right = nil   then
      begin
      x := nptr^.parent;
      if  x = nil   then   destroy_tree(b)
```

```
          else  if  x^.left = nptr   then
             destroy_tree(x^.left)
          else  destroy_tree(x^.right);
          end
       else
          { nptr^.left = nil,   nptr^.right <> nil }
          begin
          x := nptr^.parent;
          if  x^.left = nptr   then
             { splice and destroy }
             x^.left := nptr^.right
          else    { x^.right = nptr }
             x^.right := nptr^.right;
          nptr^.right := nil;
          destroy_tree(nptr);
          end;
       end    { nptr^.left = nil }
    else  if  nptr^.right = nil   then
       begin
       { nptr^.right = nil,   nptr^.left <> nil }
       x := nptr^.parent;
       if  x^.left = nptr   then
          { splice and destroy }
          x^.left := nptr^.left
       else    { x^.right = nptr }
          x^.right := nptr^.left;
       nptr^.left := nil;
       destroy_tree(nptr);
       end
    else
       begin
       { nptr^.left <> nil,   nptr^.right <> nil }
       y := successor(nptr);
       { y^.left = nil; splice out y }
       x := y^.parent;
       if  x^.left = y   then   x^.left := y^.right
       else   x^.right := y^.right;
       nptr^.data := y^.data;
       y^.right := nil;
       destroy_tree(y);
       end;
    end;    { remove }

end.    { bin_tree }
```

Exercises

1. An algebraic system has one operation, multiplication, denoted by juxtaposition. An *expression* consists of a string of letters and fences: ( ) [ ] { }. Write a program that tests for correct pairing of the fences. For instance, it should veto

   (ab]   a(b[c)d].

2. Remember the child's game of sequences of (left, right) or (0, 1):

   s4 = 1001, s16 = s4 s4' s4' s4 = 1001011001101001,
   s64 = s16 s16' s16' s16

   etc, where ' reverses a sequence? Write a program to determine the n-th term of the sequence.

3. Solve the "Tower of Hanoi" (disks on three pegs) puzzle.

4. The Ackerman function is defined by

   $$A(0, y) = y + 1 \qquad A(x, 0) = A(x - 1, 1)$$
   $$A(x, y) = A(x - 1, A(x, y - 1))$$

   It grows so fast that it soon blows any computer out of the water. Define a "modular Ackerman function" to be A modulo m, where m is input. Be sure to measure the depth of nesting when you use it to generate a table of values.

5. Write a program for the function p(n, k), the number of partitions of n with highest part at most k. This, unlike the program "restricted_partitions" in Section 1, should stand alone: no other function involved, either on the same level and **forward** or as a subroutine of p. Note that p(n, n) is the standard partition function p(n).

6. Write a program for the function q(n, k), the number of partitions of n into *distinct* parts, with highest part at most k. For instance, q(5, 4) = 2 because 5 = 4+1 = 3+2 shows all possibilities.

7. Write an iterative program for the difference equation in program recursive_solution_diff_eq.

8. Implement "max_node", which searches a binary tree for its right-most node.

9. The *successor* of a branch of a binary tree is defined as follows. If the node the branch points to has a right child, then the successor is the left-most branch of that right child. If not, retreat up the tree as far as possible along *right* branches; the final one is the successor. Program function "successor".

10. (Cont.) Similarly the *predecessor* of a branch is as the right-most branch of the left child if there is one, etc. Program it too.

Section 3.   Backtracking

Backtracking algorithms are efficient iterative programs for traversing labyrinths. Suppose at each node we can continue on either left, center, or right forks, or we have reached a dead end. The backtracking method will start by choosing left until we reach a dead end. The general step is this: Whenever we reach a dead end, we retreat one step. If we had gone left, we next try center; if we had done center, we next try right. But if we had gone right, we retreat another step, etc.

Let us apply this idea to the following problem. We want to find a long sequence of n elements 0, 1, 2 with the following property. For each k between 2 and n **div 2**, no two consecutive runs of k terms are the same. For instance,

```
00010002
```

is acceptable, but

```
00010000  and  00010001
```

are not because of the consecutive equal runs

```
00 00  and 0001 0001
```

respectively.

The following program solves the problem. Note that if we have found an acceptable run of n–1 terms, and we adjoin an n-th trial term, then we must only check each run that ends in n against the previous run of the same length. The **function** "valid" for testing the latest trial sequence implements this idea; read it carefully.

Program 6.3.1

```
{$R+,S+,M 65520}
program  sequence_012;    { 09/22/94 }

    uses  CRT, timer;

    const  max = 400;

    type  vector = array[1..max] of Char;

    var  v: vector;

    procedure  print_sequence;

        var  j: Word;
```

```
begin
for  j := 1  to  max  do
   begin
   Write(v[j]);
   if  j mod 50 = 0  then  WriteLn;
   end;
WriteLn;
end;

function  valid(n:  Word): Boolean;
   { Assumes v[1..(n-1)] has no equal consecutive
     subsequences.  Tests v[1..n]. }

   var  i, j, k, m: Word;
             equal: Boolean;

   begin
   equal := false;    { In case  n = 1 }
   if  Odd(n)  then  m := n shr 1 + 2
   else  m := n shr 1 + 1;
   for  k := m  to  n - 1  do
      begin    { v[k..n] versus v[2k-n-1, k-1] }
      i := k - 1;
      for  j := n  downto  k  do
         begin
         equal := (v[i] = v[j]);
         if  not equal  then  Break;
         Dec(i);
         end;
      if  equal  then  Break;
      end;
   valid := not equal;
   end;    { valid }
```

Let's pause a moment; we are now coming to the heart of the program. Observe that we begin the next v[n] with 0. If that fails, we try 1; if that fails, we try 2. If that fails also, then we backtrack to the first element that is not 2, increase it, and continue. The idea is clear I hope. Note of course that procedure "backtrack" is *not* recursive.

```
procedure  backtrack;

   var   n: Word;
        OK: Boolean;
```

```
begin    { backtrack }
n := 1;   v[1] := '0';
repeat
   if  valid(n)   then
      begin
      Inc(n);
      if  n > max  then  Break
      else  v[n] := '0';    { advance }
      end
   else
      begin    { retreat }
      while  v[n] = '2'  do  Dec(n);
      Inc(v[n]);
      end;
until  false;
WriteLn('Output of backtrack algorithm:');
print_sequence;
end;   { backtrack }

begin   { sequence_012 }
ClrScr;
get_time0;  backtrack;  get_time1;  put_time;
WriteLn;
Write('Press <Enter>: ');   ReadLn;
end.
```

Our next problem is to arrange 2^n 0's and 1's around a circle so that all the 2^n possible runs of n 0's and 1's occur (each once of course). The next program solves this problem. The listing omits an output procedure, which is in the disk file of course. Procedure "initialize" sets the first n elements of "circle". The sequence (0, 0, ..., 0) must occur, so we may start to label nodes at the beginning of this sequence. The program actually finds *all* solutions of this circle problem. Whenever it has actually found a solution, it prints that solution and then backtracks again, and keeps doing so until it can go no further. This mechanism is worth some study.

Program 6.3.2

```
program  circle_01;    { 05/21/94 }

   uses  CRT;

   const      run_size = 4;
          circle_size = 16;    { 2^run_size }
            array_size = circle_size + run_size;
```

```
type   vector = array[1..array_size] of 0..1;

var    circle: vector;
         index: Word;
         count: LongInt;

procedure  initialize;

   begin
   ClrScr;
   count := 0;
   for  index := 1  to  run_size  do
      circle[index] := 0;
   index := run_size + 1;
end;   { initialize }

procedure  backtrack;
{ circle[index] = 1 and the test has failed.
  Retreat index until the first 0, then try 1. }

   begin
   if  index > circle_size  then
      index := circle_size;
   while  circle[index] = 1  do  Dec(index);
   if  index = run_size  then  Exit;
   circle[index] := 1;
   end;   { backtrack }

function  new_run: Boolean;
   { Compares the final run of run_size
     terms with all the previous runs. }

   var   j, k: Integer;
         equal: Boolean;

   begin
   for  k := 1  to  index - run_size  do
      begin
      { Test run = run0 }
      for  j := 1  to  run_size  do
         begin
         equal := (circle[j + k - 1 ]
                   = circle[index - run_size + j]);
         if  not equal  then  Break;
         end;
      if  equal  then  Break;
```

```
        end;
    new_run := not equal;
    end;    { new_run }

begin    { circle_01 }
initialize;
repeat
if  KeyPressed  then  Halt;
    repeat
        if  KeyPressed  then  Halt;
        circle[index] := 0;    { First try 0 }
        if  new_run  then  Inc(index)
        else
            begin    { 0  doesn't work }
            if  index <= circle_size   then
                begin    { Then try 1 }
                circle[index] := 1;
                while  not new_run  do
                    begin
                    if  KeyPressed  then  Halt;
                    backtrack;
                    end;
                Inc(index);
                end
            else
                begin    { index > circle_size }
                backtrack;
                while  not new_run  do
                    begin
                    if  KeyPressed  then  Halt;
                    backtrack;
                    end;
                Inc(index);
                end;    { index > circle_size }
            end;    { 0  doesn't work }
    until  (index = array_size)
          or (index = run_size);
    if  index = array_size  then
        begin
        Inc(count);
        print;
        backtrack;
        while  not new_run  do
            begin
            if  KeyPressed  then  Halt;
            backtrack;
```

```
          if   index = run_size   then   Break;
          end;
       Inc(index);
       end;
  until    index <= run_size + 1;
  WriteLn;
  WriteLn('No more solutions');
  WriteLn;
  Write('Press <Enter> to quit:  ');    ReadLn;
  end.    { circle_01 }
```

Our final example is in the "fun and games" department: the famous eight queens problem: Place 8 queens on a chessboard so that no queen attacks any other. We shall go a little further than just finding all (92) of the solutions; we shall also pick a unique one out of each class of equivalent solutions under the action of the dihedral group of the chessboard, rotations and reflections. This boils the 92 down to 12. For an extensive discussion of this and related problems, see Ball, W W R, Mathematical Recreations and Essays 11/e, revised, Macmillan, London, 1963, Chapter 6.

As is often the case, it is easier to program from 0 base, i.e, from 0 to 7 rather than from 1 to 8. Various arrays in the program have the following meanings. First, solution[col] = row if and only if there is a queen on (row, col). Next, Q_row[r] if and only if a queen is in the row r, and Q_up_diag[diag] if and only if there is a queen in the diag-th up diagonal. Similarly, Q_down_diag[diag] if and only if a queen is in the diag-th down diagonal.

The program progresses from the first to the eighth column. When the solution up to the k-th column is satisfactory, there are 8 forks ahead for the step into the next column: row 0 to row 7. Each in turn is tried until a valid placement of the queen in that column is found, else we backtrack. Got it?

Program 6.3.3

```
program   eight_queens;    { 09/22/94 }

  uses   CRT;

  const            max = 7;
                     m = 2*max;
            all_sols = true;
            symmetry = true;
```

```
type     vector = array[0..max] of Word;
     list_ptr = ^list;
          list = record
                   sol: vector;
                   next: list_ptr;
                 end;

var    solution: vector;
           Q_row: array[0..max] of Boolean;
       Q_up_diag: array[-max..max] of Boolean;
     Q_down_diag: array[0..m] of Boolean;
            done: Boolean;
         row, col,
            diag: Integer;
           total,
           count: Word;
       root, run: list_ptr;
        out_file: file of vector;

procedure  remove_queen;
   { Assumes queen on (row, col); removes queen,
     but leaves (row, col) as the test solution. }

   begin
   Q_row[row] := false;
   Q_down_diag[col + row] := false;
   Q_up_diag[col - row] := false;
   end;

procedure  backtrack;
   { Assumes (row, col) is the test solution.
     Moves the test solution to the solution of the
     queen in the previous column (now removed) and
     continues until the queen can be moved up. }

   begin
   if  KeyPressed  then  Halt;
   Dec(col);
   row := solution[col];
   while  (row = max) and (col > 0)  do
      { Back to beginning of previous column. }
      begin
      remove_queen;
      Dec(col);
      row := solution[col];
      end;
```

```
if  row < max  then
{ Remain in the column, but move up one row. }
   begin
   remove_queen;
   Inc(row);
   end
else
   begin
   if  not all_sols  then
      WriteLn('No solutions');
   done := true;
   end;
end;    { backtrack }

function  new_solution: Boolean;
{ Accepts exactly one solution
  from each symmetry class. }

   var  transpose: vector;
            j, k: Word;
            test: array[1..8] of vector;
                { The symmetries }
         known: Boolean;

   begin    { new_solution }
   { General strategy.  Make a linked list of known
     solutions.  Test a new solution against these
     positions.  However, in general, each known
     solution has 8 symmetric solutions, so the new
     solution must be tested against all of these. }
   known := false;
   run := root;
   test[1] := solution;    { Identity }
   for  k := 0  to  max  do
      transpose[solution[k]] := k;
   test[2] := transpose;
   { Reflection in main diagonal }
   for  k := 0  to  max  do
      begin
      test[3, k] := solution[max - k];
         { Reflection in vertical axis }
      test[4, k] := transpose[max - k];
         { Rotation  -90 deg }
      test[5, k] := max - solution[k];
         { Reflect in horizontal axis }
```

```
      test[6, k] := max - transpose[k];
         { Rotation +90 deg }
      test[7, k] := max - solution[max - k];
         { Rotation 180 deg }
      test[8, k] := max - transpose[max - k];
         { Reflection in off diagonal }
   end;
  while  run^.next <> nil  do
    begin
    with  run^  do
       begin
       for  j := 1  to  8  do
          begin
          for  k := 0  to  max  do
             begin    { sol = test[j] ? }
             if  sol[k] <> test[j, k]  then
               Break;
               if  k = max  then    { Yes }
                 begin
                 known := true;  Break;
                 end;
             end;    { for k }
          end;    { for j }
       end;    { with }
    if  known  then  Break;
    run := run^.next;
    end;    { while }
  if  known  then  new_solution := false
  else
     begin
     new_solution := true;
     run := root;
     New(root);
     root^.sol := solution;
     root^.next := run;
     end;
  end;    { new_solution }

procedure  print;

  begin
  if  all_sols then
     Write('Solution ', count:2, ':   ');
```

```
      Write('[');
      for  col := 0  to   max  do
         Write(solution[col]:3);
      WriteLn(']':3);
      Write(out_file, solution);
      end;

begin    { eight_queens }
Assign(out_file, '8queens.dat');
Rewrite(out_file);
ClrScr;
total := 0;
if  all_sols  then
   begin
   Write('All solutions');
   count := 0;
   end
else  Write('A solution');
WriteLn(' of the ', max + 1, ' queens problem');
if  symmetry  then
   begin
   WriteLn('with one representative only ',
           'from each equivalence');
   WriteLn('class under the symmetry group ',
           'of the square.');
   end;
WriteLn;
{ Initialize }
done := false;
New(root);  root^.next := nil;
for row := 0  to   max  do  Q_row[row] := false;
for diag :=  -max  to   max  do
   Q_up_diag[diag] := false;
for diag := 0  to  m  do
   Q_down_diag[diag] := false;
row := 0;  col := 0;
repeat
   if  Q_row[row] or
       Q_down_diag[col + row] or
       Q_up_diag[col - row]  then
      begin
      if  row = max  then  backtrack
      else  Inc(row);
      end
   else    { Add a queen at current (row, col) }
```

```
        begin
        solution[col] := row;
        Q_row[row] := true;
        Q_down_diag[col + row] := true;
        Q_up_diag[col - row] := true;
        { Test it }
        if  col = max  then    { Done! }
            begin
            Inc(total);
            if  all_sols then
                begin
                if  not symmetry  then
                    begin
                    Inc(count);  print;
                    remove_queen;  backtrack;
                    end
                else
                    begin
                    if  new_solution  then
                        begin  Inc(count);  print;  end;
                    remove_queen;  backtrack;
                    end;
                end
            else
                begin  print;  ReadLn;  Exit;  end;
            end    { if  col = max }
        else
            begin  { Start with next column }
            Inc(col);  row := 0;
            end;
        end;    { Add a queen at current (row, col) }
    until  done;
    WriteLn;
    WriteLn('Total number of solutions: ', total,
            '    Number of fundamental solutions: ',
            count);
    WriteLn;
    Write('Press <Enter>: ');  ReadLn;
    Close(out_file);
    end.    { eight_queens }
```

Exercises

1. Write a recursive version of "sequence_012".
2. The disk file for "queens" creates a disk file "8queens.dat" of the 12 basic solutions. Its purpose is to entice you into writing a graphics program for displaying these solutions beautifully.

3. Program the placing of 5 queens on a chessboard so that each square is either occupied by a queen or attacked by a queen. Indeed, this can be done for boards up to size 11×11.

4. The classical "knight's tour" problem asks for a path on the chessboard for a knight to follow, touching each square one time. See W W R Ball, *loc cit* for a discussion of the problem. We ignore the more difficult problem of making the knight's tour reentrant, i.e, the knight should be able to move from his 64th square back to his first. (There are standard methods for transforming any solution into a reentrant solution, more art than science.) A systematic backtracking algorithm is rather inefficient, but the following idea attributed by Ball to Warnsdorff, H C works very efficiently; the knight always selects for its next square one of the squares it has not previously visited and from which it has the *least* number of possible next moves. This is counter-intuitive, and as far as I know has never been proved to work (on chess boards other than 8 by 8). There it is; program it!

Section 4. Recursive Graphical Algorithms

In this section we explore recursive generation of geometrical figures. After reading this section, you may turn to the first two sections of the next chapter for the related string generation of curves, and probabilistic generation of figures with iterated function schemes. The programs in this section and in the first section of the next chapter are based on the ideas of "turtle graphics". The "turtle" is a (bound) vector with its base point at the cursor and a vector, telling its next relative move from its base point, leaving a line (of droppings) behind. The basic turtle operations are rotating the turtle by an angle and drawing from the current turtle position by the amount of the vector, to the new turtle base point. The "Graph" unit procedure *LineRel* does this; we must write our own rotation procedures.

We start with a recursive program for the *von Koch curve*. It uses only rotations by 90° and –90°. The procedure "draw" essentially tells us the structure of the curve:

Draw, turn+, draw, turn–, draw, turn–, draw, draw, turn+, draw, turn+, draw, turn–, draw.

Program 6.4.1

```
program  von_koch_curve;    { 09/16/94 }
   { Recursive program }

   uses  CRT, Graph, graphs;

   var  ch: Char;

   procedure  rotate_plus(var  x, y: Integer);

      var  t: Integer;
```

```
    begin  t := x;   x := y;   y := -t;   end;

  procedure  rotate_minus(var  x, y: Integer);

    var  t: Integer;

    begin  t := x;   x := -y;   y := t;   end;

  procedure  draw(x, y: Integer;   s: Word);
      { (x, y) is a unit vector in the direction of
        the next move; s is the length of the move. }

    const  lim = 8;

    var  t: Word;

    begin
    if  s >= lim   then
        begin
        t := s div 4;
        draw(x, y, t);
        rotate_plus (x, y);   draw(x, y, t);
        rotate_minus(x, y);   draw(x, y, t);
        rotate_minus(x, y);
        draw(x, y, t);   draw(x, y, t);
        rotate_plus (x, y);   draw(x, y, t);
        rotate_plus (x, y);   draw(x, y, t);
        rotate_minus(x, y);   draw(x, y, t);
        end
    else  LineRel(s*x, s*y);
    end;

  begin    { von_koch_curve }
  open_graph;
  MoveTo(20, 200);
  draw(1, 0, 512);
  OutTextXY(200, 460, 'Press <Space>:');
  ch := ReadKey;
  close_graph;
  end.
```

The next program is a two dimensional *Cantor set*, first open middle squares deleted but all boundaries remain; then just the proper boundaries shown (with a very dark shadow of the rest, which you can easily remove by a color change):

Program 6.4.2

```pascal
program  twoD_Cantor_set;    { 09/16/94 }
   { See R Sedgewick, Algorithms 2/e, Addison-
   Wesley, 1988, p.59 }

   uses  CRT, Graph, graphs;

   var  ch: Char;

   procedure  draw(x, y: Integer;   size: Word);

      var  s: Word;

      procedure  solid_rectangle(x, y, size: Integer);

         begin
         Rectangle(x - size, y - size,
                   x + size, y + size);
         Bar(x - size + 1, y - size + 1,
             x + size - 1, y + size - 1);
         end;

      begin
      if  size > 1  then
         begin
         s := size div 2;
         draw(x - size, y + size, s);
         draw(x - size, y - size, s);
         draw(x + size, y + size, s);
         draw(x + size, y - size, s);
         end;
      solid_rectangle(x, y, size);
      end;

   begin    { twoD_Cantor_set }
   open_graph;
   SetFillStyle(SolidFill, Black);
   SetColor(Yellow);
   draw(GetMaxX div 2, GetMaxY div 2, GetMaxY div 4);
   OutTextXY(265, 235, 'Press <Space>');
   ch := ReadKey;
   SetColor(Blue);   { Try black too }
   SetWriteMode(XORPut);
   draw(GetMaxX div 2, GetMaxY div 2, GetMaxY div 4);
   SetColor(LightRed);
```

```
OutTextXY(265, 235, 'Press <Space>');
ch := ReadKey;
close_graph;
end.
```

The *Peano curve* offers an interesting programming challenge. The basic shape is like a squared off Z, or LCD computer numeral 2:

Draw, draw, turn+, draw, turn+, draw, draw, turn–, draw, turn–, draw, draw.

The reverse of this is a squared-off S. The curve of degree 0 is empty. The curve of degree n consists of three rows of Z's and S's connected together. We actually can include the Z and the S in a single procedure by adding a boolean variable for the direction. I hope the recursive procedure "draw_Z" is clear. You should try calling it with n = 1, 2, 3, 4.

Program 6.4.3

```
program  peano_curve;    { 09/20/94 }
   { Recursive program }

   uses   CRT, Graph, graphs;

   var       ch: Char;
           x, y: Integer;

      procedure   draw_Z(n: Word;   plus: Boolean);

         procedure   rotate(plus: Boolean);

            var   t: Integer;

            begin
            if  plus  then
               begin   t := x;   x := -y;   y := t;   end
            else
               begin   t := x;   x := y;   y := -t;   end;
            end;

      begin   { draw_Z }
      if  n > 0   then
         begin
         draw_Z(n - 1, plus);
         LineRel(x, y);   draw_Z(n - 1, not plus);
         LineRel(x, y);   draw_Z(n - 1, plus);
         rotate(plus);
         LineRel(x, y);   rotate(plus);
         draw_Z(n - 1, not plus);
```

```
      LineRel(x, y);   draw_Z(n - 1, plus);
      LineRel(x, y);   draw_Z(n - 1, not plus);
      rotate(not plus);
      LineRel(x, y);   rotate(not plus);
      draw_Z(n - 1, plus);
      LineRel(x, y);   draw_Z(n - 1, not plus);
      LineRel(x, y);   draw_Z(n - 1, plus);
      end;
   end;    { draw_Z }

begin    { peano_curve }
open_graph;
MoveTo(0, 0);
x := 5;   y := 0;
draw_Z(4, true);
OutTextXY(500, 460, 'Press <Space>:');
ch := ReadKey;
close_graph;
end.
```

The *Hilbert curve* is somewhat more complicated to program; let's take a crack at it. The basic shape is a squared-off U. The curve of order 0 is empty. The rules are different in the odd and even order cases; that is what adds a complication.

Program 6.4.4

```
program  hilbert_curve;    { 09/20/94 }
   { Draw Hilbert curves of orders 1..7 recursively }

   uses  CRT, Graph, graphs;

   const  size: array[1..7] of Integer
              = (380, 127, 57, 25, 12, 6, 3);

   var  order: Word;
          ch: Char;
        x, y: Integer;

   procedure  draw_U(n: Word;  plus: Boolean);
     { Even n preserve direction; odd n reverse. }

      procedure  rotate(plus: Boolean);

         var  t: Integer;
```

```
        begin
        if  plus   then
           begin  t := x;   x := -y;   y := t;   end
        else
           begin  t := x;   x := y;   y := -t;   end;
        end;

    begin    { draw_U }
    if  n > 0   then
       begin
       if  Odd(n)   then
          begin
          draw_U(n - 1, not plus);
          LineRel(x, y);   rotate(plus);
          draw_U(n - 1, plus);
          LineRel(x, y);
          draw_U(n - 1, plus);
          rotate(plus);   LineRel(x, y);
          draw_U(n - 1, not plus);
          end
       else
          begin
          draw_U(n - 1, not plus);
          rotate(plus);   LineRel(x, y);
          draw_U(n - 1, plus);
          rotate(not plus);   LineRel(x, y);
          rotate(not plus);
          draw_U(n - 1, plus);
          LineRel(x, y);   rotate(plus);
          draw_U(n - 1, not plus);
          end;
       end;
    end;    { draw_U }

begin    { hilbert_curve }
open_graph;
for  order := 1  to  7  do
   begin
   SetColor(8 + order);
   MoveTo(100, 50);
   x := size[order];
   y := 0;
   if  Odd(order)   then  draw_U(order, true)
   else  draw_U(order, false);
   SetColor(White);
   OutTextXY(30, 0, 'Press <Space>: ');
```

```
  ch := ReadKey;
  ClearDevice;
  end;
 close_graph;
end.
```

Exercises

1. The disk program "hilbert3" is another recursive program for generating the Hilbert plane-filling curve. Run it, then try to figure out how it works. (I won't reveal where I found it—in my view, it is a model of confusing recursive code.)

Section 5.   Permutations

A *permutation* in this section is a one–one function on the interval [0..max] of words onto itself. If max = 2, the 6 permutations are

012   021   102   120   201   210

In general there are (max+1)! permutations. We'll use the standard bracket notation for permutations, for example,

$$\sigma = \begin{bmatrix} 0 & 1 & 2 & 3 & 4 \\ 3 & 2 & 0 & 4 & 1 \end{bmatrix}$$

denotes the permutation σ defined by

$$\sigma(0) = 3 \qquad \sigma(1) = 2 \qquad \sigma(2) = 0 \qquad \sigma(3) = 4 \qquad \sigma(4) = 1$$

We start a unit of basic operations on permutations.

Program 6.5.1

```
unit  permutat;    { 09/22/94 }
{ Basic definitions and algorithms for permutations }

interface

   const   max = 10;

   type   vector = array[0..max] of Word;
          Str10 = string[20];
```

```
procedure   print(s: Str10;   var   x: vector);
procedure   random_permutation(var   p: vector);
procedure   random_perm(n, k: Word;   var   p: vector);
procedure   encode(var   P, C: vector);
procedure   decode(var   C, P: vector);
procedure   factorial_encode(var   P, C: vector);
procedure   factorial_decode(var   C, P: vector);
procedure   cyclic_perm1(var   V: vector;   r: Word);
procedure   cyclic_perm2(var   V: vector;   r: Word);
procedure   invert1(var   x, y: vector);
procedure   invert2(var   x: vector);
procedure   invert3(var   x: vector);
```

The procedure "print" implements the bracket notation above. It uses the IBM boxing characters to make the sides of the matrix:

implementation

```
procedure   print(s: Str10;   var   x: vector);

    var   k:   Word;

    begin
    WriteLn(#218:Length(s) + 4, ' ':3*max + 2, #191);

    Write(#179:Length(s) + 4);
    Write(0:2);
    for   k := 1   to   max   do   Write(k:3);
    WriteLn(#179);

    WriteLn(s, ' = '#179,   ' ':3*max + 2, #179);

    Write(#179:Length(s) + 4);
    Write(x[0]:2);
    for   k := 1   to   max   do   Write(x[k]:3);
    WriteLn(#179);

    WriteLn(#192:Length(s) + 4, ' ':3*max + 2, #217);
    end;   { print }
```

Generating a random permutation presents a challenge, and we give two algorithms. We need random permutations in order to test other procedures.

```
procedure  random_permutation(var  p: vector);

  var  j, k, t: Word;

  begin
  Randomize;
  for  j := 0  to  max  do  p[j] := j;
  for  k := max  downto  1  do
     begin
     j := Random(k + 1);      { 0 <= j <= k }
     if  j <> k  then
         begin
         t := p[k];  p[k] := p[j];  p[j] := t;
         end;
     end;     { for  k }
  end;     { random_permutation }

procedure  random_perm(n, k: Word;  var  p: vector);
  { Attributed to R W Floyd }

  var  i, j, l: Word;
           S: set of 0..255;

  begin
  if  (n > 255) or (k > n) or (k > max)  then
     Halt;
  S := [];
  l := 0;
  Randomize;
  for  i := max - k  to  max - 1  do
     begin
     j := Random(i);     { 0  j < i }
     if  j in S  then
         begin  p[l] := i;  S := S + [i];  end
     else
         begin  p[l] := j;  S := S + [j];  end;
     Inc(l);
     end;
  end;     { random_perm }
```

There are situations in which it is useful to encode permutations in various ways. There must be a way to put the integers $1 \ldots n!$ in one–one correspondence with the $n!$ permutations of $0..max$, where $max = n - 1$. Actually, the two encodings we give go to vectors rather than integers, but the idea is the same. Look particularly at the second, which uses the "factorial radix" expansion of integers: Each positive integer n has a unique expansion

$$n = c_1 1! + c_2 2! + \cdots + c_r r! \qquad 0 \leq c_j \leq j$$

With each encoding procedure there is a corresponding decoding.

```
procedure  encode(var  P, C: vector);
   { P is a permutation of 0..max
     Given k, suppose P[s] = k.
     Then C[k] = #P[j] < k, j = 0..s-1 }

   var  j, k, count: Word;

   begin
     for  k := 0  to  max  do
      begin
      count := 0;
      for  j := 0  to  max  do
         begin
         if  P[j] < k  then  Inc(count)
         else  if  P[j] = k  then  Break;
         end;    { for  j }
      C[k] := count;
      end;    { for  k }
   end;    { encode }

procedure  explain_encode(var  P, C: vector);
   { P  is a permutation of 0..max
     This explains how the coding is done. }

   var  k, count: Word;

   begin
   for  k := max  downto  0  do
      begin
      { Find count so  P[count] = k, ie, invert P }
      count := 0;
      while  P[count] < k  do  Inc(count);
      C[k] := count;
      { Delete k from P }
      if  count < k  then
         Move(P[count + 1], P[count],
              (k - count)*SizeOf(Word));
      end;    { for  k }
   end;    { explain_encode }
```

```
procedure  decode(var  C, P: vector);
   { 0 <= C[k] <= k   for all k }

   var  j, k, count: Word;

   begin
   for  k := 0  to   max  do
      begin
      count := C[k];
      for  j := k + 1  to   max  do
         begin
         if  C[j] <= count  then   Inc(count);
         end;
      P[count] := k;
      end;    { for  k }
   end;    { decode }

procedure  explain_decode(var  C, P: vector);
   { 0 <= C[k] <= k   for all k }

   var  k, count: Word;

   begin
   P[0] := 0;
   for  k := 1  to   max  do
      begin
      count := C[k];
      if  count < k  then
         Move(P[count], P[count + 1], k - count);
      P[count] := k;
      end;
   end;    { explain_decode }

procedure  factorial_encode(var  P, C: vector);
   { See Knuth, D, Art of Computer Programming
     vol 2, 2/e, Addison-Wesley, 1981, p 64.
     There is a 1-1 correspondence between P's
     and (max+1)!   P --> C, which corresponds to
        c1 1! + c2 2! + c3 3! +...   }

   var  i, k: Word;
```

```
begin
for  k := max  downto  1  do
    begin
    i := 0;
    while  P[i] <> k  do  Inc(i);
    C[k]  := i;
    P[i]  := P[k];
    P[k]  := k;
    { P ends as (0 1 2 ... max) }
    end;
end;

procedure  factorial_decode(var  C, P: vector);

var  j, k: Word;

begin
P[0] := 0;
for  k := 1  to  max  do
    begin
    j := C[k];    { j  k }
    if  j < k  then
        begin  P[k] := P[j];  P[j] := k;  end
    else  P[k] := k;
    end;    { for  k }
end;    { decode }
```

We have two procedures for cyclic permutations. You should think of a very large max, so that it is not possible to declare more than one vector variable. Then the procedures must execute "in place".

```
procedure  cyclic_perm1(var  V: vector;  r: Word);
    { Effects the cyclic permutation
        (0..max) --> (r..max, 0..r-1) }

var  j, k, t: Word;

begin
j := 0;  k := r - 1;
{ Reverse (0..r-1) }
while  j < k  do
    begin
    t := V[j];  V[j] := V[k];  V[k] := t;
    Inc(j);  Dec(k);
    end;
```

```
  { Reverse (r..max) }
  j := r;   k := max;
  while  j < k   do
     begin
     t := V[j];   V[j] := V[k];   V[k] := t;
     Inc(j);   Dec(k);
     end;
  { Reverse (0..max) }
  j := 0;   k := max;
  while  j < k   do
     begin
     t := V[j];   V[j] := V[k];   V[k] := t;
     Inc(j);   Dec(k);
     end;
  end;    { cyclic_perm1 }

procedure  cyclic_perm2(var  V: vector;   r: Word);
{ Effects the permutation
     (0..max) --> (r..max, 0..r-1).
  Analysis:  The permutation is the product of
  g = gcd(r, max+1) independent (max+1)/g cycles
  starting with 0, 1,...   Indeed the permutation is
       Tx = x + r (mod   max+1)
  so    (T^k)k x = x + kr   (mod   max+1)
  and   x + kr = x (mod   max+1)
  iff   k = 0   (mod   (max+1)/g ).  }

  var  g, i, j, k, m: Word;
                   t: Word;

  function  gcd(m, n: Word): Word;

     var  r: Word;

     begin
     while  n > 0   do
        begin
        r := m mod n;   m := n;   n := r;
        end;
     gcd := m;
     end;
```

```
begin
g := gcd(r, max + 1);
for  i := 0  to  g - 1  do
    begin
    t := V[i];
    k := i;  m := i + r;
    for  j := 1  to  (max + 1) div g - 1  do
        begin
        if  m > max  then  m := m - max - 1;
        V[k] := V[m];
        k := m;  m := m + r;
        end;
    V[k] := t;
    end;    { for  i }
end;    { cyclic_perm2 }
```

Now we take up the problem of finding the inverse of a permutation. If there is no difficulty with memory, it is very easy:

```
procedure  invert1(var  x, y: vector);

    var  j: Word;

    begin
    for  j :=  0  to  max  do  y[x[j]] := j
    end;    { invert1 }
```

However, much more sophisticated algorithms are required to invert "in place". Both of the following are worthy of study:

```
procedure  invert2(var  x: vector);
    { Inversion "in place".  Each cycle is inverted,
      and its elements are marked. }

    var  i, j, k, t: Word;

    begin
    i := 0;
    repeat
        k := i;  j := x[k];
        while  j <> i  do
            begin
            t := k + 2*max;  k := j;
            j := x[k];  x[k] := t;
            end;
```

```
        x[i] := k + 2*max;
        while  (i <= max) and (x[i] > max)    do
            Inc(i);
    until  i > max;
    for  k := 0  to   max  do  x[k] := x[k] - 2*max;
    end;    { invert2 }

  procedure  invert3(var  x: vector);
    { J. Boothroyd }

    var  i, j, k: Word;

    begin
    for  i := 0  to   max  do  x[i] := x[i] + 2*max;
    for  i := 0  to   max  do
        begin
        k := i;  j := x[i];
        while  j <= max  do
            begin
            k := j;  j := x[k];
            end;
        x[k] := x[j - 2*max];  x[j - 2*max] := i;
        end;    { for i }
    end;    { invert3 }

end.    { unit permutat }
```

Much interesting work has gone into algorithms for listing all permutations. We shall look at three programs for doing so; all share some common declarations and procedures, which we organize into a unit:

Program 6.5.2

```
unit  enumperm;    { 09/28/94 }

interface

    const  abs_max = 10;

    type  vector = array[1..abs_max] of Word;

    var  count: LongInt;
         max: Word;
```

```
procedure   get_max;
procedure   switch(var   v: vector;   j, k: Word);
procedure   print(var   x: vector);
```

implementation

```
procedure   get_max;

    begin
    WriteLn(
    'All permutations of 1..n will be listed.');
    WriteLn('Enter n with 3 <= n <= ', abs_max);
    Write('n = ');   ReadLn(max);
    if   max < 3   then   max := 3;
    if   max > abs_max   then   max := 5;
    count := 0;
    end;

procedure   switch(var   v: vector;   j, k: Word);
    { v[j] <--> v[k] }

    var   t: Word;

    begin
    t := v[j];   v[j] := v[k];   v[k] := t;
    end;

procedure   print(var   x: vector);

    var   k: Word;

    begin
    Inc(count);
    Write(count:4, '   ');
    for   k := 1   to   max   do   Write(x[k]:3);
    WriteLn;
    if   count mod 20 = 0   then
        begin
        Write('To continue, press <Enter>: ');
        ReadLn;
        end;
    end;

end.   { unit   enumperm }
```

The first problem, solved by R. J. Ord-Smith, is to generate all permutations in reverse lexicographic order.

Program 6.5.3

```
program  all_permutations_1;    { 09/28/94 }
   { R J Ord-Smith's algorithm for generating all
     permutations in reverse lexicographic order.
     The first k-1 elements are moved before the
     k-th is moved. }

   uses  enumperm;

   var       x, ind: vector;
                   j: Word;
         even, done: Boolean;

   procedure  next_perm;

      var  i, j, k: Integer;

      begin   { next_perm }
      if  even  then
         begin
         j := 2;
         Inc(ind[2]);    { Now increment  ind. }
         while  ind[j] > j  do
            begin
            ind[j] := 0;   Inc(j);   Inc(ind[j]);
            end;
         if  j = max  then  done := true
         else
            begin
            switch(x, j + 1, ind[j]);
            i := 1;   k := j;
            repeat    { List  j  x's  in order. }
               switch(x, i, k);
               Inc(i);   Dec(k);
            until  i >= k;
            end    { else }
         end    { even }
      else  switch(x, 1, 2);
      even := not even;
      end;    { next_perm }
```

```
begin    { all_permutations_1 }
{ Initialize }
get_max;
for  j := 1  to   max   do
    begin  x[j] := j;   ind[j] := 0;   end;
even := false;   done := false;
repeat
    print(x);
    next_perm;
until   done;
Write('Done: press <Enter>: ');   ReadLn;
end.    { all_permutations_1 }
```

The next permutation enumerator, due to J. Boothroyd, has each permutation in the list obtained from the previous one by interchanging two elements. Boothroyd published his algorithm in Computer Journal 10 (1967) p. 311. There are two survey articles on the subject of enumerating permutations: Ord-Smith, R J, Computer Journal 13 (1970), pp. 152-155, *ibid* 14 (1971), p. 136-139.

Program 6.5.4

```
program  all_permutations_2;    { 09/28/94 }
    { J. Boothroyd's algorithm for listing all
      permutations, where each differs from the
      previous one by one transposition. }

    uses  enumperm;

    var  x, coeff: vector;
                k: Word;

    procedure  next_coeff;

        begin
        Inc(coeff[1]);
        k := 1;
        while  coeff[k] > k  do
            begin
            coeff[k] := 0;
            Inc(k);
            if  k = max  then  Exit;
            Inc(coeff[k]);
            end;
        end;    { next_coeff }
```

```
procedure   next_permutation;

   begin
   if   not Odd(k) or (coeff[k] <= 2)   then
      switch(x, k, k + 1)
   else   switch(x, k + 1 - coeff[k], k + 1);
   end;   { next_perm }

begin    { all_permutations_2 }
get_max;
for   k := 1   to   max   do
   begin   x[k] := k;   coeff[k] := 0;   end;
k := 1;
repeat
   print(x);
   next_coeff;
   if   k = max   then   Break;
   next_permutation;
until   false;;
Write('Press <Enter>: ');   ReadLn;
end.    { all_permutations_2 }
```

Perhaps the deepest algorithm in this family is that of H. F. Trotter, published
as Algorithm 115, Collected Algorithms of the ACM (CACM 1962).   Each
permutation in the list of permutations is obtained from the previous one by
interchanging a pair of *adjacent* elements. If the elements were bells, this would
be the UK sport of "ringing changes". I have put two versions of the algorithm
on disk, "allperm3" and "allperm4"; the latter listed here:

Program 6.5.5

```
program   all_permutations_4;    { 09/28/94 }
   { Lists all permutations of 1..max in
     "change ringing" order: each permutation is
     obtained from the previous one by a single
     transposition of adjacent elements. }

uses   enumperm;

var   x, index: vector;
         Move: array[1..abs_max] of (left, right);
            m: Word;
         done: Boolean;
            k: Word;    { global, for tracing }
```

```
procedure   compute_next_perm;

    var   i: Word;

    begin
    k := 0;   m := 1;
    if   Move[m] = left   then   Dec(index[m])
    else   Inc(index[m]);
    while   (index[m] = max - m + 1) or
            (index[m] = 0)   do
       begin
       if   index[m] = 0   then     { Off left end }
          begin
          Move[m] := right;    { shift right }
          Inc(k);    { move switch point right }
          end
       else    { Off right end; shift left }
          Move[m] := left;
       if   m = max - 1   then
          begin
          switch(x, max - 1, max);
          done := true;
          Break;    { while loop }
          end
       else
          begin
          Inc(m);   { move to next element }
          if   Move[m] = left   then   Dec(index[m])
          else   Inc(index[m]);
          end;
       end;    { while }
    if   not done   then    { switch }
       begin
       i := k + index[m];
       switch(x, i, i + 1);
       end;
    end;    { compute_next_perm }

begin    { all_permutations_4 }
get_max;
{ Initialize }
for   k := 1   to   max   do
   begin
   x[k] := k;
   index[k] := 0;   Move[k] := right;
   end;
```

```
   done := false;
   m := 1;
   repeat
      print(x);
      compute_next_perm;
   until  done;
   Write('Press <Enter>: ');   ReadLn;
   end.    { all_permutations_4 }
```

A permutation of 0..max has been a vector of 0..max. Given such a vector, we can use it to permute a vector of, say, reals indexed from 0 to max. If max is large, it is essential to effect the permutation in place. For instance a vector of 5000 extended reals occupies 50,000 bytes, so there is no possibility of another such vector. The following program does this in two ways. The first method assumes that we can mark those reals that have been processed already. For instance if they are all positive, then changing signs marks the processed ones. The second method assumes that we have at least enough memory for an array of bytes, so we can start with 0's and mark with 1's the processed elements. If we were really tight for space, we could use individual bits, which of course would require more code and run time.

Program 6.5.6

```
program  permute;    { 09/23/94 }
  { To permute "in place". }

  const   max = 4999;

  type        Real = Extended;
            vector = array[0..max] of Word;
       real_vector = array[0..max] of Real;

  procedure  permute_in_place(var  A: real_vector;
                              var  P: vector);
     { Assumed A[k] > 0.0 }

  {  P is a permutation on  0..max-1.   The problem
     is to permute the elements of  A   according
     to  P, "in place".   Assumed: the elements of  A
     can be marked, e.g, by changing their signs. }

     var    hold: Real;
            run, k: Word;
```

```
begin
k := 0;
while  k < max   do
   begin     { Process a cycle }
   hold := A[k];   run := k;
   while  P[run] <> k  do
      begin
      a[run] := - A[P[run]];
      run := P[run];
      end;
   A[run] := -hold;
   repeat
      Inc(k)
   until  (A[k] > 0) or (k = max);
   end;
for  k := 0  to  max  do  A[k] := - A[k];
end;    { permute_in_place }

procedure  permute_in_place2(var  A: real_vector;
                             var  P: vector);

var     hold: Real;
      run, k: Word;
      control: array[1..max] of Byte;

begin
FillChar(control, SizeOf(control), 0);
k := 0;
while  k < max   do
   begin     { Process a cycle }
   hold := A[k];   run := k;
   while  P[run] <> k  do
      begin
      a[run] := A[P[run]];
      control[run] := 1;
      run := P[run];
      end;
   A[run] := hold;
   repeat
      Inc(k)
   until  (control[k] = 0) or (k = max);
   end;
end;    { permute_in_place2 }
```

```
begin
end.
```

Sometimes it is useful to list "permutations with duplicates". For instance we can list all 6 arrangements of a, a, b, b:

aabb   abab   abba   baab   baba   bbaa

The following program does this in general. You might have some fun testing it, and computing exactly how many arrangements there are in various cases.

Program 6.5.7

```
program  permutations_with_duplicates;    { 09/26/94 }
   { J P N Phillips algorithm, following M K Shen
     Lists all distinct permutations of a sequence
     with duplications. }

   uses   CRT;

   const   abs_max = 4;

   type   vector = array[1..abs_max] of Char;

   var              x: vector;
         max, count: Word;
         decreasing: Boolean;

   procedure  initialize;

      var   ch: Char;
            k: Word;

      begin
      max := abs_max;    { var needed }
      count := 0;
      Randomize;
      ch := 'a';
      for  k := 1  to  max  do
        begin
        x[k] := ch;
        if  Random(2) = 0  then  Inc(ch);
        end;
      end;
```

```
procedure  print(var  x: vector);

    var    k: Word;

    begin
    for  k := 1  to   max  do   Write(x[k]:3);
    WriteLn;
    end;

procedure  find_next_perm(var x: vector;
                          var  decreasing: Boolean);

    var   i, j, k: Word;
              t: Char;

    procedure  transpose(j, k: Word);

       var   t: Char;

       begin
       t := x[j];   x[j] := x[k];   x[k] := t;
       end;

    procedure  three_cycle(i, j, k: Word);

       var   t: Char;

       begin
       t := x[i];   x[i] := x[j];
       x[j] := x[k];   x[k] := t;
       end;

    procedure  search;

       begin
       if   max in [2, 3]   then
          begin
          decreasing := true;   i := 1;   j := max;
          end
       else
          begin    { max >= 4 }
          i := max - 3;
          { Find largest i: x[i] < x[i + 1] }
          while  (i >= 1) and (x[i + 1] <= x[i])   do
             Dec(i);
```

```
      if  i >= 1   then    { x[i + 1] <= x[i] }
          begin   { Find largest k: x[i] < x[k] }
          k := max;   t := x[i];
          while  x[k] <= t  do  Dec(k);
          transpose(i, k);
          Inc(i);   j := max;
          end   { Find largest k }
       else
          begin
          decreasing := true;   i := 1;   j:= max;
          end;
       end;    { max >= 4 }
    repeat
       transpose(i, j);   Inc(i);   Dec(j);
    until  j <= i;
    end;    { search }

  begin   { find_next_perm }
  decreasing := false;
  if  x[max - 1] < x[max]   then
     transpose(max - 1, max)
  else  if  max = 2   then   search
  else  if  x[max - 2] < x[max]   then
     { x[max - 2] < x[max] <= x[max - 1]; }
     three_cycle(max, max - 1, max - 2)
  else  if  x[max - 2] < x[max - 1]   then
     { x[max] <= x[max - 2] < x[max - 1];   }
     three_cycle(max, max - 2, max - 1)
  else    { x[max] <= x[max - 1] <= x[max - 2] }
     search;
  end;    { find_next_perm }

begin   { permutations_with_duplicates }
ClrScr;
initialize;
repeat
   print(x);   find_next_perm(x, decreasing);
   Inc(count);
   if  count mod 20 = 0   then
      begin
      Write('Press <Enter>: ');   ReadLn;
      end;
until  decreasing;
WriteLn(count, ' permutations');
Write('Press <Enter>: ');   ReadLn;
end.
```

Each permutation can be expressed in a unique way, up to order, as a product (composition) of independent cycles. On your disk you will find a program "factor_perm" which takes as input any product of cycles and returns the same permutation factored into independent cycles.

This brings us to fun and games. A magic square may be thought of a permutation of $1..n^2$. There is a rather easy algorithm for generating odd order magic squares; even order is quite hard, so we'll leave it for an exercise.

Program 6.5.8

```pascal
program  magic_square;    { 06/03/94 }
   { D M Collison, Algorithm 118, CACM, 1962 }

   uses   CRT;

   const   m_2 = 6;
           max = 2*m_2;
          even = not Odd(m_2);

   type  matrix = array[1..max - 1, 1..max - 1]
                     of Word;

   var    y: matrix;
       j, k: Word;
        sum: Word;

    procedure  odd_square(var  y: matrix);
       { De la Loubre method, 1693 }

       var  i, j, k: Integer;

       begin
       for  j := 1  to max - 1  do
           for  k := 1  to  max - 1  do  y[j, k] := 0;
       i := m_2;  j := max - 1;
       for  k := 1  to  Sqr(max - 1)  do
           begin
           if  y[i, j] <> 0  then
               begin
               Dec(i);  Dec(j, 2);
               if  i = 0  then  i := i + max - 1;
               if  j <= 0  then  j := j + max - 1;
               end;   { if }
           y[i, j] := k;
           Inc(i);
```

```
         if  i = max  then  i := i - max + 1;
         Inc(j);
         if  j = max  then  j := j - max + 1;
         end;    { for k }
      end;    { odd_square }

  begin    { magic_square }
  odd_square(y);
  WriteLn('Magic square of order ', max - 1);
  WriteLn;
  for  j := 1  to  max - 1  do
     begin
     for  k := 1  to  max - 1  do  Write(y[j, k]:4);
     WriteLn;
     end;
  ReadLn;
  for  j := 1  to  max - 1  do
     begin
     sum := 0;
     for  k := 1  to  max - 1  do
        sum := sum + y[k, j];
     Write( 'row sum ', j:2,   ': ', sum);
     sum := 0;
     for  k := 1  to  max - 1  do
        sum := sum + y[j, k];
     GotoXY(30, WhereY);
     WriteLn('column sum ', j:2,   ': ', sum);
     end;
  sum := 0;
  for  j := 1  to  max - 1  do
     sum := sum + y[j, j];
  Write('main diagonal sum: ', sum);
  GotoXY(30, WhereY);
  sum := 0;
  for  j := 1  to  max - 1  do
     sum := sum + y[j, max - j];
  WriteLn('other diagonal sum: ', sum);
  ReadLn;
  end.    { magic_square }
```

Exercises

1. Test the coding and decoding procedures in the unit "permutat".
2. Test the procedures for inverting permutations in the unit "permutat".
3. Test the procedures for cyclic permutations in the unit "permutat".

4. Look up a method for generating even order magic squares and program it. The method of Devedec is given in Kraitchik. M, Mathematical Recreations 2/e. Dover, 1953, pp. 150–152.

Section 6. Combinations

"Combination" is an old fashioned word for subset. Everyone knows the Pascal triangle, a table of the number of k–subsets of an n–set, called "n choose k", and written

$$C_{n,k} = \binom{n}{k} = \frac{n!}{k!(n-k!)}$$

The formula follows from the iteration formula

$$\binom{n}{k} = \binom{n-1}{k-1} + \binom{n-1}{k} \qquad \binom{n}{0} = 1$$

A subset of k elements either contains n (there are $C_{n-1,\,k-1}$ of these) or doesn't contain n (there are $C_{n-1,\,k}$ of these); that explains the formula. Our first program uses the iteration formula, but recursively; its speed should impress (rather depress) you.

Program 6.6.1

```
program  Pascal_triangle;    { 09/28/94 }

   var   n, k: Word;

   function  n_choose_k(n, k: Word): LongInt;

      begin
      if   k = 0   then   n_choose_k := 1
      else   if   k = n   then   n_choose_k := 1
      else   n_choose_k := n_choose_k(n - 1, k - 1)
                             + n_choose_k(n - 1, k);
      end;    { n_choose_k }
```

```
begin    { Pascal_triangle }
for   n := 0  to   12   do
   begin
   Write(n_choose_k(n, 0):(40 - 3*n));
   for   k := 1   to   n   do
      Write(n_choose_k(n, k):6);
   WriteLn;
   end;
WriteLn;
n := 18;   k := 10;
WriteLn('Be patient;   computing ',
         n, ' choose ', k);
WriteLn(n, ' choose ', k, ' = ', n_choose_k(n, k));
WriteLn;
Write('Press <Enter> to quit: ');   ReadLn;
end.    { Pascal_triangle }
```

Much more interesting than counting subsets is listing them. We start with a rather simple algorithm for listing all k–subsets of 1..n in lexicographic order. Although the program is very brief, its recursive call is worth study.

Program 6.6.2

```
program  all_k_subsets;    { 09/28/94 }
   { Generates all k-subsets of 1..n in
     lexicographic order }

   const   max = 20;

   var      set_size,
         subset_size,
               delta: Word;
                  S: array[0..max] of Word;

   procedure  fill_subset(m: Word);
      { Entry: S[0..m-1] known, strictly increasing
        Exit: S[m..k] known and S printed, m fixed }

      var  j: Word;

      procedure  print_comb;

         var   k: Word;
```

```pascal
      begin
      Write('{');
      for  k := 1  to  subset_size - 1  do
         Write(S[k], ', ');
      WriteLn(S[subset_size], '}');
      end;    { print_comb }

   begin    { fill_subset }
   for  j := S[m - 1] + 1  to  delta + m  do
      begin
      S[m] := j;
      if  m = subset_size  then  print_comb
      else  fill_subset(m + 1);
      end;
   end;    { fill_subset }

begin    { all_k_subsets }
WriteLn;
WriteLn('All k-subsets of  1..n  enumerated.');
WriteLn;
Write   ('Enter n: ');   ReadLn(set_size);
Write   ('Enter k: ');   ReadLn(subset_size);
WriteLn;
if  (set_size > max) or
    (subset_size > set_size)  then  Halt;
delta := set_size - subset_size;
S[0] := 0;   fill_subset(1);
WriteLn;
Write('Press <Enter>: ');   ReadLn;
end.
```

We shall inspect three more programs for generating subsets, all recursive and interesting in different ways. They all use a common declarations, initialization, and output procedures in a unit "subsets". We omit the routine body of "print" in this listing:

Program 6.6.3

```pascal
unit   subsets;    { 09/28/94 }

interface

   const   max = 5;

   var     S: array[1..max] of 0..1;
        count: Word;
```

```
procedure   initialize;
procedure   print(direct: Boolean);

implementation

   procedure   initialize;

      var   k: Word;

      begin
      for   k := 1   to   max   do   S[k]  := 0;
      WriteLn('count',  ' ':4,  'code':max,
                    ' ':4,  'subset');
      count := 0;
      end;    { initialize }

   procedure   print(direct: Boolean);
         { Prints S[k] from 1 to max if direct,
           from max downto 1 if not direct }

      begin .... end;    { print }

end.    { unit   subsets }
```

The array S determines a subset; k is in the subset if and only if S[k] = 1. The following programs list all subsets of 1..max. Here is the first; its recursion is pretty transparent: First list the subsets that contain n, then those that do not.

Program 6.6.4

```
program   all_subsets1;    { 09/28/94 }
{ Takes each possibility for S[n] and then recursion }

   uses   subsets;

   procedure   next_subset(n: Word);

      begin
      if   n = 0   then   print(false)
      else
         begin
         S[n]  := 1;
         next_subset(n - 1);
         S[n]  := 0;
         next_subset(n - 1);
         end;
      end;    { next_subset }
```

```
begin    { all_subsets1 }
initialize;
next_subset(max);
Write('Press <Enter>: ');    ReadLn;
end.
```

The next subset generator is a variation of this program, using a double induction on n and k.

Program 6.6.5

```
program  all_subsets2;    { 09/28/94 }
{ Goes by rule for generating combinations }

  uses   subsets;

  var   k: Word;

  procedure   next_k_subset(n,  k: Word);

      var   i: Word;

      begin
      if   n = 0   then   print(true)
      else
          begin
          S[n]  := 0;
          if   k < n   then   next_k_subset(n - 1, k);
          S[n]  := 1;
          if   k > 0   then   next_k_subset(n - 1, k - 1);
          end;
      end;

  begin    { all_subsets2 }
  initialize;
  for  k := 0  to   max  do   next_k_subset(max, k);
  Write('Press <Enter>: ');    ReadLn;
  end.
```

The final program for listing subsets is based on "Gray codes". In this listing, each subset differs from the previous one by changing the status of a single element. Its recursion is quite elegant.

Program 6.6.6

```
program  all_subsets3;    { 09/08/94 }
{ Nijenhuis, A and H.S. Wilf, Combinatorial Algorithms,
  Academic Press, New York, 1978, pp. 14 ff.
  Modified by R. M. Dieffenbach }

  uses   subsets;

  procedure  gray_code(n: Word);

     begin
     if   n = 0   then   print(false)
     else
        begin
        gray_code(n - 1);
        S[n] := 1 - S[n];
        gray_code(n - 1);
        end;
     end;

  begin    { all_subsets3 }
  initialize;
  gray_code(max);
  Write('Press <Enter>: ');   ReadLn;
  end.
```

Chapter 7 More Algorithms

Section 1. String Generation of Curves

We begin this chapter with a fresh look at recursive curves—from an entirely different viewpoint. The instructions for drawing a curve can be in the form of a string. For instance the basic Z shape can be written "LL+L+LL–L–LL", where L means "line segment", + means "turn forward", and – means "turn backward". The idea, called "string rewriting system", of using a string of characters to represent a recursive graphics figure was worked out by P. Prusinkiewicz, based on earlier work on modeling organism growth. See Saupe, D, Appendix C in Peitgen H-O, et al, Fractal Images, Springer-Verlag, 1988.

We'll start with a simpler version of the von Koch curve: a closed curve whose basic configuration is a square. The seed string is "L+L+L+L", and there is only one replacement rule; it replaces each instance of L in the string by fixed string, called the "rule". This is applied enough times to reach the desired resolution, and the final string is read to plot the curve. Of course 255 characters is not adequate, so we use a character array for the "string". If you will draw the rule on paper, its meaning will be obvious.

Program 7.1.1

```pascal
program  von_koch_island;    { 09/17/94 }
   { Generates von Koch fractal island by
      "string rewriting system". }

   uses  CRT, Graph, graphs;

   var      Str: array[1..4000] of Char;
         no_chars,   { 7 + 52(8^3 - 1)/7 = 3803 enough }
            depth: Word;
              ch: Char;
          c, s,
          size: Word;

   procedure  generate_string;

      const  seed = 'L+L+L+L';
             rule = 'L+L-L-LL+L+L-L';
              min = Length(rule);

      var  j, k: Word;
             uu: string[min];
             { Needed because Move requires var's  }
```

312

```
begin
no_chars := Length(seed);
uu := seed;
for  j := 1  to  no_chars  do
    Str[j] := uu[j];
uu := rule;
for  j := 1  to  depth  do
    begin
    for  k := no_chars  downto  1  do
        if  Str[k] = 'L'  then
            begin
            Move(Str[k + 1], Str[k + min],
                no_chars - k);
            Move(uu[1], Str[k], min);
            no_chars := no_chars + min - 1;
            end;
    end;
end;    { generate_string }

procedure  draw;
    { (x, y) is a vector in the direction of the
      next move; size is the length of the move. }

    var  x, y: Integer;
         k: Word;

    procedure  rotate_plus(var  x, y: Integer);
        { Rotation + 90 deg }

        var  t: Integer;

        begin  t := x;  x := -y;  y := t;  end;

    procedure  rotate_minus(var  x, y: Integer);
        { Rotation -90 deg }

        var  t: Integer;

        begin  t := x;  x := y;  y := -t;  end;

    begin    { draw }
    x := size;  y := 0;
    MoveTo(150, 120);
```

```
for  k := 1  to   no_chars   do
      case   Str[k]   of
            'L':  LineRel(Round(x), Round(y));
            '+':  rotate_plus (x, y);
            '-':  rotate_minus(x, y);
         end;
   end;    { draw }

begin    { von_koch_island }
open_graph;
size := 256;
for   depth := 0  to  3  do
      begin
      generate_string;
      draw;
      OutTextXY(500, 460, 'Press <Space>:');
      ch := ReadKey;
      ClearDevice;
      size := size div 4;
      end;
close_graph;
end.
```

There can be several rules for more complicated graphs. The following program again draws the Peano curve. In addition to the turns +, − and L, there are two other characters in the string, A and B. They *draw nothing*. Their purpose is to expedite the two replacement rules. If you ignore the A's and B's in the rules (which don't draw) you see that the rules are the Z and the S.

Program 7.1.2

```
program  peano_curve2;    { 09/17/94 }
   { Peano curve by "string rewriting system". }

   uses   CRT, Graph, graphs;

   var       Str: array[1..17000] of Char;
         no_chars,    { 16401 needed }
            depth: Word;
               ch: Char;
             size: Integer;
```

```
procedure  generate_string;

   const    seed = 'A';
   { A = LL+L+LL-L-LL   B = LL-L-LL+L+LL }
            rule1 = 'ALBLA+L+BLALB-L-ALBLA';
            rule2 = 'BLALB-L-ALBLA+L+BLALB';
            min1 = Length(rule1);
            min2 = Length(rule2);

   var  j, k: Word;
        uu1: string[min1];
        uu2: string[min2];
        { Needed because Move requires var's }

   begin
   Str[1] := seed;
   no_chars := 1;
   uu1 := rule1;   uu2 := rule2;
   for  j := 1  to  depth  do
      begin
      for  k := no_chars  downto  1  do
         if  Str[k] = 'A'  then
            begin
            Move(Str[k + 1], Str[k + min1],
                 no_chars - k);
            Move(uu1[1], Str[k], min1);
            no_chars := no_chars + min1 - 1;
            end
         else  if  Str[k] = 'B'  then
            begin
            Move(Str[k + 1], Str[k + min2],
                 no_chars - k);
            Move(uu2[1], Str[k], min2);
            no_chars := no_chars + min2 - 1;
            end;
      end;
   end;   { generate_string }

procedure  draw;
   { (x, y) is a vector in the direction of the
     next move; size is the length of the move. }

   var  x, y: Integer;
        k: Word;
```

```
procedure  rotate_plus(var  x, y: Integer);
   { Rotation + 90 deg }

   var  t: Integer;

   begin  t := x;  x := -y;  y := t;  end;

procedure  rotate_minus(var  x, y: Integer);
   { Rotation -90 deg }

   var  t: Integer;

   begin  t := x;  x := y;  y := -t;  end;

begin    { draw }
x := size;  y := 0;
MoveTo(100, 0);
for  k := 1  to  no_chars  do
   case  Str[k]  of
      'L': LineRel(x, y);
      '+': rotate_plus (x, y);
      '-': rotate_minus(x, y);
      end;
end;    { draw }

begin    { peano_curve2 }
open_graph;
size := 200;
for  depth := 1  to  4  do
   begin
   generate_string;
   draw;
   OutTextXY(200, 450, 'Press <Space>:');
   ch := ReadKey;
   ClearDevice;
   size := size div 4 + 2;
   end;
close_graph;
end.
```

The disk program "peano3" is another example of a string generating system for the same Peano curve, but with four rules instead of two. The disk program "hilbert2" does the same thing for the Hilbert curve.

We now come to the program "bush_fractal". It shows how to handle the "string" if it is even longer than 64K, so cannot be represented by a character array—use a file instead.   You may find the device of alternating two files instructive. Note that the program uses untyped files for speed. You should try **file of** Char for comparison. You will of course compare the speed of "hilbert2" with the recursive version "hilbert" above.

   "Bush_fractal" also illustrates another important tool. When drawing recursive ferns, trees, and the like, the turtle often wants to go out on one fork of a limb, draw something, return, then go out on the other fork. This requires saving the state of the turtle before it forks and retrieving that state after completion of the drawing on the fork. The state of the turtle, a bound vector, consists of its base point and vector. We store turtle states in a stack, a linked list. The string symbol [means push a state onto the list, that is, add it to the head of the list. The symbol] means pop the state off the list. Remember that a stack is a LIFO (last in, first out) structure.

Program 7.1.3

```
{$I+,S+}
program  bush_fractal;    { 09/18/94 }
   { bush fractal by "string rewriting system". }

   uses   CRT, Graph, graphs;

   const   seed = 'L';
           rule = 'LL+[+L-L-L]-[-L+L+L]';
            min = Length(rule);
            max = 50;

   type   Real = Extended;
          node_ptr = ^node;
          node = record
                        xx, yy: Integer;
                        dx, dy: Real;
                          next: node_ptr;
                    end;

   var   f0, f1: file;
          depth: Word;
             ch: Char;
          c, s: Real;
         buffer: array[1..max] of Char;
         result,
       k, size: Word;
            root,
            temp: node_ptr;
              uu: string[min];
{ Note, even though uu is not referred to outside
```

of generate_string, it is assumed to be unchanged when generate_string is reentered, so it MUST be a global variable. }

```
procedure  generate_string;

    var    j, k: Word;
              ch: Char;

    begin
    if  depth = 0  then
        begin    { initialize }
        Assign(f0, 'temp0');   Assign(f1, 'temp1');
        Rewrite(f0, 1);
        uu := seed;    { temporary use }
        BlockWrite(f0, uu[1], 1);
        uu := rule;
        end
    else
        begin
        Reset(f0, 1);   Rewrite(f1, 1);
        while  not Eof(f0)   do
            begin
            BlockRead(f0, buffer, max, result);
            for  k := 1  to  result  do
                begin
                ch := buffer[k];
                if  ch = 'L'  then
                    BlockWrite(f1, uu[1], min)
                else  BlockWrite(f1, ch, 1);
                end;
            end;
        Close(f0);   Close(f1);
        if  Odd(depth)   then
            begin
            Assign(f0, 'temp1');   Assign(f1, 'temp0');
            end
        else
            begin
            Assign(f1, 'temp1');   Assign(f0, 'temp0');
            end;
        end;
    end;    { generate_string }
```

```pascal
procedure  draw;
  { (x, y) is a vector in the direction of the
    next move; size is the length of the move. }

  var  x, y: Real;
         k: Word;

  procedure  rotate_plus(var  x, y: Real);
    { Rotation  +pi/8 }

    var  t: Real;

    begin
    t := x;
    x := c*x + s*y;
    y := -s*t + c*y;
    end;

  procedure  rotate_minus(var  x, y: Real);
    { Rotation  -pi/8 }

    var  t: Real;

    begin
    t := x;
    x := c*x - s*y;
    y := s*t + c*y;
    end;

  begin   { draw }
  x := size;  y := 0;
  MoveTo(0, 300);
  while  not Eof(f0)  do
    begin
    BlockRead(f0, buffer, max, result);
    for  k := 1  to  result  do
        case  buffer[k]  of
            'L': LineRel(Round(x), Round(y));
            '[': begin   { push }
                    temp := root;
                    New(root);
                    with  root^  do
                        begin
                        next := temp;
                        xx := GetX;  yy := GetY;
```

```
                                dx := x;   dy := y;
                                end;
                        end;
                ']': begin      { pop }
                        with  root^  do
                            begin
                            MoveTo(xx, yy);
                            x := dx;   y := dy;
                            end;
                        temp := root;
                        root := root^.next;
                        Dispose(temp);
                        end;
                    '+': rotate_plus (x, y);
                    '-': rotate_minus(x, y);
                end;     { case }
            end;     { while not eof(f0) }
        end;     { draw }

begin     { bush_fractal }
c := Cos(Pi/8);   s := Sin(Pi/8);
open_graph;
size := 128;
for  depth := 0  to  5  do
    begin
    if  depth = 5  then
        OutTextXY(0, 460, 'Be patient!');
    generate_string;
    Reset(f0, 1);
    draw;
    OutTextXY(500, 460, 'Press <Space>:');
    ch := ReadKey;
    ClearDevice;
    size := size div 2 + 1;
    end;
close_graph;
Close(f0);
Erase(f0);   Erase(f1);
end.
```

Exercises

1. Here is part of the disk program "dragon". The angle is 90°, and A and B are replaced by rule1 and rule2 resp. Try to figure out what it does before you run it.

```
    program  dragon_curve;
        const    seed = 'A';
                 rule1 = 'A+BL+';   rule2 = '-LA-B';
```

2. Here is part of the disk program "pine". The angle is 22.5°, and X are replaced by rule[X], where X = A...E. Try to figure out what it does before you run it.

```
program   pine;
    const    seed = 'AELLL';
             rule: array['A'..'E'] of string[14]
                 = ('[+++B][---B]DA',
                    '+C[-B]E',
                    '-B[+C]E',
                    'DE',
                    'L[-LLL][+LLL]L');
```

3. Here is part of the disk program "ragweed". The angle is 22.5°, and A and B are replaced by rule1 and rule2 resp. Try to figure out what it does before you run it.

```
program   ragweed;
    const    seed = 'B';
             rule1 = 'A[-LLL][+LLL]LA';
             rule2 = 'BLA[+B][-B]';
```

4. Here is part of the disk program "snowflak". The angles are +60° and −120°, and L is replaced by rule. Try to figure out what it does before you run it.

```
program   snowflake;
      const    seed = 'L';
               rule = 'L+L-L+L';
```

5. Here is part of the disk program "squared". The angle is 90°, and L is replaced by rule. Try to figure out what it does before you run it.

```
program   Sierpinski_square;
      const    seed = 'L+L+L+L';
               rule = 'LL+L+L+L+LL';
```

6. Here is part of the disk program "willow". The angle is π/7, and by L is replaced by rule. Try to figure out what it does before you run it.

```
program   willow;
      const    seed = 'L';
               rule = 'L[+L]L[-L]L';
```

Section 2. Probabilistic Graphical Algorithms

The Sierpinski curve is like a Cantor set. It begins with a solid triangle. The sides of the triangle are bisected, and these points connected, making four similar triangles. The open middle triangle is removed, etc. The disk program "sierpnsk" is a string generating system for this plane set. Now we shall generate this set by another program with other ideas.

For a technical account of the generation of fractals by iterated function schemes (IFS), see Falconer, K, Fractal Geometry, Wiley, 1990, pp 113 ff. For a less technical, more promotional account, see Barnsley, M, Fractals Everywhere, Academic Press, 1988. See also Barnsley, M, Fractal Modelling of the Real World, Chapter 5 in H-O Peitgen, et al, Fractal Images, Springer-Verlag, 1988, p. 224.

   The idea is best applied to sets that can be covered with a finite number
of smaller sets similar to the original set. Generally, affine transformations are
used which contract, i.e., reduce Euclidean distance. If you can find a few such
transformations whose images of the desired set cover that set, then you can
generate the set by applying these transformations in random order to any initial
point. A few of the initial iterations will produce a little noise outside of the
desired set, but that can be fixed easily. Let us now look at an example.

Program 7.2.1

```pascal
program   triangle;

   uses   Graph, graphs, CRT;

   type   Real = Extended;

   var  ch: Char;

   procedure   draw;

      const   mid_x = 640 div 2;
              mid_y = 480 div 2;
              radius = 275;

      var   x0, y0,
            x1, y1,
            x2, y2,
              x, y: Real;
           color: Integer;
               k: LongInt;
               p: Word;

      procedure   initialize;

         var   c, s: Real;

         begin
         c := Cos(2*Pi/3);   s := Sin(2*Pi/3);
         x0 := mid_x + radius;
         y0 := mid_y;
         x1 := mid_x + c*radius;
         y1 := mid_y + s*radius;
         x2 := mid_x + c*radius;
         y2 := mid_y - s*radius;
         end;
```

```
begin    { draw }
initialize;
Randomize;
x := mid_x + 100.0*Random;
y := mid_y + 100.0*Random;
for  k := 1  to  50000  do
    begin
    p := Random(3);
    case  p  of
        0: begin
            x := 0.5*(x0 + x);
            y := 0.5*(y0 + y);
            color:= Yellow;
            end;
        1: begin
            x := 0.5*(x1 + x);
            y := 0.5*(y1 + y);
            color  := LightGreen;
            end;
        2: begin
            x := 0.5*(x2 + x);
            y := 0.5*(y2 + y);
            color := LightRed;
            end;
        end;
    PutPixel(Round(x), Round(y), color);
    end;
end;    { draw }

begin    { triangle }
open_graph;
draw;
OutTextXY(0, 465, 'Press <Space>:');
ch := ReadKey;
close_graph;
end.
```

The three affine transformations are in the **case** statement. They each contract the original triangle by a factor of 2 towards one of its vertices.

The following example draws a fractal tree where each limb is similar to the whole tree. The trunk of the tree is generated as follows; with small probability, the next iterate is simply a random point in the trunk, which in this case is a thin rectangle.

In general, it doesn't matter how you pick the probabilities of going any direction. For a uniform spread of point density, you pick probabilities as follows.

For a fixed set, like the tree trunk just mentioned (which cannot be covered with similar smaller copies of the whole tree), you simply assign a small probability of picking a random one of its points. Otherwise you are dealing with nonsingular affine transformations, and you assign to each a probability proportional to the absolute value of its determinant. This is after deducting the small probability for a possible fixed set.

Program 7.2.2

```pascal
program  tree;    { 09/21/94 }

   uses   Graph,  graphs, CRT;

   type   Real = Extended;

   var   ch: Char;

   procedure  draw;

      const       x0 = 640 div 2;
                  y0 = 450;
                   r = 0.5;
              radius = (1.0 - r)*y0;
               theta = 0.25*Pi;

      var   t, x, y,
            c, s: Real;
               k: LongInt;
               p: Real;

      begin    { draw }
      Randomize;
      c := Cos(theta);   s := Sin(theta);
      x := 1.0;
      y := 0.0;
      for  k := 1  to  50000  do
         begin
         p := Random;
         if  p <= 0.02   then
            begin
            x := 0.04*Random - 0.02;
            y := Random;
            end
         else  if  p <= 0.51   then
            begin
            x := r*x;  y := r*y;
```

```
      t := x;
      x := c*x + s*y;   y := -s*t + c*y + 1.0;
      end
   else
      begin
      x := r*x;   y := r*y;
      t := x;
      x := c*x - s*y;   y := s*t + c*y + 1.0;
      end;
   PutPixel(x0 + Round(radius*x),
            y0 - Round(radius*y), Yellow);
      end;
   end;    { draw }

begin     { tree }
open_graph;
draw;
OutTextXY(0, 465,  'Press <Space>:');
ch := ReadKey;
close_graph;
end.
```

Exercises

1. Examine the disk program "fern", and example of an IFS with four affine transformations, one of which is singular. Try to analyze what it does before you run it.
2. Examine the disk program "wheel", an example of an IFS with six affine transformations. Try to analyze what it does before you run it.
3. Program a vortex: about 20 some arms spiraling slowly towards the center, each arm made up of infinitely many copies of the whole, size proportional to distance from the center. (See Peitgen &Saupe, p. 168 for the general idea.)

Section 3. Searching

In this section we discuss some standard and nonstandard searching problems. This is a topic that occupies large portions of most texts on programming and on algorithms, and it is well worth a section here. We begin with searching linear arrays, which we usually refer to as vectors. A *binary search* is very efficient, and we start with it. One warning though: in all of these programs that involve extensive referencing of array elements, it is a good practice to test with range checking {$R+} on.

Binary search locates an element between adjacent members of an *ordered* vector. Its run time is O(log m), where m is the length of the vector. As you can see from the code of function "trap" below, binary search is very similar to the bisection method of solving $F(x) = 0$; both use a "divide and conquer" strategy.

Program 7.3.1

```pascal
program  binary_search;

  uses    CRT;

  const   max = 1000;

  type    vector = array[1..max] of Real;

  var     v: vector;
          j, k: Word;

  function  trap(var  v: vector;   x: Real): Word;
     { Assumes v is in increasing order.
       Locates j:   v[j] <= x < v[j + 1]  }

     var   first, mean, last: Word;

     begin
     if  x < v[1]   then
        begin  trap := 0;   Exit;   end;
     if  x > v[max]   then
        begin  trap := max;  Exit;   end;
     first := 1;   last := max;
     while  last - first > 1  do
        begin    { Divide and conquer }
        mean := (first + last) shr 1;
        if  x < v[mean]   then   last := mean
        else  first := mean;
        if  KeyPressed  then   Halt;
        end;
     trap := first;
     end;    { trap }

  begin
  for  j := 1  to  20  do
     begin
     Randomize;
     v[1]  := Random;
     for  k := 2  to   max  do
        v[k]  := v[k - 1] + Random;
     WriteLn('Test ', j:2, ':', trap(v, 490.0):6);
     end;
  ReadLn;
  end.
```

Now we consider some problems dealing with unsorted vectors. Our first is finding the largest element. The most obvious, direct way of doing so follows.

Program 7.3.2

```
program  find_max_element1;    { 10/01/94 }

   const   n = 10000;

   type   vector = array[1..n] of Real;

   var   k, c: Word;
           x: Real;
           V: vector;

   procedure   find_max(var   V: vector;
                          var   k: Word;   var   x: Real);
      { Returns   x = V[k], the max in V[1..n] }

      var   j: Word;

      begin
      x := V[1];   k := 1;
      j := 1;
      while   j < n   do
          begin
          Inc(c, 2);
          Inc(j);
          if   V[j] > x   then
              begin   k := j;   x := V[j];   end;
          end;
      end;

   begin    { find_max_element1 }
   Randomize;
   for   k := 1   to   n   do   V[k] := 100.0*Random;
   c := 0;
   find_max(V, k, x);
   WriteLn('V[',    k, '] = ', x);
   WriteLn(c, ' comparisons');
   Write('Press <Enter>: ');   ReadLn;
   end.
```

I used a **while** loop instead of a **for** loop in "find_max" so that its cost of doing business would be apparent. There are two tests in each traversal of the loop: $j < n$ and $V[j] > x$, for a total of $2n - 2$ comparisons. This can be cut in half by slightly more precise programming. This program uses an idea we have not seen up to now: a *sentinel*. This is an additional term of a vector whose purpose is to diminish the number of tests. In this case the sentinel is V[n+1]. Its use is not completely free because assignments of the sentinel must be made. If we know *a priori* an upper bound for the terms of V, then V[n+1] could be assigned once and for all, and the program speeded. Anyhow, its cost is n comparisons.

Program 7.3.3

```
program   find_max_element2;    { 10/01/94 }

   const   n = 10000;

   type   vector = array[1..n + 1] of Real;

   var     V: vector;
           x: Real;
           k: Word;
        c, d: Word;    { Counts comparisons }

   procedure   find_max(var   V: vector;   var   k: Word;
                           var   x: Real);
      { Returns   x = V[k], the max in V[1..n] }

      var   j: Word;

      begin
      j := 1;
      while   j <= n   do
         begin
         Inc(c);
         x := V[j];   k := j;
         V[n + 1] := x + 1;    { sentinel }
         Inc(j);
         while   V[j] <= x   do
            begin
            Inc(d);
            Inc(j);
            end;
         { Now either (j <= n) and (V[j] > x)
            or   j = n + 1. }
         end;
      end;    { find_max }
```

```
begin    { find_max_element2 }
. . . . . . . . . . . . . . . . . . . . . . . . . . . . . .
end.
```

Similarly, we can find the smallest element of an unordered array. A much more challenging problem is to find the second smallest, third smallest, ..., k-th smallest. It is not at all obvious how to do this in a reasonably efficient manner, and the following elegant O(n) algorithm of A. Hoare should be surprising.

The algorithm is rather subtle; its idea is to sort partially the vector V so that its first k–1 elements are at most its k-th element, and all elements beyond are at least as large. See the second comment in the listing. Note that elements of V are repeatedly switched during the program, so V will never be the same (well, hardly ever); it has the same elements, but in another order. We'll take the program in pieces.

Program 7.3.4

```
program   k_th_smallest;    { 10/03/94 }
    { Find the k-th smallest element in a vector
      of some ordered type.   Iterative version.
      See Bentley, J, Programming Pearls,
      Addison-Wesley 1986, p. 179. }

    uses   CRT;

    const   max = 180;

    type   my_type = Word;
           vector = array[1..max] of my_type;

    var      V: vector;
          j, k: Word;

    procedure   find_kth_smallest(k: Word);
    { Algorithm of A. Hoare: Partially sort V so that
        V[1..k-1] <= V[k] <= V[k+1..max] }

        var   j0, j1,
              j, p: Word;

        procedure   transpose(j, k: Word);

            var   t: my_type;
```

```
begin
t := V[j];   V[j] := V[k];   V[k] := t;
end;
```

This is routine so far; [j0, j1] will be a subinterval of [1..max], initially the whole interval, and j is the control variable for a **for** loop from j0 +1 to j1. The subroutine "transpose" switches two members of V, but not their indices (j and k are value parameters). Read the first comment that follows, explaining what is invariant in each traversal of the **while** loop. The index p in [j0, j1] is determined by the **for** loop so that all elements of V for the interval [j0+1, p] are less than V[j0] and all V[j] for j in [p+1, j1] are greater than or equal to V[j0]. Then the values of V[p] and V[j0] are switched and the relation of p to k (we are seeking the k-th smallest) determines the next subinterval [j0, j1]. Let's get on with the listing:

```
begin    { find_kth_smallest }
j0 := 1;   j1 := max;
while   true   do
    begin
    { The while loop preserves the condition
        V[1...j0-1] <= V[j0..j1] <= V[j1+1..max]
        Each traversal decreases [j0, j1]. }
    p := j0;
    for  j := j0 + 1  to  j1  do
        if  V[j] < V[j0]   then
            begin  Inc(p);   transpose(p, j);   end;
    { After the for loop,
        V[j0+1..p] < V[j0],   V[p+1..j1] >= V[j0] }
    transpose(j0, p);
    { Now  V[j0..p-1] < V[p] <= V[p+1..j1] }
    if  k < p  then   j1 := p - 1
    else  if  k > p  then   j0 := p + 1
    else   Exit;
    transpose(j0, j1);
    end;    { while }
end;    { find_kth_smallest }

begin    { k_th_smallest }
ClrScr;
Randomize;
for  j := 1  to  max  do  V[j] := Random(100);
WriteLn('Raw data:');
for  j := 1  to  max  do  Write(V[j]:4);
WriteLn;
```

```
k := 25;
find_kth_smallest(k);
WriteLn('Processed data:');
for  j := 1  to   max   do
   begin
   if  j = k  then   TextColor(White)
   else   TextColor(LightGray);
   Write(V[j]:4);
   end;
WriteLn;
WriteLn(max,  ' elements   ',
         k,  '-th smallest:  ', V[k]);
Write('Press <Enter>:  ');   ReadLn;
end.   { k_th_smallest }
```

Note how the answer is highlighted in the final display. We'll use this technique in more programs. The disk listing of "k_th_smallest" is more complete, and measures some statistics.

The next problem is to maximize the sum of a run of consecutive elements of a real vector. We take the convention that if all the terms of the vector are negative, then the maximum sum equals 0, taken for the vacuous run. Of course, if all terms are nonnegative, the maximum is the sum of all elements; the interesting situation is when some elements are positive, some negative.

One's first guess is to compute the sum for each subinterval [j0, j1] of [1, max] and take the largest. This is obviously an $O(max^2)$ algorithm, which is hopelessly slow if max is large, say the 5000 we'll test. The following very nice $O(max)$ algorithm of David Gries is the sort of obvious mathematical reasoning that seems hard to find when you are programming.

Suppose we have solved the problem for interval [1, k-1], and s is the largest sum of a run of consecutive terms in that subinterval. Then either s equals the largest sum of a run in [1, k] or the largest run for V[1, k] ends at k. So, in addition to keeping track of the largest sum s for [1, k] we also keep track of the largest sum for [1, k] ending at k, and upgrade these two variables simultaneously. That is the gist of the mathematical induction.

Program 7.3.5

```
program  max_run_sum;
{ Algorithm of David Gries.   See Bentley, J,
  Programming Pearls, Addison-Wesley 1986,
  pp. 73-74. }

   const   max = 5000;
```

```
var            V: array[1..max] of Real;
               k: Word;
          max_sum: Real;
```

{ To find the largest possible sum V[k0..k1 of
  consecutive terms. }

```
   procedure   find_max_sum(var   max_sum: Real);

     var   sum_to_k: Real;
                  k: Word;

     begin
     max_sum := 0.0;   sum_to_k := 0.0;
     for  k := 1  to   max  do
        begin
        { When k increases, max_sum is the max sum
          in V[1..k-1].   It can only be exceeded by
          a sum ending at k, the largest which
          is sum_to_k. }
        sum_to_k := sum_to_k + V[k];
        if   sum_to_k < 0.0   then   sum_to_k := 0.0;
        if   max_sum < sum_to_k   then
           max_sum := sum_to_k;
        end;
     end;    { find_max_sum }

  begin     { max_run_sum }
  Randomize;
  for  k := 1  to   max  do
     V[k] := Random - 0.5;
  find_max_sum(max_sum);
  WriteLn('Maximum run sum: ', max_sum:6:6);
  Write('Press <Enter>: ');   ReadLn;
  end.
```

The following program deals with a vector of 0's and 1's. We seek the longest run of 1's. Again, at first glance this looks like an $O(max^2)$ problem, but it does have a nice $O(max)$ solution. The idea is similar to that of the previous algorithm: the longest run in [1, k] is either the longest run in [1, k−1] or a run ending at k. We omit details of the "display" subroutine, which are in the disk file, however.

Program 7.3.6

```
program  longest_1_run;    { 10/02/94 }
   { Find the longest run of 1's in a 01 array. }
```

```pascal
uses   CRT;

const   max = 1600;

var   B: array[1..max] of 0..1;
        j, longest,
      first, last: Word;

procedure  display;
. . . . . . . . . . . . . . . . . .

procedure  find(var  longest, first, last: Word);

   var   k, longest2k: Word;

   begin
   longest := 0;   last := 0;   longest2k := 0;
   for  k := 1   to   max   do
      begin
      if   B[k] = 1   then
         begin
         Inc(longest2k);
         if   longest2k > longest   then
            begin
            longest := longest2k;   last := k;
            end;
         end
      else   longest2k := 0;
      end;
   first := last - longest + 1;
   end;   { find }

begin   { longest_1_run }
Randomize;
for  j := 1   to   max   do   B[j] := Random(2);
find(longest, first, last);
display;
WriteLn('Longest run of 1''s: A[', first, '..',
         last, ']; length = ', longest);
Write('Press <enter>: ');   ReadLn;
end.
```

Now we consider some search problems in matrices. We begin with a matrix
A with strictly increasing rows and strictly increasing columns and an element z

which we know to be in A somewhere; where? Our first attack is a generalization of binary search in vectors, and is a pretty routine recursive algorithm.

Program 7.3.7

```
program  search_array;
    { A  is an array of words,  strictly increasing rows
      and columns.   To find the location of z in A. }

    const  m0 = 100;   n0 = 100;

    type   matrix = array[1..m0, 1..n0] of Word;

    var         A: matrix;
        xm, ym: Word;    { mean values }
         z, w: Word;     { array values }
         found: Boolean;

    procedure  initialize(var  A: matrix);

        var  i, j: Word;

        begin
        for  i := 1  to  m0  do
            for  j := 1  to  n0  do
                A[i, j] := 2*(i + (j - 1)*m0);
        end;

    procedure  search(x0, y0, x1, y1, z: Word);

        begin
        xm := (x0 + x1) div 2;
        ym := (y0 + y1) div 2;
        w := A[xm, ym];
        if  w < z  then
            begin
            if  xm < x1  then
                search(xm + 1, y0, x1, ym, z);
            if  found  then  Exit;
            if  ym < y1  then
                search(x0, ym + 1, x1, y1, z);
            if  found  then  Exit;
            end
        else  if  w > z  then
            begin
            if  x0 < xm  then
                search(x0, ym, xm - 1, y1, z);
```

```
      if  found  then  Exit;
      if  y0 < ym  then
          search(x0, y0, x1, ym - 1, z);
      if  found  then  Exit;
      end
   else  found := true;
   end;    { search }

begin    { search_array }
initialize(A);
Write('Enter  z:  ');  ReadLn(z);
found := false;
search(1, 1, m0, n0, z);
if  found  then
    WriteLn('A[', xm, ', ', ym, '] = ', z)
else  WriteLn(z, '  not in  A');
WriteLn;
Write('Press <Enter>: ');  ReadLn;
end.    { search_array }
```

The problem has a much more efficient solution, due to David Gries, a model
of algorithm simplicity.

Program 7.3.8

```
program  array_search;    { 10/01/94 }
   { A[i, j]  is increasing in both i and j.
    Given z in A,  find (one of) its locations.
    See Gries, D, Science of Programming,
    Springer-Verlag, 1981, pp. 215, 346-347. }

const  m0 = 50;  n0 = 50;

type  matrix = array[1..m0, 1..n0] of Word;

var         A: matrix;
       m, n, c: Word;

procedure  saddleback_search(z: Word;
                             var  m, n: Word);
```

```
    begin
    m := 1;   n := n0;
    while   true   do
        begin
        Inc(c);
        if   A[m, n] < z   then   Inc(m)
        else   if   A[m, n] > z   then   Dec(n)
        else   Break;
        end;
    end;    { saddleback_search }

begin    { array_search }
c := 0;
for   m := 1   to   m0   do
    for   n := 1   to   n0   do
        A[m, n] := (m - 1)*n0 + n - 1;
{ worst case }
saddleback_search((m0 -1)*n0, m, n);
WriteLn('A[', m, ', ', n, '] = ', A[m, n]);
WriteLn(c, ' comparisons');
WriteLn;
ReadLn;
end.    { array_search }
```

The final problem of this section asks for the largest square of 1's in a 0-1 matrix. Obviously we have to look at each element of the matrix, so the algorithm's run time must be at least $O(\text{rows} * \text{cols})$. Surprisingly, we can do this well, however, at the cost of some space. Index[x, y] holds the size of the largest 1 square with lower right corner (x, y). As it is a matrix of words, it takes more than twice the space of the data matrix A of bytes. An exercise suggests what to do about this. As usual, we omit here the details of the output routine. To understand the clever computation of t in the nested **for** loop, please draw a careful figure on graph paper.

Program 7.3.9

```
program   largest_1_square;    { 10/02/94 }

    uses   CRT;

    const   rows = 20;
            cols = 79;
```

```
var          A: array[1..rows, 1..cols] of Byte;
          x, y,
        x0, y0,
        x1, y1,
          max: Word;

procedure  display;
. . . . . . . . . . . . . . . . .

function  minimum(x, y, z: Word): Word;

   var   t: Word;

   begin
   if  x < y  then  t := x  else  t := y;
   if  z < t  then  t := z;
   minimum := t;
   end;

procedure  find_square;
  { Returns  upper left (x0, y0),
             lower right (x1, y1) }

   var    index: array[0..rows, 0..cols] of Word;
          { Side of the largest square with
            lower right (x, y) }
        t, x, y: Word;

   begin    { find_square }
   for  x := 0  to  rows  do
      for  y := 0  to  cols  do  index[x, y] := 0;
   max := 0;   x1 := 0;   y1 := 0;
   for  x := 1  to  rows   do
      for  y := 1  to  cols   do
         begin
         if  A[x, y] = 1  then
            begin
            t := minimum(index[x - 1, y],
                         index[x, y - 1],
                         index[x - 1, y - 1] ) + 1;
            index[x, y] := t;
            if  t > max  then
               begin
               max := t;
               x1 := x;  y1 := y;
               end;
```

```
            end;    { if   A[x, y] }
        end;    { for  y,  x }
    if   max = 0   then
        begin   x0 := 0;   y0 := 0;   end
    else
        begin
        x0 := x1 - (max - 1);
        y0 := y1 - (max - 1);
        end;
    end;    { find_square }

{ Cost:   Time: O(M*N);    Space: O(M*N) }

begin    { largest_1_square }
Randomize;
for  x := 1  to  rows  do
    for  y := 1  to  cols  do
        begin
        if  Random(10) = 0  then  A[x, y] := 0
        else   A[x, y] := 1;
        end;
find_square;
display;
WriteLn('Largest 1 square: (',
        x0, ', ', y0,
        ') to (', x1, ', ', y1,
        ').   Side length: ', max, '.');
WriteLn;
Write('Press <Enter> to quit: ');   ReadLn;
end.    { largest_1_square }
```

Exercises

1. Let's start with an easy one. Let V be an unordered vector of words. Write a Boolean-valued function of x that returns whether x is in V.

2. A large sparce matrix is stored linearly. There are max nonzero elements. Given: value[k] is the element at (row[k], col[k]). Define a function "element" that returns the element in the (r, c) position of the matrix. (see Kernighan, B W and Plauger, P J, Elements of Programming Style 2/e, McGraw-Hill, 1978, p. 22.)

3. Generalize "search_array" to 3-dimensional ordered tensors.

4. Find the longest constant run in a non-decreasing vector.

5. Modify "max_run_sum" so that it reports not only the maximum sum of a run, but the bounds of the run too.

6. Given a vector of words, there may two runs that are the same, possibly overlapping. Write a program to find the longest consecutive subsequence (run) that occurs again later. If possible, the output should show the runs in color.

7. Suppose that the matrix A in "largest_1_square" is large, say 1000×1000. It could not be in memory, but might be stored as a disk file, in row major order. Now the roadblock is the matrix "index" which cannot possibly be in memory. If you look at the code where "index" is referenced, you see that only three elements of "index" are referenced, two in the previous row, one in the current row. Rewrite the program using only a single row and an extra element for auxiliary storage, in this case, 1001 words. You will learn a very useful technique from this exercise.

Section 4. Sorting

Sorting, like searching, is a big deal in computer science. In this section we'll cover a number of sorting procedures, both standard and not so standard. You will need to sort arrays sometimes, even in scientific programs. For instance, to graph a parametric surface with hidden line removal, you must first compute all the little rectangles that approximate the surface, then sort them from back to front before you plot them, so that pieces closer to the observer cover pieces farther back.

The following unit contains many of the by now classical sorting algorithms for one-dimensional arrays. The interface section will show you what we are up to. Note that we are sorting vectors of "Word", but this could be any integer or real type.

Program 7.4.1

```
unit   sorting;     { 09/14/94 }

interface

    uses   CRT;

    const   max = 1000;

    type   vector = array[0..max] of Word;

    procedure   init_vector(var   v: vector);
    procedure   display_vector(var   v: vector);
    procedure   selection_sort(var   v: vector);
    procedure   insertion_sort_1(var   v: vector);
    procedure   insertion_sort_2(var   v: vector);
    procedure   insertion_sort_3(var   v: vector);
    procedure   shell_sort(var   v: vector);
    procedure   heap_sort(var   v: vector);
    procedure   quick_sort(var   v: vector);
```

We start the implementation with routines for displaying a vector:

implementation

```
procedure  display_vector(var  v: vector);

    var  k: Word;

    begin
    for  k := 0  to  max  do  Write(v[k]:5);
    WriteLn;
    Write('Press <Enter>: '); ReadLn;
    end;   { display_vector }

procedure  init_vector(var  v: vector);

    const  max_2 = max div 2;

    var  k: Word;

    begin
    Randomize;
    for  k := 0  to  max  do  v[k] := Random(max_2);
    end;   { init_vector }
```

Selection sort searches the array from its first to its last element. Whenever it finds an element smaller than the zero-th element, it exchanges that element with the zero-th element. Then it repeats the process, seeking elements smaller than the first element, etc.

```
procedure  selection_sort(var  v: vector);
    { Searches for the smallest, then the next
      smallest, etc. }

    var  j, k, t: Word;

    begin
    for  j := 0  to  max - 1  do
       { v[0..j-1] the lowest j elements, sorted }
       for  k := j + 1  to  max  do
          begin
          if  v[k] < v[j]  then
             begin
             t := v[j];  v[j] := v[k];  v[k] := t;
             end;
          end;   { Now v[j] = successor of v[j-1] }
    end;   { selection_sort }
```

Insertion sort (also called *bubble sort*) is similar. It moves down the line, interchanging adjacent terms that are out of order. If this is done enough times, the vector becomes sorted. We include three versions of this; the third one, which uses "Move" is much faster than the other two, although still an O(n²) algorithm. It's just that the multiplier of n² is smaller by about a factor of 2.5. The second version is from Bentley, J, Programming Pearls, Addison Wesley, 1986, p. 108.

```
procedure  insertion_sort_1(var   v: vector);
     { Induction on j: v[0..j] sorted }

   var   j, k, t: Word;

   begin
   for  j := 1   to   max   do
       begin
       k := j;
       while   (k > 0) and (v[k] < v[k - 1])   do
           begin
           t := v[k];   v[k] := v[k - 1];
           v[k - 1] := t;
           Dec(k);
           end;
       end;
   end;    { insertion_sort_1 }

procedure  insertion_sort_2(var   v: vector);
     { Induction on j: v[0..j] sorted. }

   var   j, k, t: Word;

   begin
   for  j := 1   to   max   do
       begin
       t := v[j];   k := j;
       while   (k > 0) and (t < v[k - 1])   do
           begin
           v[k] := v[k - 1];
           Dec(k);
           end;
       v[k] := t;
       end;
   end;    { insertion_sort_2 }
```

```
procedure   insertion_sort_3(var   v: vector);
   { Induction on j: v[0..j-1] sorted.
     Speeded by using "move". }

   var   k, t: Word;
             j: LongInt;

   begin
   for  k := 1  to   max   do
       begin    { v[0..k-1] is sorted }
       t := v[k];
       j := k;   Dec(j);
       while   (j >= 0) and (v[j] > t)   do   Dec(j);
       Inc(j);
       { Now   0 <= j <= k,   v[j] <= t,
          and  (j = k   or   v[j + 1] > t) }
       if  j < k   then
           begin    { Shift k - j elements one right }
           Move(v[j], v[j + 1], (k - j)*SizeOf(Word));
           { Fill the gap }
           v[j] := t;
           end;
       end;    { for  k }
   end;    { insertion_sort_3 }
```

Shell sort is a clever modification of insertion sort, named after its inventor, D. L. Shell. Apparently it is at least as fast as $O(n^{3/2})$.

```
procedure   shell_sort(var   v: vector);

   var h, j, k, t: Word;

   begin
   h := 1;
   repeat   h := 3*h + 1   until   h > max;
   repeat
       h := (h - 1) div 3;
       for  j := h  to   max   do
       { Insertion sort arithmetic progressions
         with difference h }
           begin
           t := v[j];   k := j;
           while   (k >= h) and (v[k - h] > t)   do
               begin
               v[k] := v[k - h];
```

```
            k := k - h;
            end;
        v[k] := t;
        end;
    until  h = 0;
    end;    { shell_sort }
```

Heap sort is one of the winners in the race for sorting speed; it sorts an n–element vector in time O(n log n). It is conceptually quite complicated, and I am not going to attempt an explanation, just refer you to R. Sedgewick, Algorithms 2/e, Addison-Wesley, 1988, 152-156.

```
    procedure  heap_sort(var  v: vector);

        var  k, t, last: Word;

        procedure  sift(first: Word);

            var  j, last2, t: Word;

            begin    { sift }
            t := v[first];   last2 := last div 2;
            while  first <= last2   do
                begin
                j := 2*first;
                if  (j < last) and (v[j] < v[j + 1])   then
                    Inc(j);
                if  v[j] <= t   then   Break;
                v[first] := v[j];   first := j;
                end;
            v[first] := t;
            end;    { sift }

        begin    { heap_sort }
        last := max;
        for  k := last div 2   downto   0  do   sift(k);
        repeat
            t := v[0];   v[0] := v[last];   v[last] := t;
            Dec(last);   sift(0);
        until  last = 0;
        end;    { heap_sort }
```

Anthony Hoare's *quick sort* is my favorite because of its speed, O(n log n), and its clarity. In brief, you pick the point in the center of the vector. Then you compare terms to the right of that point to terms to the left, and switch them if out of order. After this first pass, every element left of the pivot point is smaller than every element to the right. Then you quick sort the left and right sides of the pivot point by recursion. Note that "quick_sort" is a driver to get "qs" started. The latter calls itself recursively.

```
procedure  quick_sort(var  v: vector);

   procedure  qs(left, right: Integer);

      var  i, j, t: Integer;

      var  pivot: Real;

      begin
      if  left < right  then
         begin
         i := left;  j := right;
         pivot := v[(left + right) shr 1];
         repeat
            while  v[i] < pivot  do  Inc(i);
            while  pivot < v[j]  do  Dec(j);
            if  i <= j  then
               begin
               if  i < j  then
                  begin    { Switch }
                  t := v[i];
                  v[i] := v[j];  v[j] := t;
                  end;
               Inc(i);  Dec(j);
               end;
         until  i > j;
         qs(left, j);
         qs(i, right);
         end;
      end;   { qs }

   begin    { quick_sort }
   qs(0, max);
   end;

end.   { unit  sorting }
```

Here are results of a series of experiments:

	500 terms	1000 terms	2000 terms	4000 terms
Selection	1.21 sec	4.83 sec	19.34 sec	77.61 sec
Insertion 1	1.10 sec	4.56 sec	18.51 sec	73.22 sec
Insertion 2	0.55 sec	2.31 sec	9.45 sec	37.18 sec
Insertion 3	0.22 sec	0.94 sec	3.79 sec	14.77 sec
Shell	0.11 sec	0.16 sec	0.33 sec	0.77 sec
Heap	0.06 sec	0.17 sec	0.39 sec	0.82 sec
Quick	0.06 sec	0.16 sec	0.33 sec	0.66 sec

If you wish to pursue this subject, you will find an enormous amount of material in Knuth, D E, Art of Computer Programming, vol 3, Chapter 5, Addison-Wesley, 1973.

The following example is a very special sorting problem: the vector only contains three different elements. This situation is so special that an $O(n)$ algorithm is possible. The problem is given in Dijkstra, E, Discipline of Programming, Prentice Hall, 1976, pp 111-116 and in Gries, D, Science of Programming, Springer-Verlag, 1981, p. 214.

Program 7.4.2

```pascal
program  sort_3_items;    { 09/14/94 }
   { To sort an array of 0..2 as quickly as possible. }

   uses  CRT;

   const  max = 600;

   type  three = 0..2;
         vector = array[1..max] of three;

   var      a: vector;
         count: LongInt;
          freq: array[three] of Word;

   procedure  fill_vector(var  a: vector);

      var  k: Word;
           t: three;
```

```
begin
for  t := 0  to  2  do  freq[t] := 0;
Randomize;
for  k := 1  to  max  do
    begin
    a[k] := Random(3);  Inc(freq[a[k]]);
    end;
end;

procedure  print_vector(var  a: vector);

var  k: Word;

begin
for  k := 1  to  max  do  Write(Byte(a[k]));
WriteLn;
end;

procedure  sort(var  a: vector);

var  i, j, k: Integer;

procedure  switch(var  x, y: three);

    var  t: three;

    begin
    t := x;  x := y;  y := t;
    Inc(count);
    end;

begin    { sort }
i := 1;  k := max;  j := i;
{ a[1..i-1] = 0, a[i..j-1] = 1, a[k+1..max] = 2
  i <= j <= k }
while  j <= k  do
    case  a[j] of
        0:  begin
            if  i < j  then  switch(a[i], a[j]);
            Inc(i);  Inc(j);
            end;
        1:  Inc(j);
        2:  begin
            if  j < k  then  switch(a[j], a[k]);
            Dec(k);
            end;
```

```
      end;    { case }
    end;    { sort }

  begin    { sort_3_items }
  fill_vector(a);
  ClrScr;
  WriteLn('Unsorted A = ');
  print_vector(a);
  WriteLn;
  count := 0;
  sort(a);
  WriteLn('Sorted A = ');
  print_vector(a);
  WriteLn;
  WriteLn(freq[0], ' 0''s  ',
          freq[1], ' 1''s  ',
          freq[2], ' 2''s');
  WriteLn(count, ' transpositions');
  WriteLn;
  Write ('Press <Enter> to halt: ');
  ReadLn;
  end.    { sort_3_items }
```

Now we look at a two dimensional problem: to sort a sequence of points of the plane in lexicographic order. The program is fairly routine, but requires some care.

Program 7.4.3

```
program  lexicographic_order;    { 06/27/94 }

  const   max = 100;

  type    point = record  x, y: Integer;   end;
          vector = array[1..max] of point;

  var  V: vector;
       k: Word;

  procedure  lexord(var  V: vector;
                         first, last: Word);
  { Changes V[first..last] to lexicographic order }

      var  j, k: Word;
           p, q: point;
```

```
  begin
  p.x := V[first].x;
  p.y := V[first].y;
  j := first - 1;   k := last + 1;
  while   true   do
     begin
     repeat
        Dec(k)
     until (V[k].x < p.x) or
           ( (V[k].x = p.x) and (V[k].y <= p.y) );
     { Now V[k] <= p and V[l] > p for all l > k }
     repeat
        Inc(j)
     until   (V[j].x > p.x)
           or ( (V[j].x = p.x) and (V[j].y >= p.y) );
     { Now V[j] >= p and V[i] < p for all i < j }
     if  j < k   then
        begin
        q := V[j];   V[j] := V[k];   V[k] := q;
        end
     else   Break;
     end;     { while }
  if   first < k   then   lexord(V, first, k);
  if   k + 1 < last   then   lexord(V, k + 1, last);
  end;     { lexord }

begin     { lexicographic_order }
Randomize;
for   k :=   1   to   max   do
   begin
   V[k].x := -10 + Random(20);
   V[k].y := -10 + Random(20);
   end;
WriteLn;
WriteLn('In lex order:');
lexord(v, 1, max);
for   k :=   1   to   max   do
   begin
   Write('(':3, V[k].x:3, ', ', V[k].y:3, ')');
   if   k mod 6 = 0   then   WriteLn;
   end;
ReadLn;
end.     { lexicographic_order }
```

Sorting a very large amount of data is another subject in itself. Assume the data is in a file that is far to large to hold in computer memory. Then what? Such data can be and is sorted, as you know from everyday life. You must use a number of auxiliary files, and sort part of the data at a time. We'll look at an interesting example; the input is a file of records; the output is a file of the original in reverse order. In the program, we take the records to be words, but they could be anything.

The program uses five auxiliary files. File "buf0" is just used to pad the original file to length a power of 2. Then the file is alternately split between pairs of the other four files, each time taking chunks of size the next lower power of two from the two files that are now input, and writing these chunks to the two files that are output. Play the game on paper with small examples to see the idea.

Program 7.4.4

```pascal
program   reverse_file;    { 10/09/94 }

   const    in_name = 'source';
            out_name = 'sink';
            hold_len = 16;    { power of 2 }

   type   my_type = Word;    { or whatever }

   var    in_file,
          out_file,
               buf0: file of my_type;
               hold: array[1..hold_len] of my_type;
            passes,
          k,  count: Word;
               two2e,
          pow2,  j: LongInt;
               x: my_type;
            short: Boolean;

   procedure   open_files;

      begin
      {$I-}
      Assign(in_file, in_name);   Reset(in_file);
      if   IOResult <> 0   then
          begin
          WriteLn('File ', in_name,
                  ' cannot be opened.');
          Write('Press <Enter>: ');   ReadLn;   Halt;
          end;
```

```pascal
   Assign(out_file, out_name);
   Rewrite(out_file);
   if  IOResult <> 0  then
      begin
      WriteLn('File ', out_name,
               ' cannot be opened.');
      Write('Press <Enter>: ');   ReadLn;   Halt;
      end;
   {$I+}
   end;    { open_files }

procedure  initialize;

   begin
   { Copy in_file into buf0 and pad it to
     make length the next power of 2 }
   Assign(buf0, 'buffer0.xyz');
   Rewrite(buf0);
   count := 0;
   passes := 1;
   while  not Eof(in_file)  do
      begin
      Read(in_file, x);
      Inc(count);
      Write(buf0, x);
      end;
   Close(in_file);
   WriteLn(in_name, ' has ', count, ' records');
   short := (count <= hold_len);
   if  not  short  then
      begin
      two2e := 1;
      while  two2e < count  do  two2e := 2*two2e;
      pow2 := two2e div 2;
      { pow2 < count <= two2e }
      for  j := count + 1  to  two2e  do
         Write(buf0, x);
      end;
   Reset(buf0);
   end;    { initialize }

procedure  reverse_short;

   var  j: Word;
        x: my_type;
```

```
begin
for  j := 1  to    count    do
    begin
    Read(buf0, x);   hold[j] := x;
    end;
passes := 2;
Close(buf0);   Erase(buf0);
for  j := count   downto   1   do
    Write(out_file, hold[j]);
end;    { reverse_short }

procedure   reverse_long;

   var   buf1, buf2,
         buf3, buf4: file of my_type;
                  x: my_type;
                  j: LongInt;

   procedure   split_file;

      var  j: LongInt;

      begin
      Assign(buf1, 'buffer1.xyz');
      Rewrite(buf1);
      Assign(buf2, 'buffer2.xyz');
      Rewrite(buf2);
      for  j := 1  to   pow2   do
         begin
         Read(buf0, x);   Write(buf1, x);
         end;
      for j := 1  to   pow2   do
         begin
         Read(buf0, x);   Write(buf2, x);
         end;
      Close(buf0);   Erase(buf0);
      Close(buf1);   Close(buf2);
      end;    { split_file }


   procedure   finish;

      var   j: LongInt;
```

```
      begin
      Reset(buf1);   Reset(buf2);
      Rewrite(buf0);
      while  not Eof(buf1)  do
         begin
         for  j := 1  to  hold_len  do
            begin
            Read(buf2, x);   hold[j] := x;
            end;
         for  j := hold_len  downto  1  do
            Write(buf0, hold[j]);
         for  j := 1  to  hold_len  do
            begin
            Read(buf1, x);   hold[j] := x;
            end;
         for  j := hold_len  downto  1  do
            Write(buf0, hold[j]);
         end;    { while }
      Close(buf1);   Close(buf2);
      Erase(buf1);   Erase(buf2);
      { Remove the padding }
      Reset(buf0);
      for  j := count + 1  to  two2e  do
         Read(buf0, x);
      { Write the rest into out_file }
      while  not Eof(buf0)  do
         begin
         Read(buf0, x);   Write(out_file, x);
         end;
      Close(buf0);   Erase(buf0);
      Inc(passes, 2);
      end;    { finish }

   begin    { reverse_long }
   split_file;
   { Alternate splitting: buf3 & buf4 input,
     buf1 & buf2 output }
   while  pow2 > hold_len  do
      begin
      Inc(passes);
      Assign(buf3, 'buffer1.xyz');
      Rename(buf3, 'buffer3.xyz');
      Reset(buf3);
      Assign(buf4, 'buffer2.xyz');
      Rename(buf4, 'buffer4.xyz');
      Reset(buf4);
```

```
      Assign(buf1, 'buffer1.xyz');
      Rewrite(buf1);
      Assign(buf2, 'buffer2.xyz');
      Rewrite(buf2);
      pow2 := pow2 div 2;
      while  not Eof(buf4)   do
         begin
         for j := 1  to  pow2  do
            begin
            Read(buf4, x);   Write(buf1, x);
            end;
         for j := 1  to  pow2  do
            begin
            Read(buf4, x);   Write(buf2, x);
            end;
         for j := 1  to  pow2  do
            begin
            Read(buf3, x);   Write(buf1, x);
            end;
         for j := 1  to  pow2  do
            begin
            Read(buf3, x);   Write(buf2, x);
            end;
         end;    { while  not eof(buf3) }
      Close(buf3);   Close(buf4);
      Erase(buf3);   Erase(buf4);
      Close(buf1);   Close(buf2);
      end;    { while  pow2 > hold_len }
   finish;
   end;   { reverse_long }

begin    { reverse_file }
{ Test }
Assign(in_file, in_name);
Rewrite(in_file);
for  k := 1  to  265  do  Write(in_file, k);
{ Replace 265 by others }
Close(in_file);

open_files;
initialize;
if  short  then  reverse_short
else  reverse_long;
WriteLn(passes, ' passes through the data');
Close(out_file);
```

```
{ Test }
Assign(out_file, out_name);
Reset(out_file);
while   not Eof(out_file)   do
   begin   Read(out_file, x);   Write(x:4);   end;
ReadLn;
Close(out_file);
end.    { reverse_file }
```

We next look at two programs for transposing matrices. The first program deals with a rectangular matrix small enough to hold in memory. Let's start with the program's declarations and its output procedure, which is interesting because of its use of type casting to display both the input matrix and its transpose.

Program 7.4.5

```
program   transpose_matrix;    { 10/09/94 }

   uses    CRT;

   const   no_rows = 7;   no_cols = 9;

   type       Str4 = string[4];
            my_type = Word;
             linear = array[1..no_rows*no_cols]
                           of my_type;
            matrix1 = array[1..no_rows, 1..no_cols]
                           of my_type;
            matrix2 = array[1..no_cols, 1..no_rows]
                           of my_type;

   var   A: matrix1;
         B: matrix2;
         L: linear;
         { The problem is to transpose A into B }
         j, k: Word;

   procedure   display(rows, cols: Word;
                       var   V: linear;
                       name: Str4 );
      { Type casting is used so this procedure
        can display both A and its transpose B }

      var   j, k, l: Word;
```

```
begin
GotoXY(1, 1 + (rows + 1) div 2);
Write(name, ' = ');
l := Length(name) + 4;
GotoXY(1, 1);   Write(#218);
GotoXY(l + 4*cols + 1, 1);    Write(#191);
for  j := 1  to  rows  do
    begin
    GotoXY(1, 1 + j);
    Write(#179);
    for  k := 1  to  cols   do
        Write(V[(j - 1)*cols + k]:4);
    Write(#179);
    end;
GotoXY(1, rows + 2);   Write(#192);
GotoXY(l + 4*cols + 1, rows + 2);   WriteLn(#217);
end;   { display }
```

We shall give two different approaches to transposing a rectangular matrix and discuss a third. The first two assume there is enough memory to hold both the matrix and its transpose. The first is the absolutely direct and obvious transpose method; just write it down.

```
procedure   transpose1(var  A: matrix1;
                       var  B: matrix2);

    var  j, k: Word;

    begin
    for  j := 1  to  no_rows  do
        for  k := 1  to  no_cols  do
            B[k, j] := A[j, k];
    end;
```

The second is more subtle. It is based on the linear storage of a matrix in row major order. It uses a nice number-theoretic idea for permuting to the row-major order of the transpose. For instance

```
1 2 3 4 5 6  and 1 4 2 5 3 6
```

are transposes.

```
procedure   transpose2(var   A: matrix1;
                        var   B: matrix2);

    var   L, LP: linear;
          x: Word;

    function  psi(x: Word): Word;
        { This is a permutation of [1, mn] which
          is transposition when applied to the
          elements of A in the linear order in
          which they are stored.}

        begin
        psi := ((x - 1) mod no_rows)*no_cols
                 + ((x - 1) div no_rows) + 1;
        end;

    begin
    L := linear(A);    { Type cast! }
    for  x := 1  to  no_rows*no_cols  do
        LP[x] := L[psi(x)];
    B := matrix2(LP);    { Cast back! }
    end;
```

If there is enough memory to hold the matrix, but not enough to hold its transpose too, then we have to transpose "in place". This is a highly complicated algorithm, and I will only give a hint and a reference: Brenner, N, Matrix transposition in place, Algorithm 467, CACM 16, 1973, pp 692-694.

```
procedure   transpose_in_place(var   L: linear);
    { Suppose L represents A:
         A[j, k] = L[(j - 1)*no_cols + k].
       Then  B[j, k] = L[(j - 1)*no_rows + k]
                     = A[k, j]
                     = L[(k - 1)*no_cols + j]. }
    begin
    end;

begin    { transpose_matrix }
ClrScr;
Randomize;
for  j := 1  to  no_rows  do
    for  k := 1  to  no_cols  do
        A[j, k] := Random(1000);
display(no_rows, no_cols, linear(A), 'A');
```

```
transpose2(A, B);
Window(1, no_rows + 3, 80, 25);
display(no_cols, no_rows, linear(B), 'B');
ReadLn;
end.    { transpose_matrix }
```

Transposing a very large matrix that is stored on disk in row-major order is an entirely different problem. The following algorithm, from Knuth, D E, loc. cit, p. 573, is similar to the program above for reversing a file. Again, the file is first padded with additional rows, to make the next power of two rows. Then the program works with two pairs of files and alternately puts rows from one pair into the other pair. We'll leave the details of the input and output procedures to the disk listings, and get to the main algorithm.

Program 7.4.6

```
program  transpose_stored_matrix;    { 10/09/94 }
    { Input: a disk file of  m X n  records, storing
      an m X n matrix A in row order:  First n records,
      first row of A; next n records, second row
      of A, etc.

      Output: a disk file for the transpose  A'.

      Source: D Knuth, vol 3, p. 573. }

    const  no_rows = 5;   no_cols = 3;

    type  my_type = string[2];    { or whatever }

    var  in_file, out_file: file of my_type;
                     j, k: Word;
                 c1, c2: Char;

    procedure  initialize_source;
    procedure  print_source;
    procedure  print_sink;

    procedure  sort;

        var  F1, F2, F3, F4: file of my_type;
                  rec, dummy: my_type;
              i, j, k, pow2: Word;
                     even: Boolean;
```

```
begin    { sort }
dummy := '--';
pow2 := 1;
while  pow2 < no_rows  do  pow2 := 2*pow2;
{ Add rows of dummy to replace no_rows
  by the next power of 2. }
Assign(F1, 'source.dat');
Reset(F1);
Assign(in_file, 'buffer0.dat');
Rewrite(in_file);
while  not  Eof(F1)  do
   begin
   Read(F1, rec);  Write(in_file, rec);
   end;
for  j := 1  to  pow2 - no_rows  do
     for  k := 1  to  no_cols  do
        Write(in_file, dummy);
Close(F1);
Reset(in_file);
Assign(F1, 'buffer1.dat');  Rewrite(F1);
Assign(F2, 'buffer2.dat');  Rewrite(F2);
Assign(F3, 'buffer3.dat');  Rewrite(F3);
Assign(F4, 'buffer4.dat');  Rewrite(F4);
while  not Eof(in_file)  do
   begin
   { Put odd numbered rows in F3, even in F4 }
   for  k := 1  to  no_cols  do
      begin
      Read(in_file, rec);  Write(F3, rec);
      end;
   for  k := 1  to  no_cols  do
      begin
      Read(in_file, rec);  Write(F4, rec);
      end;
   end;    { while }
Close(in_file);  Erase(in_file);
j := 1;
{ From now on, sources and sinks will alternate.
  F3, F4 will be sources, F1, F2 sinks. }
even := true;
repeat
   even := not even;
   Close(F1);  Close(F2);  Close(F3);  Close(F4);
   if  even  then
      begin
      Assign(F3, 'buffer1.dat');
```

```
      Assign(F4,  'buffer2.dat');
      Assign(F1,  'buffer3.dat');
      Assign(F2,  'buffer4.dat');
      end
  else
      begin
      Assign(F3,  'buffer3.dat');
      Assign(F4,  'buffer4.dat');
      Assign(F1,  'buffer1.dat');
      Assign(F2,  'buffer2.dat');
      end;
  Reset(F3);   Reset(F4);
  Rewrite(F1);   Rewrite(F2);
  while  true  do
      begin
      if  Eof(F3)  then  Break;    { while loop }
      for  i := 1  to  no_cols  do
          begin
          for  k := 1  to  j  do
              begin
              Read(F3, rec);   Write(F1, rec);
              end;
          for  k := 1  to  j  do
              begin
              Read(F4, rec);   Write(F1, rec);
              end;
          end;    { for i }
      if  Eof(F3)  then  Break;    { while loop }
      for  i := 1  to  no_cols  do
          begin
          for  k := 1  to  j  do
              begin
              Read(F3, rec);   Write(F2, rec);
              end;
          for  k := 1  to  j  do
              begin
              Read(F4, rec);   Write(F2, rec);
              end;
          end;    { for i }
      end;    { while  true }
  j := 2*j;
until  j = pow2;
Close(F2);   Close(F3);   Close(F4);
Erase(F2);   Erase(F3);   Erase(F4);
{ Filter out the dummies. }
Reset(F1);
```

```
Assign(out_file, 'sink.dat');
Rewrite(out_file);
while   not   Eof(F1)   do
    begin
    Read(F1, rec);
    if   rec <> dummy   then   Write(out_file, rec);
    end;
Close(F1);   Erase(F1);
Close(out_file);
end;   { sort }

begin   { transpose_stored_matrix }
initialize_source;
ClrScr;
print_source;
sort;
print_sink;
ReadLn;
end.
```

Exercises

1. Write a program to test the procedures of the unit "sorting".

Section 5. Genetic Algorithms

Genetic algorithms are programs for optimization based on ideas from evolution: survival of the fittest, inheritance of acquired characteristics and genetic material, and mutations. This is a relatively recent subject, started about 25 years ago by John H. Holland. I recommend a recent article, Srinivas, M and Patnaik, L M, Genetic algorithms: a survey, IEEE Computer 27 (1994), 17–26 for an introduction and references to the literature.

Suppose the problem is to maximize $F(x)$ on an interval $[a, b]$. Start with a first generation colony of say 50 random real numbers in the interval. Pass from one generation to the next as follows. Colony members x with higher *fitness* $F(x)$ have a chance of more descendents in the next generation. Some of the least fit members are replaced by fresh random reals from the interval. Form breeding pairs of members of the current generation; some members of the current generation may appear in several pairs because they are more fit (higher F). The two members of a pair exchange (*crossover*) some genetic information (switch some bits) to create their two descendents. Then *mutate* a small number of bits in these descendents, i.e, change the bits to their opposites.

Genetic algorithms have been applied to a very wide range of optimization problems such as the travelling salesman problem, Hamiltonian graphs, multivariable functions, nonlinear equations, population migration, machine learning, knapsack problem, VLSA circuit layout, classroom scheduling, and on and on.

We'll set up a program to handle optimization of real functions of several variables. Our test function will be one with many local maxima and an absolute maximum at the origin:

$$F(x,\, y) = \left(2 - x^2 - y^2\right)\cos 25x \cos 25y$$

The program on disk contains some other functions as comments.

The first problem in applying genetic algorithms to a problem is to represent members of the population as strings of bits. In our case it is convenient to use the real type "Comp" which, I remind you, goes with the 80x87 coprocessor and which is represented by 8 bytes, or 64 bits. The high bit is the sign bit, the other 63 bits represent an "integer", so a "Comp" x seems to have domain $\left[-(2^{63} - 1),\, (2^{63} - 1)\right]$, which we easily scale into any domain. Borland Pascal's type casting allows easy access to the individual bits of a "Comp" and we use a type "vector" to view a "Comp" as an array of 8 bytes.

Let's look at the opening of the program. Among the constants are two probabilities, for *crossover* (exchange of genetic information) and for *mutation*. "No_bits" is the number of bits in each point, that is, dimension times 64 bits per "Comp". The four lines following the declaration of "no_bits" compute the largest "Comp", as a real number, for scaling.

Program 7.5.1

```
program  genetic_extrema;    { 10/06/94 }

   uses  CRT;

   const      colony_size = 100;    { 30 -- 200, even }
           crossover_prob = 0.60;
           mutation_prob = 0.05;
                      dim = 2;    { dimension }
                  no_bits = 64*dim;
                   two_15 = 32768.0;   two_16 = 65536.0;
                   two_32 = two_16*two_16;
                   two_63 = two_15*two_16*two_32;
                 max_comp = two_63 - 1.0;
```

```
type           Real = Extended;
              point = array[1..dim] of Real;
             member = array[1..dim] of Comp;
         colony_vec = array[1..colony_size] of member;
        permutation = array[1..colony_size] of Word;
          value_vec = array[1..colony_size] of Real;
             vector = array[0..7] of Byte;
              Str30 = string[30];
```

The next two points hold the interval bounds; in this case $[-1, 1] \times [-1, 1]$.

```
const   upp_bd: point = (1.0, 1.0);
        low_bd: point = (-1.0, -1.0);
```

```
var           colony: colony_vec;
        member_value,
       member_fitness: value_vec;
                 bit: array[0..7] of Byte;
               count: Word;
             max_val: Real;
          max_member: member;
             a, b: point;
                ch: Char;
```

```
function  F(p: point): Real;

    var   x, y: Real;

    begin
    x := p[1];   y := p[2];
    F := (2.0 - Sqr(x) - Sqr(y))
            *Cos(25.0*x)*Cos(25.0*y);
    end;    { F }
```

"Max_val" and "max_member" are the largest fitness (value of F) in the current generation and the member with that fitness. The points a and b are the constants for scaling; the next procedure is an affine map from "Comp"s into our intervals. The procedure "value" computes the value of F at a member of the colony, that is, a pair of "Comp"s.

```
procedure   member2point(mem: member;
                              var   p: point);

    var   j: Word;
```

```
begin
for  j := 1  to  dim  do
    p[j] := a[j]*mem[j] + b[j];
end;

function  value(var  mem: member): Real;

    var  p: point;

    begin
    member2point(mem, p);
    value := F(p);
    end;    { value }
```

The next procedure generates a random vector of 8 random bytes and type-casts it into a random "Comp" to generate each component of a point. The procedure "initialize" initializes an array "bit" of bytes needed later to pick out individual bits for mutation, initializes the a[] and b[] coefficients of scaling affine maps, initializes the colony randomly, and finally sets the initial generation count to 0.

```
procedure  get_random_member(var mem: member);

    var  i, j: Word;
            v: vector;

    begin
    for  j := 1  to  dim  do
        begin
        for  i := 0  to  7  do  v[i] := Random(256);
        mem[j] := Comp(v);
        end;
    end;    { get_random_member }

procedure  initialize;
    { Initializes:
            bit[]   a[]   b[]
            colony[]
            member_value[]
            max_val
            count to 0 }

    var   i, j, k: Word;
                t: Real;
                v: vector;
```

```
begin    { initialize }
{ Initialize  bit[] }
j := 1;
for  k := 0  to  7  do
   { bit[k] = 2^k; e.g., bit[3] = 00001000 }
   begin
   bit[k] := j;
   j := j shl 1;
   end;
{ Initialize the affine map:
  x --> a[j]x + b[j] maps [-max_comp, max_comp]
     onto [ low_bd[j], up_bd[j] ]. }
for  j := 1  to  dim  do
   begin
   a[j] := 0.5*(upp_bd[j] - low_bd[j])/max_comp;
   b[j] := 0.5*(upp_bd[j] + low_bd[j]);
   end;
{ Initialize the colony }
max_val := -1.0e4932;
Randomize;
for  k := 1  to  colony_size  do
   begin
   get_random_member(colony[k]);
   t := value(colony[k]);
   member_value[k] := t;
   if  t > max_val  then
      begin
      max_val := t;
      max_member := colony[k];
      end;
   end;
count := 0;
end;    { initialize }
```

In order to handle the fitness of individuals systematically, the next procedure maps the values linearly into the unit interval, with the minimum of F mapping to 0 and the maximum to 1.

```
procedure  compute_fitness;
   { Maps member_value linearly into
     member_fitness, with values in [0, 1],
     including the endpoints. }
```

```
var    min, max,
          a,  b,  t: Real;
                k: Word;

begin
min := member_value[1];    max := member_value[1];
for  k := 2  to colony_size  do
     begin
     t := member_value[k];
     if  t > max  then    max := t
     else  if  t < min  then    min := t;
     end;
a := 1.0/(max - min);    b := - a*min;
for  k := 1  to colony_size  do
     member_fitness[k] := a*member_value[k] + b;
end;    { compute_fitness }
```

This brings us to evolutionary procedure "select_fittest". First it replaces the least fit members with random replacements. (My idea—omit it if you please.) Next it divides up the unit interval, giving each member of the colony a subinterval of length proportional to its fitness. (Note that fitnesses have not been computed for the replaced low fitness members.) If a random real falls into the interval of a member, that member is selected as a parent for the next generation.

```
procedure  select_fittest;
   { Survival of the fittest:
     The number of offspring of an individual is
     proportional to its member_fittness. }

var              x,
        total_fit: Real;
        High, Low,
           mid, k: Word;
           chance: array[0..colony_size] of Real;
      hold_colony: colony_vec;

begin
total_fit := 0.0;
for  k := 1  to  colony_size  do
   begin    { Introduce new genetic material }
   if  member_fitness[k] <= 0.20  then
        get_random_member(colony[k]);
   total_fit := total_fit + member_fitness[k];
   end;
chance[0] := 0.0;
```

```
for  k := 1  to  colony_size  do
   chance[k] := chance[k - 1]
                    + member_fitness[k]/total_fit;
{ chance[colony_size] = 1.0.
  Each member has a number of offspring
  in the next generation proportional
  to its member_fitness.
  Upgrade colony members simultaneously. }
hold_colony := colony;
for  k := 1  to  colony_size  do
   begin
   x := Random;
   Low := 0;
   High := colony_size;
   repeat    { binary search }
      mid := (Low + High) div 2;
      if  x < chance[mid]  then  High := mid
      else  Low := mid;
   until  High <= Low + 1;
   colony[k] := hold_colony[High];
   end;
end;    { select_fittest }
```

The next evolutionary process pairs off the parents randomly and switches some of their genes. For pairing, we need a random permutation generated by an algorithm we have seen earlier.

```
procedure  random_perm(var  perm: permutation);
   { Makes perm a random permutation
     of 1..colony_size }

   var  j, k, t: Word;

   begin
   for  j := 1  to  colony_size  do  perm[j] := j;
   for  k := colony_size  downto  2  do
      begin
      j := 1 + Random(k);    { 1 <= j <= k }
      if  j < k  then
         begin
         t := perm[k];
         perm[k] := perm[j];  perm[j] := t;
         end;
      end;    { for  k }
   end;    { random_perm }
```

Two (of several known) crossover procedures are given. *Uniform crossover*, means any bit whatever may be switched between the two parents to create their two children. *Single point crossover* means all bits to the left of a particular bit are switched. The idea is to keep together good genetic material. (Keep a grain of salt handy.)

```
procedure  uniform_crossover;
   { Mating pairs may switch some genes
     for their offspring }

   var          perm: permutation;
            i, j, k,
              m, n: Word;
      bi, bj, mask: Byte;
            vi, vj: vector;

   begin
   random_perm(perm);
   k := 1;
   repeat
      { Random pairing }
      i := perm[k];   Inc(k);
      j := perm[k];   Inc(k);
      if  Random < crossover_prob   then
      { Cross some genes over }
         begin
         { Select components to uniform_crossover }
         for  m := 1  to  dim  do
            begin    { Convert type }
            vi := vector(colony[i, m]);
            vj := vector(colony[j, m]);
            { Select the bytes to crossover }
            for  n := 0  to  7  do
               begin
               { Select bits to crossover }
               mask := Random(256);
               { Effect the crossover in byte n
                 of vi and vj. }
               bi := (not mask and vi[n])
                      or (mask and vj[n]);
               bj := (mask and vi[n])
                      or (not mask and vj[n]);
               vi[n] := bi;  vj[n] := bj;
               end;    { Convert back }
```

```
              colony[i, m] := Comp(vi);
              colony[j, m] := Comp(vj);
              end;
          end;    { if random < crossover_prob }
      until  k > colony_size;
      end;    { uniform_crossover }

   procedure  single_point_crossover;

      var     perm: permutation;
          i, j, k,
         m, n, p: Word;
              b: Byte;
           vi, vj: vector;

      begin
      random_perm(perm);
      k := 1;
      repeat
         { Random pairing }
         i := perm[k];   Inc(k);
         j := perm[k];   Inc(k);
         if  Random < crossover_prob   then
         { Cross some genes over }
            begin
            m := Random(dim) + 1;    { 1..dim }
            vi := vector(colony[i, m]);
            vj := vector(colony[j, m]);
            n := Random(8);    { 0..7 }
            for  p := 0  to  n  do
               begin
               b := vi[p];
               vi[p] := vj[p];   vj[p] := b;
               end;
            colony[i, m] := Comp(vi);
            colony[j, m] := Comp(vj);
            end;    { if  random < crossover_prob }
      until  k > colony_size;
      end;    { single_point_crossover }
```

The final evolutionary process is *mutation*.

```
   procedure  mutate;
      { Random flipping of genes }
```

```
var   i, j, k, m: Word;
                 v: vector;

begin
for   i := 1   to   colony_size   do
    for   j := 1   to   dim   do
        begin     { Convert type }
        v := vector(colony[i, j]);
        for   k := 0   to   7   do
            for   m := 0   to   7   do
                if   Random <= mutation_prob   then
                    v[k] := v[k] xor bit[m];
        { Convert back }
        colony[i, j] := Comp(v);
        end;
end;     { mutate }
```

"Update_max" locates the member "max_member" with highest value and its corresponding value "max_val".

```
procedure   update_max;

var   j, k: Word;
           t: Real;

begin
for   k := 1   to   colony_size   do
    begin
    t := value(colony[k]);
    member_value[k] := t;
    if   t > max_val   then
        begin
        max_val := t;   max_member := colony[k];
        end;
    end;
end;     { update_max }
```

The remaining procedure is for screen output, and the main program action follows. Good luck testing the program and possibly adapting it to other optimization problems.

```
procedure   display;

var   p: point;
          k: Word;
```

```pascal
begin    { display }
Write('Generation', count:5);
WriteLn('  Max F(x) to date = ', max_val:8:6);
member2point(max_member, p);
Write('  at (');
for  k := 1  to  dim  do
   begin
   Write(p[k]:8:6);
   if  k = dim  then  WriteLn(')')
   else  Write(', ');
   end;
end;    { display }

begin    { genetic_extrema }
ClrScr;
initialize;
WriteLn('Colony size: ', colony_size);
WriteLn('Crossover probability: ',
        crossover_prob:6:4);
WriteLn('Mutation probability: ',
        mutation_prob:6:4);
WriteLn;
WriteLn('To suspend: <Pause> -- ',
        'To resume: any key.');
WriteLn('To halt: <Space>');
Window(1, 9, 80, 25);
repeat
   Inc(count);
   compute_fitness;
   select_fitest;
   single_point_crossover;
   mutate;
   update_max;
   display;
   if  KeyPressed  then  Break;
until  false;
WriteLn;
display;
Write('Press <Enter>: ');   ReadLn;
end.    { genetic_extrema }
```

Section 6.   Miscellaneous Algorithms

We begin with a program on partially ordered sets (posets). A *poset* is a set P
with a relation "$<<$" satisfying

1. x << x is never true.
2. (transitivity) If x << y and y << z, then x << z.

Suppose we are given a finite set P and a relation << on that set. Suppose we extend that relation to other pairs in the set by applying transitivity as far as we can. Then we either have a poset, or find x << x for some element x. Our problem is to determine which, and to compute the extended relation.

The problem can be put this way. The finite set P can be 1..max. The relation "<<" is given by a max × max Boolean matrix. When we extend this relation by transitivity, the result is a new Boolean matrix, and it is the matrix of a poset if its diagonal elements all equal False. To make the problem more challenging, we insist on computing the extended matrix "in place".

The program *poset* is based on an ingenious algorithm of S Warshall, Jour ACM, 9 (1962), 11-12 which makes the computation much faster than you might expect. See also R W Floyd, CACM 96, (1962), and P Z Ingerman, CACM 141 (1962). (Recall that CACM means "Collected algorithms of the ACM".)

Let's start the program with basic declarations and procedures for generating a random relation and for printing a relation. The latter uses IBM boxing characters as usual.

Program 7.6.1

```
program  poset;    { 05/23/94 }

   uses  CRT;

   const  max = 10;

   type  relation = array[1..max, 1..max]  of  Boolean;

   var  bad_data: Boolean;
               A: relation;
               k: Word;

   procedure  random_relation;

      var  j, k: Word;

      begin
      for  j := 1  to  max  do
         for  k := 1  to  max  do  A[j, k] := false;
      Randomize;
      for  j := 1  to  max  do
         A[1 + Random(max), 1 + Random(max)] := true;
      end;    { random_relation }
```

```
procedure  print_relation(var  A: relation;
                               col, row: Word);

   var  j, k: Word;

   begin
   GotoXY(col, row);
   Write(#218);
   Write(#191 : 2*max);
   Inc(row);
   for  j := 1  to  max  do
      begin
      GotoXY(col, row);
      Write(#179);
      for  k := 1  to  max - 1  do
         if  A[j, k]  then  Write('T ')
         else  Write('F ');
      if  A[j, max]  then  Write('T')
         else  Write('F');
      Write(#179);
      Inc(row);
      end;
   GotoXY(col, row);
   Write($192);
   Write(#217 : 2*max);
   WriteLn;
   end;    { print_relation }
```

Now we come to the Warshall algorithm. Its three nested **for** loops means that it is an $O(max^3)$ algorithm, which should be surprising. The comments should explain how it works.

```
procedure  generate_poset(var  A: relation;
                               var  bad_data: Boolean);

   var  i, j, k: Word;

   begin
   for  j := 1  to  max  do
      begin
      for  i := 1  to  max  do
         if  A[i, j]  then
            for  k := 1  to  max  do
               { The next line cannot change  A[i, k]
                 from true to false. }
```

```
                A[i, k] := A[i, k] or A[j, k];
          bad_data := A[j, j];
          if  bad_data  then  Break;
          { At this point, for any i, j, k,
              if  A[i, j]  and  A[j, k],  then  A[i, k] }
        end;
      end;   { generate_poset }

  begin    { posets }
  random_relation;
  ClrScr;
  GotoXY(1, 1 + max div 2);
           { 123456789 }
  Write   ('Input  = ');
  print_relation(A, 10, 1);
  generate_poset(A, bad_data);
  if  bad_data  then  WriteLn('Inconsistent input')
  else
      begin
      k := WhereY;
      GotoXY(1, k + max div 2);
               { 123456789 }
      Write   ('Output = ');
      print_relation(A, 10, k);
      end;
  Write('Press <Enter> to quit: ');   ReadLn;
  end.    { posets }
```

Our next problem is concerned with minimizing the work in multiplying matrices. Specifically, we want to group the matrices in a product

$$P = A_1 A_2 A_3 \cdots A_p$$

so that the number of scalar multiplications is as small as possible. For instance, if A is $l \times m$, B is $m \times n$, and C is $n \times p$, then $(AB)C$ and $A(BC)$ involve respectively

$$lmn + lnp = ln(m + p) \quad \text{and} \quad lmp + mnp = mp(l + n)$$

multiplications. For the case

$$(l, \quad m, \quad n, \quad p) = (1, \quad 10, \quad 1, \quad 10)$$

this is 20 *vs* 200. So the way the product is computed can make a substantial difference.

You will find this "Matrix chain product" problem discussed in Sedgewick, R, Algorithms 2/e, Addison-Wesley, 1988, 598–602, and in Cormen, T, et al, Introduction to Algorithms, MIT Press, 1990, 302–309.

Our program is based on the following analysis. We are given a sequence

$$n_0 \quad n_1 \quad \cdots \quad n_p$$

of positive integers. The matrix A_j has size $n_{j-1} \times n_j$ so that the matrices have compatible sizes for multiplying. The main idea here is to look not only at the product of all the matrices, but at all possible products of contiguous factors. Let M_{ik} denote the *minimum* number of scalar multiplications for the product $A_i \cdots A_k$. Then $M_{ii} = 0$ and we have the relation

$$M_{ik} = \min_{i<j<k} \left(M_{ij} + M_{j+1,k} + n_{i-1}n_jn_k \right)$$

Think about it; in any association of all the factors for multiplication, there is a *last* multiplication of two factors,

$$(A_i \cdots A_j)(A_{j+1} \cdots A_k)$$

It is the product of an $n_{i-1} \times n_j$ matrix with an $n_j \times n_k$ matrix, and this final product involves $n_{i-1}n_jn_k$ scalar multiplications. The two bracketed products must be optimally associated of course to achieve the least number of scalar multiplications.

In addition to the array M, we also keep track of the best choice of the j that yields the minimum. Thus we need a second array $S = [S_{ik}]$, where S_{ik} will be the least (to break ties) j that gives the minimum in the recursion formula above.

The data to the problem is the vector of sizes of the matrices, which I take as a typed **const**. The output procedure "print" offers a recursive algorithm for showing the actual association of factors; I'll let you figure it out as an exercise.

Program 7.6.2

```
program  matrix_multiplication;    { 05/22/94 }

   uses  CRT;

   const  p = 10;
          N: array[0..p] of Word
             = (9, 10, 5, 3, 20, 25, 30, 4, 14, 7, 5);

   var      p1, min,
            i, j, k,
          no_factors: Word;
                     M: array[1..p, 1..p] of LongInt;
                     S: array[1..p, 1..p] of Word;

   procedure  print(i, k: Word);
      . . . . . . . . . . . . . . . . . . . . . . . .
```

```
begin
p1 := p - 1;
for  i := 1  to  p  do  M[i, i] := 0;
for  no_factors := 2  to  p  do
   begin
   for  i := 1  to  p - no_factors + 1  do
   { Compute optimal cost for factors
        i, ... ,(i + no_factors - 1) }
      begin
      k := i + no_factors - 1;
      M[i, k] := MaxLongInt;
      for  j := i  to  k - 1  do
         begin
         min := M[i, j] + M[j + 1, k]
                       + N[i - 1]*N[j]*N[k];
         if  min < M[i, k]  then
            begin
            M[i, k] := min;
            S[i, k] := j;
            end;
         end;    { for  j }
      end;    { for  i }
   end;    { for  no_factors }
ClrScr;
WriteLn('M matrix:');
{ Output of M and S here. }
Write('Optimal Product = ');   print(1, p);
WriteLn;
Write('Press <Enter> to quit: ');   ReadLn;
end.    { matrix_multiplication }
```

The next program is a rather unusual, but very good random number generator that Knuth attributes to G J Mitchell and D P Moore. See Knuth, D E, The Art of Computer Programming 2, 2/e, Addison-Wesley 1981, pages below. We include the additive and subtractive versions of the algorithm.

Program 7.6.3

```
program  random_no_generator;    { 09/30/94 }
   { See Knuth, D., vol. 2 , pp 26 ff;
     Exercise 23, p.36; Solution, p. 530. }
```

```
var   cycle55: array[1..55] of Real;
              { cyclic array }
      index_j,
      index_k,
            k: Word;
         sum: Real;
       chi_sq,
      current: Real;
         freq: array[0..19] of Integer;

function   Random: Real;

   var   x: Real;

   begin
   (*    { Subtractive generator }
   x := cycle55[index_k] - cycle55[index_j];
   if   x < 0.0   then   x := x + 1.0;   *)
   { Addivite generator }
   x := cycle55[index_k] + cycle55[index_j];
   if   x >= 1.0   then   x := x - 1.0;
   cycle55[index_k] := x;
   Dec(index_j);   Dec(index_k);
   if   index_j = 0   then   index_j := 55
   else   if   index_k = 0   then   index_k := 55;
   Random := x;
   end;

function   flip_coin(bias: Real): Boolean;

   begin
   if   bias = 1.0   then   flip_coin := true
   else   flip_coin := Random <= bias;
   end;

procedure   Randomize;
{ Get seeds and initialize cycle55 }

   var   seed0, seed1,
            current,
            previous: Real;
                OK: Boolean;
             j, k: Word;
```

```
begin    { randomize }
repeat
   WriteLn('Enter distinct seeds');
   repeat
      Write(
      'Enter first seed s0, 0.0 < s0 < 1.0: ');
      ReadLn(seed0);
      OK := (seed0 > 0.0) and (seed0 < 1.0);
      if  not OK  then  WriteLn('Out of range');
   until  OK;
   repeat
      Write(
      'Enter second seed s1, 0.0 < s1 < 1.0: ');
      ReadLn(seed1);
      OK := (seed1 > 0.0) and (seed1 < 1.0);
      if  not OK  then  WriteLn('Out of range');
   until  OK;
   OK := Abs(seed1 - seed0) > 1.0e-8;
   if  not OK  then
      WriteLn('The seeds are too close');
until  OK;
cycle55[55] := seed1;
current := seed0;
previous := seed1;
k := 0;
for  j := 1  to  54  do
   begin
   k := k + 21;
   if  k >= 55  then  Dec(k, 55);
   cycle55[k] := current;
   current := previous - current;
   if  current < 0.0  then
      current := current + 1.0;
   previous := cycle55[k]
   end;
index_j := 24;
index_k := 55;
{ Flush out some initial values }
for  j := 1  to  1000  do  seed0 := Random;
end;    { randomize }
```

```
begin    { random_no_generator }
for  k := 0  to  19  do  freq[k]  := 0;
Randomize;
sum := 0.0;
for k := 1  to  20000  do
    begin
    current := Random;
    sum := sum + current;
    Inc( freq[Trunc(20.0*current)] );
    end;
WriteLn('20,000 sample random numbers in [0, 1)');
WriteLn('Frequencies in intervals [0.0, 0.05],',
    '  ... ,  [0.95, 1.00]');
for  k := 0  to  19  do  WriteLn(k:4, freq[k]:8);
chi_sq := 0.0;
for  k := 0  to  19  do
    begin
    chi_sq := chi_sq + Sqr(freq[k] - 1000);
    end;
chi_sq := 0.001*chi_sq;
WriteLn('average = ', sum/20000.0:6:4);
WriteLn('chi squared = ', chi_sq:5:2);
ReadLn;
end.    { random_no_generator }
```

Our next example is one of efficiency in computing. We are given a vector

$$\mathbf{z} = [z_1, z_2, \cdots, z_m]$$

and subintervals

$$I_1 = [a_1, b_1] \quad I_2 = [a_2, b_2] \quad \cdots \quad I_m = [a_m, b_m]$$

of [0, m]. The problem is to compute all sums

$$x_j = \sum_{\{k \,|\, j \in I_k\}} z_k$$

There is an obvious $O(m^2)$ algorithm for doing this: initialize the vector **x** to 0 and sum over each of the m intervals. This is programmed below as "slow_summation", mainly to provide a test for the better algorithm that follows.

It is not at all obvious that there is an $O(m)$ algorithm; the device is to work with the first differences of the **x** array. As is often the case in mathematics, cleverly introducing additional quantities is the key to a solution of a problem.

Program 7.6.4

```
program  interval_sums;    { 10/11/94 }
  { Array z[k] is a given vector.
    Intervals  I_k = [ a[k], b[k] ] <= [0, max]
    are given.

    The problem is to compute
       x[j] = sum(z[k]) over all k for which j in I_k.

    Obvious time = O(max^2) algorithm:
       x := 0.0;
       for  k := 0  to   max   do
           for  j := a[k]  to  b[k]   do
                x[j] := x[j] + z[k];

    Do it in time = O(max). }

  const   max = 20;

  type    my_type = Word;
          vector = array[0..max] of my_type;
          endpoint = array[0..max] of Word;

   var   x, y, z: vector;
            a, b: endpoint;
               k: Word;

  procedure  slow_summation(var   a, b: endpoint;
                            var   z, x: vector);

     var  j, k: Word;

     begin    { O(max^2) algorithm }
     for  k := 0  to   max   do   x[k] := 0;
     for  k := 0  to   max   do
        for  j := a[k]   to  b[k]   do
             x[j] := x[j] + z[k];
     end;
```

```
procedure  fast_summation(var   a, b: endpoint;
                          var   z, x: vector);
   { Analysis:
      d is the first difference of x:
          d[j] = x[j] - x[j+1]
      x[j] = sum z[k],   a[k] <= j <= b[k]
      x[j] =    sum   z[k]    -    sum   z[k]
            0 <= j <= b[k]        0 <= j < a[k]
      d[j] =    sum   z[k]    -    sum   z[k]
             j = b[k]              j = a[k] - 1 }

   var      d: vector;
         j, k: Word;

   begin    { fast_summation }
   for  k := 0  to   max  do  d[k] := 0;
   for  k :=  0  to   max  do
      begin    { Compute d }
      j := b[k];
      d[j] := d[j] + z[k];
      j := a[k];
      if  j > 0   then
         begin
         Dec(j);   d[j] := d[j] - z[k];
         end;
      end;    { for k }
   { Solve the difference equation
         x[j] - x[j+1] = d[j] }
   x[max] := d[max];
   for  k := max - 1  downto  0  do
      x[k] :=   x[k + 1] + d[k];
   end;    { fast_summation }

begin    { interval_sums }
{ Initialize z, a, b }
Randomize;
for  k := 0  to   max  do   z[k] := Random(21);
Randomize;
for  k := 0  to   max  do
   begin
   a[k] := Random(max + 1);
   b[k] := a[k] + Random(max - a[k] + 1);
   end;
{ Compute }
fast_summation(a, b, z, x);
slow_summation(a, b, z, y);
```

```
{ Print }
WriteLn('k':2, '    [ a,   b]   ',
        'z[k]', 'x[k]':8, 'y[k]':8);
WriteLn;
for  k := 0  to  max  do
    begin
    Write(k:2, '   [', a[k]:2, ', ',
                     b[k]:2, ']  ', z[k]:4);
    WriteLn(x[k]:8, y[k]:8);
    end;
ReadLn;
end.    { interval_sums }
```

The final program for this section solves the following problem: Display and count all m×n matrices of elements precisely 1, ..., mn, where each row and each column is strictly increasing.

The algorithm itself is quite interesting, and I suggest that you trace through it in a low case, as indicated, before trying larger matrices. Also draw a few pictures to show the staircase of corners. Notice that this is essentially a backtracking algorithm. To this end, I have inserted several tracing statements which follow the flow of the program. These statements are compiled conditionally. They are compiled only if the symbol "trace" is defined. To define it change the comment { $define trace} into a compiler directive by deleting the space in "{ $". (There is a discussion of conditional compilation in Appendix 2.)

Program 7.6.5

```
program  list_all_matrices;    { 10/11/94 }
    { Scans all m * n matrices with elements  1..mn
      and strictly increasing rows and columns.
      Recursive algorithm.  }

    { $define  trace }
    uses   CRT;

    const  m0 = 2;  n0 = 4;
           file_name = 'con';

    type   matrix = array[1..m0, 1..n0] of Word;
```

```pascal
var     count, k,
            value,
        nest_level: Word;
                a: matrix;
            corner: array[1..m0] of 0..n0;
            { n0 >= corner[1] >= corner[2] >=...>= 0 }
            out_file: Text;

procedure   print_matrix(var   a: matrix);

    var   i, j: Word;

    begin
    Inc(count);
    WriteLn(out_file, #218, ' ':4*n0, #191);
        for   i := 1   to   m0   do
            begin
            Write(out_file, #179);
            for   j := 1   to   n0   do
                Write(out_file, a[i, j] : 4);
            WriteLn(out_file, #179);
        end;
    WriteLn(out_file, #192, ' ':4*n0, #217);
    Write('Press <Enter>: ');   ReadLn;
    end;    { print_matrix }

procedure   get_next_corner(var   i, j: Word);
    { Input: i      Output: (i, j) }

    begin
    {$ifdef   trace }
    WriteLn('Entering get_next_corner: i = ', i);
    {$endif }
    while   i <= m0   do
        begin
        j := corner[i] + 1;
        if   (j <= n0) and
            ( (i = 1) or (corner[i - 1] >= j) )   then
                Break;
        Inc(i);
        end;
    {$ifdef   trace }
    WriteLn('Exiting get_next_corner: ',
            '(i, j) = (', i, ', ', j, ')');
    {$endif }
    end;    { get_next_corner }
```

```
procedure  survey;
{ On entry, the matrix elements
    a[1, 1],  ... ,  a[1, corner[1]]
    a[2, 1],  ... ,  a[2, corner[2]]
    .........................
    a[m, 1],  ... ,  a[m, corner[m]]
  have been determined, and
  a[i, corner[i]] = value is the last assigned. }

  var  i, j: Word;

  begin    { survey }
  if  KeyPressed  then  Halt;
  Inc(nest_level);
  i := 1;
  get_next_corner(i, j);
  while  i <= m0  do
     begin
     { Move forward }
     corner[i] := j;
     {$ifdef  trace }
     Write('Level = ', nest_level,
           '  Assigned; corner = [',
     corner[1], ', ', corner[2], ']''    ');
     {$endif }
     Inc(value);
     a[i, j] := value;
     {$ifdef  trace }
     WriteLn('Assigned: a[', i, ', ', j,
             '] = ', a[i, j]);
     {$endif }
     if  (i = m0) and (j = n0)   then
         print_matrix(a)
     else  survey;
     { Backtrack }
     corner[i] := j - 1;
     Dec(value);
     Inc(i);
     if  i <= m0  then  get_next_corner(i, j);
     end;   { while }
  Dec(nest_level);
  end;    { survey }
```

```
begin     { list_all_matrices }
Assign(out_file, file_name);
Rewrite(out_file);
count := 0;
nest_level := 0;
for  k := 1  to  m0  do  corner[k] := 0;
value := 0;
survey;
WriteLn(out_file,
         'Count(', m0, ', ', n0, ') = ', count);
ReadLn;
end.    { list_all_matrices }
```

Exercises

1. Write a program to compute and store the smallest 10,000 elements of the set S of positive integers generated by 1 and closed under a finite set of linear polynomials with positive integer coefficients. An example of such a set of linear polynomials is {2x+1, 7x+3}, in which case S = {1, 3, 7, 10, 15, ...}.

2. Program the following two-dimensional initial value problem for a difference equation, in the form of a matrix of a finite number of columns and infinitely many rows. The first row is all 1's as are the first and last columns. Any element elsewhere equals the sum of the two elements immediately diagonally above it.

3. The "Problem of Josephus": Arrange n people around a circle. Execute each third person, and keep going until one person remains. Which one? Program this.

4. Write a program to score bowling. First visit the local bowling alley to learn the scoring rules if you have never bowled yourself.

5. Program the game of roulette as played at Monte Carlo and elsewhere.

6. Develop an output procedure "print" for the program "matrix_multiplication" that will show the actual optimal association of factors.

Chapter 8 Discrete Mathematics

Section 1. Number Theory

Elementary number theory offers many opportunities to experiment, look for patterns, and test conjectures, often requiring only fairly simple programs. In this section we look at some easy examples.

Our first problem is to count the integers in the interval $[m, n]$ that are divisible by 3 or by 7, but not by both. We offer two solutions, and you might study their relative efficiencies.

Program 8.1.1

```pascal
program  counting_methods;    { 04/23/95 }

   function  count1(m, n: Word): Word;

      var  c, j: Word;

      begin
      c := 0;
      for  j := m  to  n  do
          if  (j mod 3 = 0) xor
              (j mod 7 = 0)  then  Inc(c);
      count := c;
      end;

   function  count2(m, n: Word): Word;

      begin
      count2 := n div 7 + n div 3 - 2*(n div 21)
                  - (m-1) div 7 - (m-1) div 3
                  + 2*((m-1) div 21);
      end;

   begin    { counting_methods }
   WriteLn(count1(1000, 40000));
   WriteLn(count2(1000, 40000));
   ReadLn;
   end.
```

The next program sets up the greatest common divisor of two nonnegative integers, not both 0, in three different ways. Recall that $\gcd(m, n)$ is the largest integer that divides both m and n.

The GCD may be defined recursively as follows:

$$gcd(m, n) = \begin{cases} m & \text{if } n = 0 \\ gcd(n, m) & \text{if } m < n \\ gcd(m - n, n) & \text{otherwise} \end{cases}$$

Our first gcd function simply translates the definition into Pascal.   It is an interesting exercise to prove that the other two functions give the same result.

Program 8.1.2

```pascal
program  gcd;     { 04/24/95 }

   function  gcd1(m, n: Word): Word;
      { Recursive }

      begin
      if  n = 0  then   gcd1 := m
      else  if  m < n   then   gcd1 := gcd1(n, m)
      else   gcd1 := gcd1(m - n, n)
      end;

   function  gcd2(m, n: Word): Word;
      { Recursive -- much faster }

      begin
      if  n = 0  then   gcd2 := m
      else  if  m < n   then   gcd2 := gcd2(n, m)
      else   gcd2 := gcd2(n, m mod n);
      end;

   function  gcd3(m, n: Word): Word;
      { iterative -- faster still }

      var   r: Word;

      begin
      if  m < n  then
         begin   r := n;   n := m;   m := r;   end;
      { Now   n < m }
      while   n > 0 do
         begin
         r := m mod n;
         m := n;   n := r;
         end;
      gcd3 := m;
      end;
```

```
begin    { gcd }
WriteLn(gcd1(29008, 36368));
WriteLn(gcd2(29008, 36368));
WriteLn(gcd3(29008, 36368));
ReadLn;
end.    { gcd }
```

Sometimes we need the gcd of many integers. The following program finds it fairly efficiently.

Program 8.1.3

```
program  gcd_vector;    { 04/24/95 }

    const   abs_max = 1000;

    type   vector = array[1..abs_max] of LongInt;

    var   dim: Word;     { dim  abs_max }
            v: vector;
            k: Word;

    function  gcd(v: vector): LongInt;

        var    j, j0, k: Word;
               hold, min: LongInt;

        begin    { First locate first nonzero v[j0] }
        for  j := 1  to   dim  do   v[j] := Abs(v[j]);
        j0 := 1;
        while   (j0 <= dim) and (v[j0] = 0)   do   Inc(j0);
        if   j0 = dim + 1   then
            begin   gcd := 0;   Exit;   end;
        { Now locate smallest min = v[k]. }
        min := v[j0];   k := j0;
        for  j := j0 + 1  to   dim   do
            begin
            if   (v[j] > 0) and (v[j] < min)   then
                begin  k := j;   min := v[j]   end;
            end;
        repeat
            hold := min;
            for  j := j0  to   k - 1   do
                v[j] := v[j]  mod   min;
            for  j := k + 1  to   dim   do
                v[j] := v[j]  mod   min;
            for  j := j0  to   dim   do
```

```
        if   (v[j] > 0) and (v[j] < min)   then
             begin   k := j;   min := v[j]   end;
   until   min = hold;
   gcd := min;
   end;    { gcd }

begin    { gcd_vector }
dim := 5;
Randomize;
for  k := 1  to   dim  do
    begin
    v[k] := Random(4999) + 1;
    Write(v[k]:8);
    end;
WriteLn;
WriteLn('gcd = ', gcd(v));
Write('Press <Enter>: ');   ReadLn;
end.
```

Pascal's m **mod** n produces a result in the interval $[0, n-1]$. Sometimes a result in the interval $(-n/2, n/2]$ or in the interval $[1, n]$ is needed. The following program handles these possibilities.

Program 8.1.4

```
program   residues;    { 04/24/95 }

   function   modulo0(m, n: Integer): Integer;
      { m = r (modulo n),  -n/2 < r <= n/2 }

      begin
      m := m mod n;
      if   2*m <= n   then modulo0 := m
      else   modulo0 := m - n;
      end;

   function   modulo1(m, n: Integer): Integer;
      { m = r (modulo n),  1 <= r <= n }

      begin
      m := m mod n;
      if   m = 0   then   modulo1 := n
      else   modulo1 := m;
      end;
```

```
begin    { residues }
WriteLn(modulo0(66, 10));
WriteLn(modulo1(66, 10));
ReadLn;
end.
```

Prime numbers and the factoring of integers into primes forms a big part of number theory. In the following section, we handle sieve methods for constructing tables of primes. The unit "primes" contains a fairly elementary sieve for small primes. The disk file is with the programs for this section, and I do not think it worthwhile to display it here. The disk program "prime_factors" uses "primes" plus some fairly sophisticated tools to do factoring of longint's. It comes from an article by Blair, W D, Lacampagne, C B, and Selfridge, J.

Quadratic residues is another big topic in number theory. Again, for this, I include a program on disk, but do not want to go into the theory of Legendre symbols and quadratic reciprocity. The program is "square_root_mod_p" and you should have some fun experimenting with it.

The final program of this section solves the problem of representing integers as the sum of two perfect squares.

Program 8.1.5

```
program  sums_2_squares;    { 04/23/95 }
    { Compile a list of all numbers <= max that are
      sums of two squares, sqr(x) + sqr(y), with all
      possible (x, y). }

    const  top = 150;
           max = (top*top) div 2;

    type  pair_ptr = ^pair_rec;
          pair_rec = record
                            x, y: LongInt;
                            next: pair_ptr;
                     end;

    var      square: array[1..top] of LongInt;
               list: array[2..max] of pair_ptr;
          i, j, k, t,
               count: Word;
                temp: pair_ptr;
            out_file: Text;
```

```
begin    { two_squares }
Assign(out_file, 'con');  Rewrite(out_file);
for  i := 2  to  max  do  list[i] := nil;
for  j := 1  to  top  do  square[j] := Sqr(j);
for  k := 2  to  top  do
   begin
   i := k div 2;
   repeat
      t := square[i] + square[k - i];
      if  t <= max  then
         begin
         temp := list[t];
         New(list[t]);
         with  list[t]^  do
            begin
            next := temp;  x := i;  y := k - i;
            end
         end;
      Dec(i);
   until  (i = 0) or (t > max);
   end;
count := 0;
WriteLn(out_file, '    i  Pairs (x, y)');
for  i := 2  to  max  do
   begin
   temp := list[i];
   if  temp <> nil  then
      if  temp^.next <> nil  then
         begin
         Inc(count);
         Write(out_file, i : 5, ':');
         while  temp <> nil  do
            with  temp^  do
               begin
               Write(out_file,
                     '(':2, x, ', ', y, ')');
               temp := next;
               end;
         WriteLn(out_file);
         end;
   if  count = 20  then
      begin
      ReadLn;
      count := 0;
      end;
   end;    { for i }
```

```
WriteLn(out_file);
Write('Press <Enter>: ');   ReadLn;
end.    { two_squares }
```

Exercises

1. Let m and n be positive constant Words. Define functions

$$\Phi(x) = (x \bmod n)m + x \operatorname{div} n \qquad \Psi(x) = (x \bmod m)n + x \operatorname{div} m$$

Write a program to verify experimentally that these functions are inverse to each other.

2. Write a program that produces the smallest positive integer that is the sum of two perfect cubes in two different ways; ditto fourth powers.

3. Write a program to count integers that are the sum of three squares.

4. A *Dirichlet series* is a sum of the form $\sum_{n=1}^{\infty} a_n/n^s$, where $s > 1$. Write a program for multiplying two (truncated) Dirichlet series.

5. Write a program to test the procedure "find_next_prime" of the unit "primes".

Section 2. Prime Sieves

A *prime* is an integer p greater than 1 having no positive integer divisors except 1 and p. To find all primes up to a bound *max*, we first cast out all multiples of the first prime, 2. The smallest remaining number, 3, is the second prime. We next cast out all of its multiples; the smallest remaining number, 5, is the next prime. We then cast out all of its multiples, . . . This method is the *Sieve of Eratosthenes* in its most primitive form. Note that only multiples of primes $p \le \sqrt{max}$ must be cast out because the least prime factor of any integer in our range is at most this large.

Program 8.2.1

```
program  sieve_1;    { 04/23/95 }
    { Basic sieve of Eratosthenes }

    uses   CRT;

    const   max = 1000;

    var   b, j, k: Word;
            flag: array[1..max] of Boolean;
```

```
begin    { sieve_1 }
flag[1] := false;
for  j := 2  to   max  do   flag[j] := true;
b := Trunc(Sqrt(max));
k := 0;
while  k <= b  do
   begin
   repeat  Inc(k)  until  flag[k];
   j := 2*k;
   while  j <= max  do
      begin
      flag[j] := false;
      j := j + k;
      end;
   end;    { while }
for  j := 1  to   max  do
   if  flag[j]  then   Write(j:8);
WriteLn;  WriteLn;
Write('Press <Enter>:  ');   ReadLn;
end.    { sieve_1 }
```

If we need to find a lot of primes, memory becomes a consideration. We will probably want to store the primes in a file as fast as we find them, rather than in an array. The following program can be used to find very many primes; its space requirement is $O(\sqrt{m})$. This is paid for in time; the running time is $O\left(m^{3/2}\right)$. The input to this program is $\sqrt{m}$. Its output is a list of the first m − 1 primes.

The theory behind "sieve_2" goes as follows. Let

$$p_1 < p_2 < p_3 < \cdots$$

denote the sequence of primes. It is known that for each $n \geq 2$ we have $p_{n^2-1} < (p_n)^2$. Thus to find the first $n^2 - 1$ primes, we must only cast out multiples of the primes $p_1 \cdots p_n$. Actually, for any one of these primes p, we may start with its multiple p^2 because any smaller multiple is actually divisible by some smaller prime, hence cast out already. (No need to duplicate work.) For each of these early primes, its highest relevant multiple is stored in the array "mult", which is the only real drain on memory. The array "prime" is not needed if we opt for file storage. Of course, if we go for larger primes, many of the variables will have to be changed from Word to Longint type. (I have omitted some formatting in the printed listing; it will be found in the disk file.)

Program 8.2.2

```pascal
program  sieve_2;     { 04/23/95 }
   { Finds the first (m - 1) primes. Note that
     n = sqrt(m) is the actual input. }

   uses   CRT;

   const   n = 20;   m = n*n;

   var        p, s,
         { p is the next number tested for primality
           s is a place holder for squares of primes. }
        i, j, k: Word;
         accept: Boolean;
          prime: array[1..m - 1] of Word;
           mult: array[1..n - 1] of Word;
            { mult[k] holds multiples of prime[k],
               starting with s = prime[k]^2. }

   begin    { sieve_2 }
   ClrScr;
   prime[1] := 2;
   p := 2;   k := 1;    s := Sqr(prime[1]);
   for  i := 2  to   m - 1  do
      begin
      repeat    { Sieve; compute prime[i] }
         Inc(p);
         if  s <= p   then
            begin
            { Executed once for each k < n }
            mult[k] := s;    { mult[k] = prime[k]^2 }
            Inc(k);
            s := Sqr(prime[k]);
            end;    { if }
         accept := true;
         for  j := 1  to    k - 1   do
            begin
            if  mult[j] < p   then
                mult[j] := mult[j] + prime[j];
            { Now p <= mult[j] }
            accept := (mult[j] > p);
            if  not accept   then   Break;
            { p is composite }
            end;
      until   accept;    { p is prime }
```

```
        prime[i] := p;
      end;    { for i }
  WriteLn;   WriteLn;
  for  i := 1  to  10  do  Write(i:8);   WriteLn;
  for  i := 1  to  m - 1  do  Write(prime[i]:8);
  WriteLn;   WriteLn;
  Write('Press <Enter>: ');   ReadLn;
  end.    { sieve_2 }
```

The final sieve program here is a very sophisticated one, originating with T A Peng, One Million Primes Through the Sieve, BYTE, Fall 1985, 243–244. It should be apparent in the two programs above that small changes will cut the work in half: we know that no even integer is prime except 2, so the sieving can be restricted to odd numbers; that is done in sieve_3.

We want a program that will find, say, the millionth prime, efficiently and without running out of memory. By some experimenting, we find that the millionth prime is a little less than 16,000,000. To sieve up to that number, we need the primes up to its square root, 4000. By other experiments, we find that there are exactly 549 odd primes less than 4000. To conserve memory, we keep track of the largest odd multiple to date of each of these early primes in the array "multiple".

Peng's clever idea is to sieve the odd numbers less than 16,000,000 in blocks of size 10,000. Actually, blocks of 4000 could be used with a very slight increase in run time (17 sec vs 16 sec on a Pentium 100).

The array "prime" contains the odd primes < block_size obtained by sieving in "initialize". At the end of processing the n-th block, k (= mult) = multiple[j] is the least k > block_size such that $2k - 1$ is a multiple of p = prime[j]. When the next block is sieved for multiples of p, the sieve starts at k.

The listing here omits some useful numerical comments and code for storage of the primes in a disk file. Please see the source file on disk.

Program 8.2.3

```
program  sieve_3;    { 04/22/95 }

  uses   CRT;

  const      sqrt_max = 4000;
                  max = sqrt_max * sqrt_max;
            odd_p_no = 549;
          block_size = 10000;
          no_blocks = max div (2*block_size);
```

```
var       p, offset,
              mult,
      count, block: LongInt;
              flag: array[1..block_size] of Boolean;
              { flag[i] = (2i-1 prime) }
              prime,
          multiple: array[1..odd_p_no] of LongInt;

procedure  initialize;

   var  i, j: LongInt;

   begin
   ClrScr;
   for  i := 1  to  block_size  do
      flag[i] := true;
   flag[1] := false;    { 2*1-1 is not prime }
   j := 0;
   for  i := 2  to  block_size  do
      begin
      if  flag[i]  then
         begin
         p := 2 * i - 1;    { next prime }
         if  p <= sqrt_max   then
            begin
            Inc(j);
            prime[j] := p;
            mult := i + p;
            { 2*mult-1 runs over 3p, 5p, 7p, ... }
            while  mult <= block_size  do
               begin
               flag[mult] := false;
               mult := mult + p;
               end;    { while }
            multiple[j] := mult;
            { Now  multiple[j] > block_size }
            end;    { if  p <= sqrt_max }
         end;    { if  flag[i] }
      end;    { for  i }
   count := 1;    { for p = 2 }
   for  i := 2  to  block_size  do
      if  flag[i]  then  Inc(count);
   offset := 1;
   end;    { initialize }
```

```
procedure   next_block;

   var   j: LongInt;

   begin
   for  j := 1  to   block_size   do   flag[j] := true;
   for  j := 1  to   odd_p_no   do
      begin
      p := prime[j];
      mult := multiple[j] - block_size;
      while   mult <= block_size   do
         begin
         flag[mult] := false;
         mult := mult + p;
         end;     { while }
      multiple[j] := mult;
      end;     { for  j }
   offset := offset + 2*block_size;
   for  j := 1  to   block_size   do
      begin
      if  flag[j]   then
         begin
         p := offset + 2 * j - 2;
         Inc(count);
         if   count = 1000000   then
            WriteLn('The ', count,
                          '-th prime = ', p);
         end;     { if   flag[i] }
      end;     { for  j }
   end;     { next_block }

begin     { sieve_3 }
initialize;
for  block := 2  to   no_blocks   do   next_block;
{ Report }
WriteLn;
WriteLn(p, ' is the ', count,
            '-th prime, largest prime < ', max);
WriteLn;
Write('Press <Enter>:');   ReadLn;
end.     { sieve_3 }
```

Exercises

1. Write a program to retrieve primes stored by, say, "sieve_3".
2. Write a program to print a table of primes read from a disk file.

3. The Boolean array flag in the program "sieve_1" fills max = 1000 bytes. If we wanted the primes up to 100,000 instead of 1000, flag would pass the 64K limit on variable size. But really, we only need a single bit to store the information of a Boolean, so 12,500 bytes are adequate. Modify "sieve_1" accordingly.

Section 3.   Multiprecision Arithmetic

Longint's are good for 9 or 10 decimal digits, but often we need integers with very many more digits than this for experiments in number theory. We'll now start a package to do arithmetic with thousands of digits. This immediately implies either arrays or linked lists to hold very long integers. The former means a decision in advance on an upper limit to how many digits; the latter implies much overhead in operations. So our first decision will be to go the route of arrays—length limited in advance.

What the *radix* (base) should be is a serious issue. A decimal digit has 10 possible values; the next largest computer unit is the byte, with 16 possible values. So if we opt for base 10, large arrays of bytes will waste a lot of memory. What is more, choosing base 10 goes against a computer's natural inclination to do binary arithmetic.

This suggests that our radix should be a power of 2. Obviously we don't need a sign attached to each digit, so we should use an unsigned integer type. This limits our choice to Byte or Word, and we go for the larger one, Word, because operations on arrays of Word will entail less work than on arrays of Byte that represent the same very long integer.

This points to radix = 2^{16}. Of course if we multiply two Words, we expect overflow, but it is not nice to handle overflow every time we add or subtract two "digits". So our final choice of radix is 2^{15}. This brings us to the first unit making up our big program MPA (Multi-Precision Arithmetic), "basics". (The listings in this section omit many comments that are included in the disk files.)

Program 8.3.1

```
unit   basics;     { 05/13/95 }

interface


const        base = $8000;    { 2^15 }
          base_1 = base - 1;
         max_deg = 1000;
         max_sci = 1510;
       text_fore = White;
       text_back = Black;
       highlight = Yellow;
```

```
type        Real = Extended;
       superint = record
                        deg: Word;
                        sign: Integer;
                        coeff: array[0..max_deg] of Word;
                     end;

    engineering = record    { base 1000 }
                        deg: Word;
                        sign: Integer;
                        coeff: array[0..max_sci] of Word;
                     end;
```

Let us examine these declarations so far. Our basic type is "superint", a record of a sign, -1, 0, 1, an array of coefficients, and the degree of the largest nonzero coefficient. You can think of a superint as a signed polynomial in the base $2\hat{}15$. The largest degree is 1000, so integers up to about $2\hat{}15000$ are allowed, which means about 4500 decimal digits. You can try larger bounds, but be a little careful as the program will use about 20 superints.

The type "engineering" is for displaying superints on the screen in groups of three decimal digits. Thus the type allows $3*1510 = 4530$ decimal digits. How to convert superint to scientific is a problem we'll face shortly. Let's move on:

```
var    zero, one, two: superint;
           parser_error: Boolean;
                   hold: array[0..10] of superint;
               out_file: Text;

procedure    full_print(const   n: Word);
procedure    approximate_print(const   n: Word);
procedure    initialize;
procedure    numlock_off;
```

We'll have much use for the superints zero...two, so we might as well name them. The superint "zero" has deg 0 and sign 0. Every other superint has sign 1 or -1. The program will deal with 11 superint variables with letter names; they are the content of the array "hold". The procedure "full_print" does the conversion of a superint to engineering form and displays the result. If the numbers in "hold" are all large, the screen cannot hold all of them so we approximate the larger ones by reals and print them; the procedure "approximate_print" does that.

We move on to the implementation. "Initialize" is easy stuff; "numlock_off" we know from a previous chapter, so we'll skip its listing. The guts of "full_print" is its subroutine "superint2engineering". It does step-by-step long division, and there does not seem to be a better way. You should study its workings. When you run the final program, you will appreciate the careful formatting in "full_print". The conversion of a superint to a real is quite routine in "approximate_print".

implementation

```
const   var_name: array[1..10] of Char
                = ('q', 'r', 's', 't', 'u',
                   'v', 'w', 'x', 'y', 'z');

procedure  full_print(const  n: Word);

   var  col0, col: Word;
            j, m: Word;
              k: Word;
              b: engineering;

   procedure  superint2engineering
                 (z: superint;   var  b: engineering);

      var  quotient, remainder, dividend: LongInt;
                                  k: Word;
      begin
      if   z.sign = 0   then
         begin  b.sign := 0;   Exit;   end;
      b.sign := z.sign;    b.deg := 0;
      repeat
         remainder := 0;
         for  k := z.deg  downto  0  do
            begin
            dividend := base*remainder + z.coeff[k];
            quotient := dividend div 1000;
            z.coeff[k] := quotient;
            remainder := dividend - 1000*quotient;
            end;    { for  k }
         b.coeff[b.deg] := remainder;
         Inc(b.deg);
         while  (z.deg <> 0) and
                (z.coeff[z.deg] = 0)   do   Dec(z.deg);
      until  (z.deg = 0) and (z.coeff[z.deg] = 0);
      Dec(b.deg);
      end;    { superint2engineering }
```

```
begin    { full_print }
Write(out_file, var_name[n], ' =');
superint2engineering(hold[n], b);
if  b.sign = 0  then  Write(out_file, '    000')
else
    begin    { b <> 0 }
    col0 := 70;   col := 9;
    m := b.deg;
    if  b.sign = -1  then  Write(out_file, ' - ')
    else  Write(out_file, ' + ');
    for  j := m  downto  0  do
        begin
        if  col >= col0  then
            begin
            WriteLn(out_file);
            col := 6;
            Write(out_file, ' ' : col);
            end;
        k := b.coeff[j];
        if  k <= 9  then  Write(out_file, '00', k)
        else  if  k < 99  then
            Write(out_file, '0', k)
        else  Write(out_file, k);
        if j <> 0  then  Write(out_file, ',');
        Inc(col, 4);
        end;    { for  j }
    end;    { b <> 0 }
WriteLn(out_file);
end;    { full_print }

procedure  approximate_print(const  n: Word);

    var  x: Real;
         k: Word;

begin    { approximate_print }
Write(out_file, var_name[n] , ' ');
if  hold[n].sign = -1  then  Write(out_file, ' -')
else  Write(out_file, ' +');
x := 0.0;
for  k := hold[n].deg  downto  0  do
    begin
    x := base*x + hold[n].coeff[k];
    end;
WriteLn(out_file, x);
end;    { approximate_print }
```

```
procedure  initialize;

    var  j: Word;

    begin
    Assign(out_file, 'con');
    Rewrite(out_file);
    zero.deg := 0;   zero.sign := 0;
    for  j := 0  to  max_deg  do  zero.coeff[j] := 0;
    one := zero;
    one.sign := 1;
    one.coeff[0]  := 1;
    two := one;   two.coeff[0]  := 2;
    for  j := 0  to  5  do  hold[j] := one;
    for  j := 6  to  10  do  hold[j] := two;
    hold[1].coeff[0]  := 100;
    hold[2].coeff[0]  := 30;
    hold[3].coeff[0]  := 100;
    end;

end.   { unit  basics }
```

The program "mpa" has three main parts: input of superints from the keyboard, output, which we have taken care of already, and the engine that does the actual computations with superints. This engine requires very careful programming to get it correct, and is not at all easy. Here is its interface:

Program 8.3.2

```
unit  engine;   { 04/07/95 }

interface

uses  basics;

type  order = (less, equal, greater);

function   rel_size(const  x, y: superint): order;
procedure  add(const  x, y: superint;
                  var  z: superint);
procedure  subtract(const  x, y: superint;
                       var  z: superint);
procedure  multiply(const  x, y: superint;
                       var  z: superint);
procedure  divide(const  x, y: superint;
                     var  q, r: superint);
```

```
procedure   gcd(const   x, y: superint;
               var   z: superint);
procedure   lcm(const   x, y: superint;
               var   z: superint);
procedure   power(const   x, y: superint;
               var   z: superint);
procedure   square_root(const   x: superint;
                        var   z: superint);
procedure   cube_root(const   x: superint;
                      var   z: superint);
```

The need for "rel_size" will be clear soon; its implementation is by case analysis:

implementation

```
var   error_flag: Boolean;

function   rel_size(const   x, y: superint): order;

    var   j: Integer;

    begin
    if   x.deg > y.deg   then   rel_size := greater
    else   if   x.deg < y.deg   then   rel_size := less
    else   if   (x.sign = 0)   then
       begin
       if   y.sign = 0   then   rel_size := equal
       else   rel_size := less;
       end
    else   if   y.sign = 0   then   rel_size := greater
    else
       begin   { same degree, x <> 0, y <> 0 }
       j := x.deg;
       while   (x.coeff[j] = y.coeff[j])   and
               (j > 0)   do   Dec(j);
       if   x.coeff[j] > y.coeff[j]   then
          rel_size := greater
       else   if   x.coeff[j] < y.coeff[j]   then
          rel_size := less
       else   rel_size := equal;
       end;   { else}
    end;    { relative_size }
```

The four basic operations of arithmetic, "add"..."divide", are the heart of this unit. The algorithms here are all based on the classical treatment in Donald

Knuth's Art of Computer Programming, vol. 2, 2/e, Addison-Wesley, 1981, pp. 250 ff. We deal first with addition and subtraction. A moment's reflection on the addition and subtraction of signed numbers should convince you each may really be the other. For instance $9+(-5) = 9-5$ and $9-(-5) = 9+5$. So both addition and subtraction of signed numbers is based on addition and subtraction of unsigned numbers, not necessarily in that order, followed by a case analysis of signs to determine which unsigned operation applies. The implementation of the unsigned operations follows. You should work through some third and fourth grade addition and subtractions with pencil and tablet (not calculator), paying careful attention to the roles of the carry in addition and the borrow in subtraction. Note that "unsigned_add" ignores signs. Particularly in subtraction, the algorithm here is not quite what you did in grade school, and should be studied carefully.

```
procedure   unsigned_add(const   x, y: superint;
                         var    z: superint);

    procedure   ordered_add(const   x, y: superint;
                            var    z: superint);
        { deg(x) >= deg(y) > 0;   z = |x| + |y|,
          z.sign undetermined }

        var   k, c, s: Word;     { c = carry, s = sum }

        begin   { ordered_add }
        z := x;
        c := 0;     { carry }
        for   k := 0  to  y.deg  do
            begin   { Add up to the high term of y. }
            s := x.coeff[k] + y.coeff[k] + c;
            if  s >= base   then
                begin
                z.coeff[k] := s - base;
                c := 1;
                end
            else
                begin
                z.coeff[k] := s;
                c := 0;
                end;
            end;    { for  k }
        { Now push the carry forward until there
          is no further carry. }
        if  c = 1   then
            for  k := y.deg + 1  to  x.deg  do
```

```
                begin
                s := x.coeff[k] + 1;
                if  s = base  then   z.coeff[k] := 0
                { Continue the loop }
                else    { Finish up }
                   begin
                   z.coeff[k] := s;
                   z.deg := x.deg;
                   c := 0;
                   Break;
                   end;
                end;    { for  k }
            if  c = 1  then    { One final carry }
               begin
               if  x.deg = max_deg  then
                  error_flag := true
               else
                  begin
                  error_flag := false;
                  z.deg := x.deg + 1;
                  z.coeff[z.deg] := 1;
                  end;
               end;
            end;   { ordered_add }

    begin   { unsigned_add }
    if  x.sign = 0  then   z := y
    else  if  y.sign = 0  then   z := x
    else  if  x.deg >= y.deg  then   ordered_add(x, y, z)
    else  ordered_add(y, x, z);
    end;   { unsigned_add }


procedure  unsigned_subtract(const  x, y: superint;
                                var  z: superint);
    { Input:  |x| > |y|.  Output:   z = |x| - |y|. }

    var  k, b, d: Word;

    begin
    z := x;   z.sign := 1;
    if  y.sign = 0  then  Exit;
    { Subtract the digits of y, right to left.
      Always borrow; check if really needed. }
    b := base;   { borrow }
```

```
for  k := 0  to  y.deg  do
   begin
   d := (x.coeff[k] + b) - y.coeff[k];
   if  d >= base  then
      begin   { borrow not needed }
      z.coeff[k] := d - base;
      b := base;
      end
   else
      begin
      z.coeff[k] := d;
      b := base_1;
      end;
   end;   { for k }
{ Now adjust for the borrow, past the digits of y. }
 if  b = base_1  then
    for  k := y.deg + 1  to  x.deg  do
       begin
       if z.coeff[k] = 0  then   z.coeff[k] := base_1
       else
          begin
          Dec(z.coeff[k]);
          Break;
          end;
       end;
  while  (z.coeff[z.deg] = 0) and (z.deg > 0)  do
     Dec(z.deg);
  if  (z.deg = 0) and (z.coeff[0] = 0)  then
     z := zero;
  end;   { unsigned_subtract }
```

As noted, "add" and "subtract" do case analyses to determine which unsigned operation to apply.

```
procedure  add(const  x, y: superint;
                 var  z: superint);
   { Input:  x, y;  Output:  z = x + y. }

   begin
   if  x.sign = y.sign  then
      begin
      unsigned_add(x, y, z);
      z.sign := x.sign;
      end
   else
```

```
   begin     { x.sign = - y.sign }
   case   rel_size(x, y)   of
          less:   begin
                    unsigned_subtract(y, x, z);
                    z.sign := y.sign
                  end;
        equal:   z := zero;
      greater:   begin
                    unsigned_subtract(x, y, z);
                    z.sign := x.sign
                  end;
      end;    { case }
    end;    { x.sign = - y.sign }
  end;    { add }

procedure   subtract(const   x, y: superint;
                     var   z: superint);
  { Input:   x, y     Output:   z = x - y. }

  begin
  if   x.sign = y.sign   then     { Same signs }
     case   rel_size(x, y)   of
            less:   begin
                      unsigned_subtract(y, x, z);
                      z.sign := -1;
                    end;
          equal:   z := zero;
        greater:   begin
                      unsigned_subtract(x, y, z);
                      z.sign := 1;
                    end;
        end    { case }
  else
     begin
     unsigned_add(x, y, z);
     z.sign := x.sign;
     end
  end;    { subtract }
```

There is never overflow when we add two Words, each less than "base", our radix. Multiplication is another matter altogether. We use a Longint to hold the product, then split it into high and low Words. While "multiply" is exactly what you would expect: multiply each coefficient of the first factor by each coefficient of the second; accumulate the results according to degree, still, much care is needed to avoid overflow.

```
procedure  multiply(const  x, y: superint;
                       var   z: superint);
  { z := x*y. }

  var   j, k, yk, c: Word;
                  p: LongInt;

  begin    { multiply }
  z := zero;
  if  (x.sign = 0) or (y.sign = 0)  then   Exit;
  z.deg := x.deg + y.deg;    { May be one short. }
  if  z.deg > max_deg  then
     begin  error_flag := true;  Exit;  end;
  z.sign := x.sign*y.sign;
  for  k := 0  to  y.deg  do
     begin
     yk := y.coeff[k];    { Avoid repeated look-up. }
     if  yk <> 0  then
        begin
        c := 0;
        for  j := 0  to  x.deg  do
           begin    { Avoid  word*word   overflow }
           p := x.coeff[j];
           p := p*yk + z.coeff[j + k] + c;
           c := p div base;
           z.coeff[j + k] := p mod base;
           end;    { for j }
        z.coeff[x.deg + k + 1] := c;
        end;
     end;    { for k }
  if  c <> 0  then
     begin
     Inc(z.deg);
     error_flag := (z.deg > max_deg);
     end;
  end;    { multiply }
```

Now we come to division with remainder, and it is a couple of orders of magnitude more complex than the previous operations. The first problem is a good choice of a trial quotient. Let us set up the division of 144 by 29, as done in 4-th grade:

$$\begin{array}{r} 7 \\ 2\,9\overline{|1\,4\,4} \\ 2\,0\,3 \end{array} \qquad \begin{array}{r} 6 \\ 2\,9\overline{|1\,4\,4} \\ 1\,7\,4 \end{array} \qquad \begin{array}{r} 5 \\ 2\,9\overline{|1\,4\,4} \\ 1\,4\,5 \end{array} \qquad \begin{array}{r} 4 \\ 2\,9\overline{|1\,4\,4} \\ 1\,1\,6 \end{array}$$

The first trial is to divide 2 into 14; result 7: too large. Try 6. Still too large, ditto 5; finally 4 is correct.

A more complicated example is (a) below, which we rewrite as (b), so that we always test divide a single digit into a two-digit number.

$$\text{(a)} \quad 1\,7\,9\overline{|9\,7\,3\,7\,4} \qquad \text{(b)} \quad 1\,7\,9\overline{|0\,9\,7\,3\,7\,4}$$

The trial quotients 09 **div** 1 = 9, 8, 7, 6 are all too large; finally 5 works, with result (c). The next trial quotients

$$\text{(c)} \quad \begin{array}{r} 5 \\ 1\,7\,9\overline{|0\,9\,7\,3\,7\,4} \\ 8\,9\,5 \\ 0\,7\,8\,7 \end{array} \qquad \text{(d)} \quad \begin{array}{r} 5\,4 \\ 1\,7\,9\overline{|0\,9\,7\,3\,7\,4} \\ 8\,9\,5 \\ 0\,7\,8\,7 \\ 7\,1\,6 \\ 7\,1\,4 \end{array} \qquad \text{(e)} \quad \begin{array}{r} 5\,4\,3 \\ 1\,7\,9\overline{|0\,9\,7\,3\,7\,4} \\ 8\,9\,5 \\ 0\,7\,8\,7 \\ 7\,1\,6 \\ 7\,1\,4 \\ 5\,3\,7 \\ 1\,7\,7 \end{array}$$

are 07 **div** 1 = 7, 6, 5, until finally 4 works, with result (d). Then we work through trial quotients 07 **div** 1 = 7, 6, 5, 4, and finally 3, with final result (e). A lot of work is involved, and this when the base is only 10. Think how many trial divisions this method would involve for our 2^{15} base!

A preliminary normalization avoids most of this work. Given a division with remainder,

$$A = BQ + R \qquad 0 \le R < B$$

multiplication by a scale factor s yields

$$sA = (sB)Q + sR \qquad 0 \le sR < sB$$

so the quotient is unchanged, and the remainder is multiplied by the scale factor s. We choose s so that the divisor has the same number of digits as before, but its first digit is at least half of the radix. In our example with divisor $B = 179$, we choose $s = 5$ with result

$$sA = 5 \cdot 97374 = 486870 \qquad sB = 5 \cdot 179 = 895 \qquad 895\overline{)486870}$$

Now the first trial quotient is $6 = 48$ **div** 8, only one too large. Work out the whole long division in this case; the second trial quotient is the correct 4, and the third trial quotient is 4, only one too large.

We can formulate the problem of finding the trial quotient this way. At each step, we must divide a number B into a number A of *one more digit:* divide

$$A = a_{n+1}r^{n+1} + a_n r^n + \cdots a_0 \quad \text{by} \quad B = b_n r^n + \cdots + b_0$$

where r is the base, and where $A < Br$. (We start with this latter condition by the device of appending an extra 0 on the left if necessary.) We seek the true quotient Q and remainder R so that

$$A = BQ + R \qquad 0 \le R < B$$

This is the typical step of a long division. Of course there will be many more digits, and we note that if a is the next dividend digit to the right, then the next trial dividend A is

$$rR + a \le r(B-1) + (r-1) = rB - 1 < rB$$

so the condition $A < Br$ persists.

Theorem 1 Set (trial quotient)

$$Q' = \min\left[(a_{n+1}r + a_n) \text{ div } b_n, \ r - 1\right]$$

Then $Q \le Q'$.

(Proofs of this and subsequent theorems are postponed until the end of this section.) The next theorem shows the benefit of normalization:

Theorem 2 If $b_n \ge r$ **div** 2, then $Q' \le Q + 2$.

The next theorem shows how to choose a scale factor s so that sB has the same number of digits as B, but its first digit is at least r **div** 2.

Theorem 3 Set $s = r$ **div** $(b_n + 1)$. Then

$$
\begin{array}{lll}
(a) & 1 \leq s \leq r - 1 & \\
(b) & sB < r^{n+1} & sA < r^{n+3} \\
(c) & sB = c_n r^n + \cdots + c_0 & c_n \geq r \textbf{ div } 2
\end{array}
$$

Our strategy so far is to normalize according to Theorem 3 and choose a trial quotient Q' as in Theorem 1. Then by Theorems 1 and 2, we have $Q \leq Q' \leq Q + 2$. Of course we hope that $Q' = Q$ but there are always the possibilities $Q' = Q + 1$ or $Q' = Q + 2$. It is good that we have at worst two wrong guesses. To find Q we test $A \geq BQ'$; if so, then $Q' = Q$; if not, then we decrement Q' and test again. If this second test succeeds, then $Q' = Q$; otherwise $Q' = Q - 1$.

Actually, even these tests are too slow for us, and we can do a lot better by checking the next terms of the expansions.

Theorem 4 Define R' by $a_{n+1}r + a_n = b_n Q' + R'$.

$$
\begin{array}{lll}
(a) & \text{If} & R'r + a_{n-1} < b_{n-1}Q' \quad \text{then} \quad Q < Q' \\
(b) & \text{If} & R'r + a_{n-1} \geq b_{n-1}Q' \quad \text{then} \quad Q \geq Q' - 1
\end{array}
$$

Theorem 4 is applied as follows. We start with the trial quotient Q', which satisfies

$$
a_{n+1}r + a_n = b_n Q' + R' \qquad 0 \leq R' < b_n
$$

If the hypothesis of (a) holds, then Q' is too large (by at most 2), so we replace Q' by $Q' - 1$, adjust R' to $R' + b_n$, and try the test again. Doing this eliminates the possibility $Q' = Q + 2$. Knuth shows that the probability of the test in Theorem 4 failing and $Q' = Q + 1$ is small, only about $2/r$. Now let's put this theory to work:

```
procedure  divide(const  x, y: superint;
                    var   q, r: superint);
   { Input:   x,   y <> 0   with   x.deg < max.
     Output:   q = x div y;   r = x mod y.
     Thus    x = y*q + r    and   0 <= r < |y|. }

   var    k, s: Word;
          r_bar: superint;    { temporary r }

   procedure  unsigned_divide(const  x, y: superint;
                                var   q, r: superint);
      { Input:   x, and y > 0,   with   x.deg <= max.
        Output:   q = |x| div y;   r = |x| mod y. }
```

```
var            yp: superint;
      k, scaler,
        yn1, q_: Word;
            yn: LongInt;

procedure   normalize;

    var   k, c: Word;    { c = carry }
            p: LongInt;

    begin    { normalize }
    scaler := base div (y.coeff[y.deg] + 1);
    if   scaler = 1   then
    { y.coeff[y.deg] >= (base div 2) }
        begin
        r := x;
        r.coeff[r.deg + 1] := 0;
        yp := y;
        end
    else
        begin
        { r := scaler * x }
        c := 0;    { carry }
        r.deg := x.deg;
        for  k := 0  to  r.deg  do
            begin    { r := scaler*x }
            p := x.coeff[k];
            p := scaler*p + c;
            c := p div base;
            r.coeff[k] := p mod base;
            end;
        r.coeff[r.deg + 1] := c;
        { yp := scaler * y }
        yp.deg := y.deg;
        c := 0;
        for  k := 0  to  y.deg  do
            begin
            p := y.coeff[k];
            p := scaler*p + c;
            c := p div base;
            yp.coeff[k] := p mod base;
            end;
        end;    { scaler > 1 }
```

```
    yn := yp.coeff[yp.deg];
    { highest coeff of yp >= base/2 }
    if  yp.deg >= 1   then
       yn1 := yp.coeff[yp.deg - 1]
    else  yn1 := 0;
    { next highest coeff }
    end;    { normalize }

 procedure  find_trial_quot(const  k: Word);
    { quotient <= q_ <= quotient + 1. }

    var  d_, r_, t0, t1: LongInt;

    begin    { find_trial_quot }
    if  r.coeff[k + 1] = yn   then   q_ := base_1
    else
       begin
       { d_ := r.coeff[k + 1] * base + r.coeff[k]
             <= base_1 * base + base_1 < base^2 }
       d_ := r.coeff[k + 1];
       d_ := d_*base + r.coeff[k];
       { d_ <= (yn - 1)*base + (base - 1)
             = yn*base - 1 }
       q_ := d_ div yn;
       { q_ < base }
       end;
    { r_ := d_ mod yn }
    r_ := q_;
    r_ := d_ - r_*yn;
    { quotient <= q_ <= quotient + 2. }
    { Test }
    if  yp.deg <> 0   then
       begin
       { t0 := yn1 * q_ }
       t0 := yn1;
       t0 := t0 * q_;
       t1 := r_ * base + r.coeff[k - 1];
       while  t0 > t1   do   { At most twice }
          begin
          Dec(q_);
          t0 := t0 - yn1;
          t1 := t1 + yn*base;
          end;
       end;    { yp.deg <> 0 }
    { Now  true q <= q_ <= true q + 1.
       q_ = quotient with probablity 1 - 2/base. }
```

```
    end;    { find_trial_quot }

procedure  find_quotient(m: Word);

    var     j, k: Word;
        s, p, d,    { sum, product, difference }
            b, c: LongInt;    { borrow, carry }

    begin    { find_quotient }
    c := 0;    { Carry for * }
    b := base;    { Borrow for - }
    j := m;
    for  k := 0  to  yp.deg  do
        begin    { j is always k + m }
        { * with carry
            p := yp.coeff[k]*q_ + c }
        p := yp.coeff[k];
        p := p*q_ + c;
        c := p div base;
        p := p mod base;

        { - with borrow }
        d := (r.coeff[j] + b) - p;
        if  d >= base  then    { No borrow }
            begin
            r.coeff[j] := d - base;
            b := base;
            end
        else
            begin    { Borrow }
            r.coeff[j] := d;
            b := base_1;
            end;
        Inc(j);
        end;    { for k }
    d := r.coeff[j];
    d := (d - c) - (base - b);
    if  d >= 0  then
        begin
        q.coeff[m] := q_;
        r.coeff[j] := d;
        end
    else
        begin    { d < 0 }
        q.coeff[m] := q_ - 1;
        c := 0;
```

```
j := m;
for  k := 0  to  yp.deg do
   begin
   s := r.coeff[j];
   s := s + yp.coeff[k] + c;
   if  s >= base  then
      begin
      r.coeff[j] := s - base;
      c := 1;
      end
   else
      begin
      r.coeff[j] := s;
      c := 0;
      end;
   Inc(j);
   end;    { for k}
r.coeff[j] := r.coeff[j] + c;
end;    { d < 0 }
end;    { find_quotient }

procedure  denormalize;
  { Replace  r by r/scaler }

  var  k: Word;
       d: LongInt;

  begin    { denormalize }
  while  (r.deg <> 0) and
         (r.coeff[r.deg] = 0)  do  Dec(r.deg);
  if  r.sign <> 0  then
     begin
     d := r.coeff[r.deg];
     for  k := r.deg  downto  1  do
        begin
        r.coeff[k] := d div scaler;
        d := d mod scaler;
        d := r.coeff[k - 1] + base*d;
        end;
     r.coeff[0] := d div scaler;
     end;
  if  (r.deg <> 0) and
      (r.coeff[r.deg] = 0)  then  Dec(r.deg);
  end;    { denormalize }
```

```
begin    { unsigned_divide }
normalize;
q := zero;
for  k := r.deg  downto  yp.deg  do
   begin
   find_trial_quot(k);
   find_quotient(k - yp.deg);
   end;
{ Now find  q.deg }
k := r.deg - yp.deg;
while  (k <> 0) and
       (q.coeff[k] = 0)  do  Dec(k);
q.deg := k;
if  (k = 0) and
    (q.coeff[0] = 0)  then  q.sign := 0
else  q.sign := 1;
r.sign := 1;
k := yp.deg;
while  (k <> 0) and
       (r.coeff[k] = 0)  do  Dec(k);
r.deg := k;
if  (k = 0) and
    (r.coeff[0] = 0)  then  r.sign := 0
else  r.sign := 1;
{ Finally divide  r  by  scaler }
if  scaler <> 1  then  denormalize;
end;    { unsigned_divide }

begin    { divide }
if  x.sign = 0  then
   begin  q := zero;   r := zero;   Exit;   end;
if  y.deg = max_deg  then
   begin  q := zero;   r := x;   Exit;   end;
if  x.sign = 1  then
   begin
   unsigned_divide(x, y, q, r);
   if  q.sign <> 0  then  q.sign := y.sign;
   end
else    { x < 0 }
   begin
   unsigned_divide(x, y, q, r_bar);
   if  q.sign <> 0  then  q.sign := - y.sign;
   if  r_bar.sign <> 0  then
      begin
      k := 0;
      repeat
```

```
                s := q.coeff[k] + 1;
                if  s = base   then
                    begin  q.coeff[k] := 0;   Inc(k);   end
                else  q.coeff[k] := s;
            until  s <> base;
            if  k > q.deg  then  q.deg := k;
            if  y.sign = 1   then    { Fix r }
                unsigned_subtract(y, r_bar, r)
            else
                begin
                add(y, r_bar, r);
                r.sign := 1;
                end;
        end    { if  r_bar.sign <> 0 }
    else  r := zero;
    end;    { else }
  end;    { divide }
```

The most difficult arithmetic is behind us. The following local variables are used by the relatively easy greatest common divisor, least common multiple, and power routines that follow.

```
var  q, r, t: superint;

procedure  gcd(const  x, y: superint;
                 var   z: superint);

    begin
    z := x;   t := y;
    if  z.deg < t.deg   then
        begin
        q := z;   z := t;   t := q;
        end;
    { Now t.deg <= z.deg }
    z.sign := 1;   t.sign := 1;
    while  t.sign <> 0   do
        begin
        divide(z, t, q, r);   z := t;   t := r;
        end;
    end;    { gcd }

procedure  lcm(const  x, y: superint;
                 var   z: superint);
```

```
begin
gcd(x, y, z);
divide(x, z, q, r);
multiply(q, y, z);
end;   { lcm }

procedure power(const  x, y: superint;
                  var  z: superint);
{ z = x^y[0] }

var       i: Integer;
      n, k: Word;
    binary: array[0..127] of Byte;

begin
n := y.coeff[0];
if  n = 0  then  z := one
else  if  n = 1  then  z := x
else
    begin    { y <> 0, 1 }
    i := -1;
    while  n > 0  do
        begin
        Inc(i);
        binary[i] := n mod 2;
        n := n div 2;
        end;
    z := x;
    for  k := i - 1  downto  0  do
        begin
        multiply(z, z, t);
        if  binary[k] = 1  then  multiply(x, t, z)
        else  z := t;
        end;
    end;   { y <> 0, 1 }
end;   { power }
```

Approximations of square root and cube root of a superint remain to be programmed. They are both based on Newton's method (Ch. 9, Sect. 4). For the square root, Newton's method replaces an approximation r to $\sqrt{a}$ by $(r + a/r)/2$. (For $\sqrt[3]{a}$, the corresponding quantity is $(2r + a/r^2)/3$.) We can use real number approximations to get a good start. The procedure "square_root" is listed here; "cube_root" is left as an exercise.

```
procedure  square_root(const  x: superint;
                         var   z: superint);

    var  j, k, n: Word;
             m: LongInt;
             w: Real;

    begin
    if  x.sign = 0  then
        begin  z := zero;   Exit;   end;
    z.sign := 1;
    n := x.deg;
    m := x.coeff[n];
    if  n <= 1  then
        begin
        z.deg := 0;
        z.coeff[0] := Round(Sqrt(m));
        Exit;
        end;
    { Find first approximation }
    if  Odd(n)  then
        begin
        m := m*base + x.coeff[n - 1];
        end;
    n := n div 2;
    z.deg := n;
    w := Sqrt(m);
    if  n = 0  then
        begin
        z.coeff[0] := Round(w);
        Exit;
        end;
    j := Trunc(w);
    z.coeff[n] := j;
    { Refine first approximation }
    w := base*(w - j);
    j := Trunc(w);
    z.coeff[n - 1] := j;
    if  n >= 2  then
        begin
        w := base*(w - j);
        z.coeff[n - 2] := Round(w);
        end;
    n := 1;  k := 1;
    while  k < z.deg  do
        begin  k := 2*k;   Inc(n);   end;
```

```
for  j := 1  to  n  do
    begin
    divide(x, z, q, r);    { q = x div z }
    add(z, q, r);     { r = z + x div z }
    m := r.coeff[r.deg];
    { Next, z := r/2 }
    z.coeff[r.deg] := m div 2;
    for  k := r.deg - 1  downto  0  do
        begin
        m := base*(m mod 2) + r.coeff[k];
        z.coeff[k] := m div 2;
        end;   { for k }
    end;    { for j }
if  z.coeff[0] >= base div 2  then
    begin
    add(z, one, r);
    z := r;
    end;
end;   { square_root }

end.   { unit engine }
```

Proof of Theorem 1: First we put bounds on A and B. Just as any 3 digit decimal number is at most $1000 - 1$, so we can bound the tail of A:

$$A = a_{n+1}r^{n+1} + a_n r^n \cdots \leq a_{n+1}r^{n+1} + a_n r^n + (r^n - 1)$$

Obviously $B \geq b_n r^n$, so

$$Q = A \text{ div } B \leq \frac{a_{n+1}r^{n+1} + a_n r^n + r^n - 1}{b_n r^n}$$

$$= \frac{a_{n+1}r + a_n + 1 - r^{-n}}{b_n} < \frac{a_{n+1}r + a_n + 1}{b_n}$$

Hence

$$Qb_n < a_{n+1}r + a_n + 1 \qquad Qb_n \leq a_{n+1}r + a_n \qquad Q \leq (a_{n+1}r + a_n) \text{ div } b_n$$

because the terms are all integers. Therefore $Q \leq Q'$.

Proof of Theorem 2: By the definition of Q':

$$Q' = \min\left[(a_{n+1}r + a_n) \text{ div } b_n, \ r - 1\right]$$

so $Q' \leq r - 1$. Also recall that we are assuming $A < Br$. If $A \geq B(r - 1)$ then $Q = A \text{ div } B \geq r - 1 \geq Q'$, so $Q' = Q$ and Theorem 2 is certainly true. Thus we may assume $A < B(r-1)$. We bound the difference $Q' - Q$ in several steps:

$$Q' \leq (a_{n+1}r + a_n) \text{ div } b_n = \left(a_{n+1}r^{n+1} + a_n r^n\right) \text{ div } (b_n r^n)$$

$$\leq A \text{ div } (b_n r^n) \leq \frac{A}{b_n r^n}$$

$$Q = A \text{ div } B \leq \frac{A}{B} < Q + 1$$

$$Q' - Q - 1 < \frac{A}{b_n r^n} - \frac{A}{B} = \frac{A}{B}\left(\frac{B - b_n r^n}{b_n r^n}\right) < (r-1)\left(\frac{B - b_n r^n}{b_n r^n}\right)$$

Now $b_n \geq r \text{ div } 2 = r/2$ by hypothesis, and $B - b_n r^n < r^n - 1$ because the difference on the left is an expansion in lower powers of r. Finally,

$$Q' - Q - 1 < (r-1)\frac{r^n - 1}{(r/2)r^n} < 2 \qquad Q' - Q < 3 \qquad Q' \leq Q + 2$$

Proof of Theorem 3: (a) Clearly, $s = r \text{ div } (b_n + 1) \geq 1$ because $b_n \leq r - 1$. Also, $b_n \geq 1$, so $s \leq r \text{ div } 2 \leq r - 1$.
  (b) First we have

$$sB = s(b_n r^n + \cdots b_0) < s(b_n r^n + r^n)$$

$$= s(b_n + 1)r^n \leq r \cdot r^n = r^{n+1}$$

By (a) and $A < r^{n+2}$, we have $sA < (r-1)r^{n+2} < r^{n+3}$.
  (c) We have $r = s(b_n + 1) + t$ with $0 \leq t \leq b_n$ by the definition of s, so

$$r \text{ div } 2 = \frac{r}{2} = sb_n + \frac{s + t - sb_n}{2}$$

The numerator on the right is at most 1:

$$s + t - sb_n \leq s + b_n - sb_n = 1 - (s-1)(b_n - 1) \leq 1$$

so

$$r \text{ div } 2 \leq sb_n + \tfrac{1}{2} \qquad r \text{ div } 2 \leq sb_n$$

We seek the leading coefficient of sB, and we have

$$sB = (sb_n)r^n \cdots + (sb_0) = c_n r^n + \cdots + c_0$$

Because of possible carries, $c_n \geq sb_n \geq r \text{ div } 2$.

Proof of Theorem 4:   (a) First suppose that $b_{n-1}Q' > R'r + a_{n-1}$. Then

$$A = a_{n+1}r^{n+1} + \cdots < a_{n+1}r^{n+1} + a_n r^n + (a_{n-1} + 1)r^{n-1}$$
$$= \left(b_n Q' + R'\right)r^n + (a_{n-1} + 1)r^{n-1}$$
$$= b_n r^n Q' + \left(R'r + a_{n-1} + 1\right)r^{n-1}$$
$$\leq b_n r^n Q' + b_{n-1}Q'r^{n-1} = \left(b_n r^n + b_{n-1}r^{n-1}\right)Q' \leq BQ'$$

We have proved $A < BQ'$, so $Q = A$ **div** $B < Q'$.
   (b) Now suppose that $b_{n-1}Q' \leq R'r + a_{n-1}$. Then

$$A \geq (a_{n+1}r + a_n)r^n + a_{n-1}r^{n-1} = \left(b_n Q' + R'\right)r^n + a_{n-1}r^{n-1}$$
$$= \left(b_n r Q' + R'r + a_{n-1}\right)r^{n-1}$$
$$\geq \left(b_n r Q' + b_{n-1}Q'\right)r^{n-1} = Q'(b_n r + b_{n-1})r^{n-1}$$

But as usual, $B \leq b_n r^n + b_{n-1}r^{n-1} + r^{n-1}$, so we have proved $A \geq Q'\left(B - r^{n-1}\right)$.
But $Q' < r$, so $Q'r^{n-1} < r^n \leq B$, hence

$$A > Q'B - B = \left(Q' - 1\right)B \qquad Q = A \text{ **div** } B \geq Q' - 1$$

Exercises

1. Implement the procedure "cube_root".
2. Write a program that displays many Fibonacci numbers.
3. Write a program that computes $n!$ for large values of n.
4. Write a program that displays many decimal digits of π. Base it on the program pi.pas in Chapter 9, Section 2, which you may look at now.

Section 4.   MPA—Input

Warning The listings in this section omit many comments that are included in the disk files. What is more, a few sections of the listings will be omitted, even on disk, but suggested as exercises. In those cases, compiled units will be found on disk. (There are some copyright considerations behind this.)

We continue with the development of the program MPA. As you have seen from the exercises in the previous section, much can be done with the unit "engine", however, we would like to have a user-friendly program for interactive experimentation. This program should display the 10 superint variables q...z, and invite you to change any one of them by entering a formula in terms of the variables themselves, constant superints, and operations, for instance,

$$x = (y \ z) + 54345767676 \ x - s \text{ gcd } t$$

The unit "in_put" has one procedure in its interface, "get_in_string". Its purpose is to read a string from the keyboard, with the usual line-editing found in editors, and do some replacements in the string, such as change all letters to lower case, change "div" to "/", "mod" to "%", "gcd" to "&", etc. As you can see, the functions of two arguments are all treated as binary operations, to avoid the difficult problem of parsing expressions like "lcm(x, y)". (Implementation of "get_in_string" is left as an exercise.)

Program 8.4.1

```
unit   in_put;      { 05/13/95 }

interface

uses   CRT, basics, engine;

const   menu_fore = Blue;
        menu_back = LightGray;
            digits = ['0'..'9'];
            alphas = ['a'..'z'];
        alpha_nums = alphas + digits;
     F1 = #59;                      F10 = #68;
     Home = #71;                 EndKey = #79;
     UpArrow = #72;           DownArrow = #80;
     PgUp = #73;                   PgDn = #81;
     LeftArrow = #75;   RightArrow = #77;

type   Str220 = string[220];

var         in_string, token: Str220;
       exit_flag, accept_flag: Boolean;

procedure   get_in_string(const  k: Word);
```

The input k of "get_in_string" ranges from 0 to 9, and the output is "in_string", a string expression for the corresponding superint q..z. Once an input string is in memory, the next job is to parse it and evaluate the superint that the string represents. So our next unit is a parser. We will translate an input string into a binary tree, called a "parse_tree". You might review the program "alg_rpn" in Chapter 6, Section 2 for the ideas that will be used. Let us start with some of the declarations in "parser". Note that the only procedure in the interface section, "parse", takes a string for its input, like the example above, and produces its value, a superint. (There is a lot of debugging code included in the disk file, but omitted in the listing here.)

Program 8.4.2

```
unit   parser;    { 05/13/95 }

interface

uses   basics, engine, in_put;

procedure   parse(var   in_string: Str220;
                  var   z: superint);
```

The implementation section starts with some declarations of constants, types, and variables. The string will be split up into a sequence of *tokens*, which are listed in the array "in_token". The final token in the list, "con", denotes any constant superint, that is, a sequence of contiguous decimal digits.

```
implementation

type   operator = (plus, mnus, tims, xdiv,
                   xmod, xgcd, xlcm, powr,
                   xneg, xpos, xsqr, xcub,
                   xsrt, xcbr, xnop, pvar,
                   qvar, rvar, svar, tvar,
                   uvar, vvar, wvar, xvar,
                   yvar, zvar,       xcon);

const  in_token: array[operator] of string[3]
                 = ('+', '-', '*', '/',
                 { +     -     *    div }
                   '%',   '&',   '#', '^',
                 { mod   gcd    lcm    ^ }
                   '-',   '+',   '$', '~',
                 { -      +     sqr   cub }
                   '\',         '`',  'nop',
                 { sqrt    cbrt           }
                   'p', 'q', 'r', 's',
                   't', 'u', 'v', 'w',
                   'x', 'y', 'z', 'con');

type   parse_tree = ^parse_node;
       parse_node = record
                    coefficient: superint;
                    case  tag: operator  of
                       plus, mnus, tims, xdiv,
                       xmod, xgcd, xlcm, powr:
                          (left, right: parse_tree);
                       xneg, xpos, xsqr, xcub,
```

```
                           xsrt, xcbr,          xnop:
                              (arg: parse_tree);
                           pvar, qvar, rvar, svar,
                           tvar, uvar, vvar, wvar,
                           xvar, yvar, zvar, xcon: ();
                     end;   { case, record }
```

```
const      binop_ops = [plus..powr];
           unop_ops = [xneg..xcbr];
           mult_sym = ['#', '$', '%', '&', '*',
                      { lcm  sqr  mod  gcd   * }
                       '/', '\', '`', '~' ];
                      { div sqrt cbrt  cub }
         factor_lead = ['(', '0'..'9', '[',
                        'p'..'z', '{'];
             fences = ['(', '[', '{'];
         plus_minus = ['+', '-'];
```

```
var      temp_ptr: parse_tree;
         temp_op: operator;
   temp1, temp2: superint;
         h, j, k: Word;
```

Now we have two housekeeping operations: the first gets rid of an unneeded parse tree; the second makes a copy of a parse tree:

```
procedure  deposit(var  p: parse_tree);

  begin
  if  p <> nil   then
     with  p^   do
        begin
        if  tag in binop_ops   then
           begin
           deposit(left);
           deposit(right);
           end
        else  if  tag in unop_ops   then
           deposit(arg);
        Dispose(p);
        p := nil;
        end;
   end;    { deposit }
```

```
function  copy_ptr(const  p: parse_tree): parse_tree;
```

```
var   s: parse_tree;

begin
if  p = nil   then   copy_ptr := nil
else
    begin
    New(s);
    with  s^   do
        begin
        coefficient := p^.coefficient;
        tag := p^.tag;
        case  tag  of
            plus..powr:
                begin
                left := copy_ptr(p^.left);
                right := copy_ptr(p^.right);
                end;
            xneg..xnop:
                arg := copy_ptr(p^.arg);
            end;
        end;
    copy_ptr := s;
    end;
  end;     { copy_ptr }
```

"Decimal_to_superint" changes a string of decimal digits into a superint; it is inverse to "superint_to_engineering" in the implementation section of "basics".

```
type   decimal = array[0..219] of Byte;

var   temp_dec: decimal;

procedure   decimal_to_superint(degree: Word;
                                var   z: superint);

    var   q, r, d: LongInt;
                k: Word;
            abs_0: Boolean;
```

```
begin
abs_0 := true;
for  k := 0  to   degree  do
    begin
    abs_0 := abs_0 and (temp_dec[k] = 0);
    if  not abs_0  then   Break;
    end;
if  abs_0  then
    begin  z := zero;  Exit;  end;
z.sign := 1;   z.deg := 0;
repeat    { loop executed at least once }
    r := 0;
    for  k := degree  downto  0  do
        begin
        d := 10*r + temp_dec[k];
        q := d div base;    { < 10 }
        temp_dec[k] := q;
        r := d - base*q;
        end;    { for  k }
    z.coeff[z.deg] := r;
    Inc(z.deg);
    while  (degree <> 0) and
            (temp_dec[degree] = 0)  do  Dec(degree);
until  (degree = 0) and (temp_dec[degree] = 0);
Dec(z.deg);
end;    { decimal_to_superint }
```

Suppose we have parsed part of the input string, and a Word variable "index" marks the beginning of what remains to be parsed. The procedure "next_token" loads the string variable "token" (declared in the unit "in_put") and advances "index" so it is ready for the subsequent token.

```
procedure  next_token(var  index: Word);

    var  ch: Char;
         j: Integer;

    procedure  extend;

        begin
        token := token + ch;
        Inc(index);
        ch := in_string[index];
        end;
```

```
begin    { next_token }
token := '';   ch := in_string[index];
if   ch in alphas   then
    repeat   extend   until   not (ch in alpha_nums)
else   if   ch in digits   then
    while   ch in digits   do   extend
else   extend;    { not alphanumeric }
while   in_string[index]  = ' '   do   Inc(index);
end;    { next_token }
```

Now we are at the heart of the matter, the procedure "expression". It must realize the formal definition in the EBNF meta-language (*Extended Backus-Naur Formalism*):

```
Expression = Term { ("+" | "-") Term }
```

"Term" will be defined later. The definition means that an *expression* is a *term* followed by an arbitrary number (none included) of + or – *terms*. For instance $2x + y - 3z$ is an expression. The function "expression" will produce a parse tree, starting from a point index in the input string. We need more formal definitions:

```
Argument = { Constant ["*"] } Primitive
Factor   = Primitive [ "^" Factor ]
```

Thus an *argument* consists of a *primitive* preceded by an arbitrary number of *constant* factors, with optional multiplication signs. A *factor* is a *primitive*, possibly to a *factor* power.
    The most detailed in this list of recursive definitions is *primitive*:

```
Primitive = Constant | "p" | "q" | ... | "z" |
            "f1" | "f2" | ... | "f10" |
            Unop Argument |
            "(" Expression ")" |
            "[" Expression "]" |
            "{" Expression "}"
```

It remains to define *term*:

```
Term = Factor { [ "*" | "/" | "%" |
                  "&" | "#" ] Factor }
```

Thus a *term* is a sequence of (at least one) *factor*'s combined with multiplicative operators *, **div**, **mod**, gcd, lcm.

Here are details of the implementation, except that "primitive" and "argument" are omitted:

```
function  expression(var  in_string: Str220;
                     var  index: Word): parse_tree;

   var  qe: parse_tree;

   procedure  find_next_non_space;

      begin
      repeat
         Inc(index)
      until  in_string[index] <> ' '
      end;

   function  argument: parse_tree;  forward;

   function  factor: parse_tree;

      var  qf: parse_tree;

      function  primitive: parse_tree;

         begin    { primitive }
         end;     { primitive }

      begin    { factor }
      qf := primitive;
      if  (not parser_error) and
          (in_string[index] = '^')   then
         begin    { Parsing: A^B^C = A^(B^C) }
         temp_ptr := qf;
         New(qf);
         with  qf^   do
            begin
            tag := powr;   left := temp_ptr;
            find_next_non_space;
            right := factor;
            end;
         end;
      if  parser_error   then
         begin  deposit(qf);   factor := nil;   end
      else   factor := qf;
      end;    { factor }
```

```
function   term: parse_tree;

   var   qt: parse_tree;

   begin   { term }
   qt := factor;
   while   (in_string[index] in mult_sym) and
           (not parser_error)   do
      begin
      temp_ptr := qt;
      New(qt);
      with   qt^   do
         begin
         case   in_string[index]   of
             '*': tag := tims;
             '/': tag := xdiv;
             '%': tag := xmod;
             '&': tag := xgcd;
             '#': tag := xlcm;
            end;
          left := temp_ptr;
          find_next_non_space;
          right := factor;
          end;   { with qt^ }
      end;   { while }
   if  parser_error  then
      begin   deposit(qt);   term := nil;   end
   else   term := qt;
   end;   { term }

function   argument;

   begin   { argument }
   end;   { argument }

begin   { expression }
qe := term;
while   (in_string[index] in plus_minus) and
        (not parser_error)   do
   begin
   temp_ptr := qe;   New(qe);
   with   qe^   do
      begin
      if   in_string[index] = '+'   then
         tag := plus
      else   tag := mnus;
```

```
            left := temp_ptr;
            find_next_non_space;
            right := term;
            end;    { with }
       end;    { while }
   if parser_error then
       begin    deposit(qe);    expression := nil;    end
   else    expression := qe;
   end;    { expression }
```

"Expression" turns an input string into a parse tree. It remains to evaluate that parse tree to produce a superint. This is done by a procedure "parse_tree_value" which, in turn, depends on two subprocedures, "binop_value" and "unop_value". The former is left as an exercise.

```
procedure    binop_value(const   x, y: superint;
                         const   op: operator;
                         var   z: superint);

   begin
   end;    { binop_value }

procedure    unop_value(const   x: superint;
                        const   op: operator;
                        var   z: superint);
   begin
   case op of
      xneg:   begin
              temp2 := x;   temp2.sign := -x.sign;
              end;
      xpos:   temp2 := x;
      xsqr:   multiply(x, x, temp2);
      xcub:   begin
              multiply(x, x, temp1);
              multiply(x, temp1, temp2);
              end;
      xsrt:   square_root(x, temp2);
      xcbr:   cube_root(x, temp2);
      end;    { case }
   z := temp2;
   end;    { unop_value }

procedure    parse_tree_value(var   p: parse_tree;
                              var   z: superint);
```

```
begin
with  p^  do
   begin
   if  tag in binop_ops  then
      begin
      parse_tree_value(left, left^.coefficient);
      parse_tree_value(right, right^.coefficient);
      binop_value(left^.coefficient,
                   right^.coefficient,
                   tag, coefficient);
      end
   else  if  tag in unop_ops  then
      begin
      parse_tree_value(arg, arg^.coefficient);
      unop_value(arg^.coefficient,
                  tag, coefficient);
      end;
   z := coefficient;
   end;
end;    { parse_tree_value }
```

We now finish the only interfaced procedure, "parse", and with it, the unit "parser":

```
procedure  parse(var  in_string: Str220;
                  var  z: superint);

   var       p: parse_tree;
        index: Word;

   begin    { parse }
   index := 1;
   p := expression(in_string, index);
   if  parser_error  then  Exit;
   parse_tree_value(p, z);
   deposit(p);
   end;    { parse }

end.    { unit  parser }
```

Our program MPA still lacks a front end that presents information on the state of the saved variables, prompts you to choose which variable to change, and requests string input. That is the purpose of the unit "menus", which you might like enough to use for some of your own programs:

Program 8.4.3

```
unit   menus;     { 05/13/95 }

interface

uses   CRT, basics, engine, in_put;

var   main_row, main_col: Word;

procedure   menu(var   col, row: Word);

implementation

const   column_map: array[0..4] of Word
                        = (2, 16, 30, 44, 58);

const   titles: array[0..1, 0..4] of string[13]
        { 1234567890123       1234567890123 }
     = ( ('Change   q       ',   'Change   r       ',
          'Change   s       ',   'Change   t       ',
                                 'Change   u       '),
         ('Change   v       ',   'Change   w       ',
          'Change   x       ',   'Change   y       ',
                                 'Change   z       ') );
const   top_line = '';       { See the disk file }
        mid_line = '';       { for these strings. }
           track = '';
     bottom_line = '';

var   row_map: array[0..1] of Word;

procedure   full_window;

   begin
   Window(1, 1, 80, 25);
   end;

procedure   reset_flags;

   begin
   exit_flag := false;
   accept_flag := false;
   parser_error := false;
   end;     { reset_flage }
```

```
procedure  menu(var  col, row: Word);

  procedure  set_up;

    var  c, r, j: Word;

    begin
    j := 0;  r := 3;
    repeat
        row_map[j] := r;
        Inc(r, 2);
        Inc(j);
    until  j = 2;
    GotoXY(1, 1);
    TextColor(menu_back);
    TextBackground(menu_fore);

    WriteLn(
        'To Select: <Home> <Up><PgDn>.    ' +
        'To Accept: <Enter>.   To Exit: <Esc>.');

    TextColor(menu_fore);
    TextBackground(menu_back);
    WriteLn(top_line);
    r := 2;
    for  j := 0  to  1  do
        begin
        Inc(r);
         GotoXY(1, r);
         Write(track);
         for  c := 0  to  4  do
            begin
            GotoXY(column_map[c], r);
            Write(titles[j, c]);
            end;
         Inc(r);
         GotoXY(1, r);
         Write(mid_line);
        end;
    GotoXY(1, r);
    Write(bottom_line);
    TextColor(menu_back);
    TextBackground(menu_fore);
    GotoXY(column_map[col], row_map[row]);
    Write(titles[row, col]);
    end;    { set_up }
```

```pascal
procedure  choose;

  var  ch0: Char;   n: Word;

  begin     { choose }
  numlock_off;
  ch0 := ReadKey;
  if  ch0 = #27  then   exit_flag := true
  else  if  ch0 = #13  then   accept_flag := true
  else  if  UpCase(ch0) in ['Q'..'Z']   then
     begin
     n := Ord(UpCase(ch0)) - Ord('Q');
     row := n div 5;
     col := n mod 5;
     accept_flag := true;
     end
  else  if  ch0 = #0   then
     begin
     ch0 := ReadKey;
     TextColor(menu_fore);
     TextBackground(menu_back);
     GotoXY(column_map[col], row_map[row]);
     Write(titles[row, col]);
     TextColor(menu_fore);
     if  ch0 in [F1..F10]   then
        begin
        n := Ord(ch0) - Ord(F1);
        row := n div 5;
        col := n mod 5;
        accept_flag := true;
        end
     else
        case  ch0   of
           Home:          begin
                          if  col = 0  then   col := 4
                          else  Dec(col);
                          if  row = 0  then   row := 1
                          else  Dec(row);
                          end;
           UpArrow:       if  row = 0  then   row := 1
                          else  Dec(row);
           PgUp:          begin
                          Inc(col);
                          if  row = 0  then   row := 1
                          else  Dec(row);
                          end;
```

```
        LeftArrow:   if  col = 0  then   col := 4
                     else  Dec(col);
        RightArrow: Inc(col);
        EndKey:      begin
                     if  col = 0  then   col := 4
                     else  Dec(col);
                     Inc(row);
                     end;
        DownArrow:   Inc(row);
        PgDn:        begin
                     Inc(col);
                     Inc(row);
                     end;
      end;    { case }
    if  col = 5  then   col := 0;
    if  row = 2  then   row := 0;
    GotoXY(column_map[col], row_map[row]);
    TextColor(menu_back);
    TextBackground(menu_fore);
    Write(titles[row, col]);
    end;    { ch0 = #0 }
  end;    { choose }

begin    { menu }
reset_flags;
set_up;
repeat  choose;  until  accept_flag or exit_flag;
TextColor(text_fore);
TextBackground(text_back);
ClrScr;
full_window;
end;    { menu }

end.    { unit  menus }
```

Finally we complete the driver program "mpa" itself—pretty brief as all of its actions and computations are in the units it **uses**:

Program 8.4.4

```
{$M 50000 }   { Protected mode stack size }
program  mpa;   { 05/13/95 }

uses  CRT, basics, in_put, parser, menus;

var  k: Word;
```

```
procedure  introduction;

    begin
    end;    { introduction }

begin    { mpa }
ClrScr;
TextColor(text_fore);
TextBackground(text_back);
introduction;
initialize;
main_row := 0;
main_col := 0;
repeat
    ClrScr;
    GotoXY(1, 10);
    for  k := 1  to  10  do
        begin
        if  hold[k].deg <= 15  then  full_print(k)
        else  approximate_print(k);
        end;
    menu(main_col, main_row);
    if  exit_flag  then  Break;
    k := 5*main_row + main_col + 1;
    repeat
        get_in_string(k);
        if  exit_flag  then  Break;
        parser_error := false;
    until  not parser_error;
    if  not  exit_flag  then
        begin
        parse(in_string, hold[k]);
        end;
until  false;
TextColor(LightGray);
Close(out_file);
ClrScr;
end.    { mpa }
```

Exercises

1. Implement the procedure "get_in_string" of the unit "in_put:".
2. Implement the procedure "argument" of the unit "parser".
3. Implement the procedure "primitive" of the unit "parser".
4. Implement the procedure "binop_value" of the unit "parser".
5. Write a program to compute the Jacobi symbol of two odd superints. (Some number theory needed.)

Section 5. Polynomials

We begin this section with a unit on classical orthogonal polynomials. We need to recall their iterative definitions. The *Chebyshev polynomials* $T_n(x)$ are defined by

$$T_0(x) = 1 \qquad T_1(x) = x \qquad T_n(x) = 2xT_{n-1}(x) - T_{n-2}(x)$$

The *Legendre polynomials* $P_n(x)$ are defined by

$$P_0(x) = 1 \qquad P_1(x) = x \qquad P_n(x) = \frac{1}{n}[(2n-1)xP_{n-1}(x) - (n-1)P_{n-2}(x)]$$

The *Hermite polynomials* $H_n(x)$ are defined by

$$H_0(x) = 1 \qquad H_1(x) = 2x \qquad H_n(x) = \frac{2}{n}[xH_{n-1}(x) - (n-1)H_{n-2}(x)]$$

Last, the *Laguerre polynomials* $L_n(x)$ are defined by

$$L_0(x) = 1 \qquad L_1(x) = x \qquad L_n(x) = \frac{1}{n}[(2n-1-x)L_{n-1}(x) - (n-1)L_{n-2}(x)]$$

In the unit "orthopoly", we'll list the implementation for Chebyshev polynomials, and leave the others, which are quite similar, as exercises.

Program 8.5.1

```
unit  orthpoly;    { 04/27/95 }

interface

type  Real = Extended;

function  Chebyshev_Poly(deg: Word;   x: Real): Real;
function  Legendre_Poly(deg: Word;   x: Real): Real;
function  Hermite_Poly(deg: Word;   x: Real): Real;
function  Laguerre_Poly(deg: Word;   x: Real): Real;

implementation

function  Chebyshev_Poly(deg: Word;   x: Real): Real;

    var   Pn, Pn_1, Pn_2: Real;
                     k: Word;
```

```
begin
if  deg = 0  then  Chebyshev_Poly := 1.0
else
   begin
   Pn_1 := 1.0;  Pn := x;
   for  k := 2  to  deg  do
      begin
      Pn_2 := Pn_1;  Pn_1 := Pn;
      Pn := 2.0*x*Pn_1 - Pn_2;
      end;
   Chebyshev_Poly := Pn;
   end;
   end;
```

. . . .

end. { unit orthpoly }

Our next unit deals with operations on power series. Of course, we can only store truncated power series in a computer, and these are polynomials, so this material is in place in a section on polynomials. "Pow_ser" contains some very basic operations on (truncated) power series, such as composition of series and inversion of series. Let us look at its interface section.

Program 8.5.2

unit pow_ser;   { 05/03/95 }

interface

```
const  max_deg = 20;
           eta = 1.0e-20;

type          Real = Extended;
     power_series = array[0..max_deg] of Real;
           matrix = array[0..max_deg, 0..max_deg]
                                  of Real;

var  Zero, One: power_series;
             k: Word;
```

```
procedure   quotient(const   F, G:   power_series;
                     var   H: power_series);
procedure   real_power(const   F: power_series;
                     p: Real;
                     var   H: power_series);
procedure   composition(var   F, G, H: power_series);
procedure   exponentiate(const   F: power_series;
                     var   H: power_series);
```

We pause here to discuss the purpose of these first four procedures. The input of "quotient" is two series with constant term 1, and the output is $H(x) = F(x)/G(x)$. "Real_power" again assumes $F(0) = 1$ and produces $H(x) = F(x)^p$. "Composition" has assumptions on F and G and produces output H:

$$F(y) = y + a_2y^2 + \cdots \qquad G(x) = x + b_2x^2 + \cdots ; \qquad H(x) = F[G(x)]$$

"Exponentiate" has no assumptions, and produces output $H(x) = \exp[G(x)]$, a special case of "composition", but with a special, must faster, algorithm.

We continue the interface:

```
procedure   invert(const   F: power_series;
                     var   H: power_series);

procedure   compose_inverse(F: power_series;
                            const   G: power_series;
                            var   H: power_series);
procedure   reversion1(F, G: power_series;
                     var   C: matrix);
procedure   reversion2(F, G: power_series;
                     var   H: power_series);
procedure   nth_power(const   F: power_series;
                     n: Word;
                     var G: power_series);
```

"Invert" has an assumption on F and output the inverse function H of F:

$$x = F(y) = y + a_2y^2 + \cdots ; \qquad y = H(x)$$

"Compose_inverse" has assumptions on G and F and produces output H:

$$G(0) = 0 \qquad x = F(y) = y + a_2y^2 + \cdots ; \qquad H(x) = G[F^{-1}(x)]$$

where F^{-1} denotes the inverse function (not the reciprocal) of F.

Both of the "reversion of series" procedures solve the same problem, in different ways. The second is probably faster than the first. Both have assumptions on F and G and produce output H:

$$F(0) = 0 \qquad G(y) = y + b_2 y^2 \cdots ; \qquad y = H(x) = G^{-1}[F(x)]$$

That is exactly what "revision2" produces. Actually, "revision1" produces a matrix C with the coefficients of all powers of y:

$$y^j = \sum_{k=1}^{\text{maxdeg}} c_{jk} x^k$$

Finally, "nth_power" has an assumption on F and produces output

$$F(0) = 0 \qquad G(x) = F(x)^n$$

We now begin the implementation. "Quotient" is pretty straightforward. If $H = F/G$, then $F = GH$. When we expand this relation and collect like powers of x, we obtain iterative formulas for the coefficients of H. (Note in this and other procedures down the line, I have put in a rather crude "Halt" on error conditions. Of course, you will want to improve this.)

implementation

```
procedure  quotient(const  F, G:  power_series;
                     var   H: power_series);

    var  r, k: Word;
         sum: Real;

    begin
    if  (F[0] <> 1.0) or (G[0] <> 1.0)    then   Halt;
    H := One;
    for  r := 1  to  max_deg  do
        begin
        sum := F[r];
        for  k := 0  to  r - 1  do
            sum := sum - H[k]*G[r - k];
        H[r]  := sum;
        end;
    end;    { quotient }
```

The implementation of "real_power" introduces a differential calculus technique that is used a great deal today in algorithms, particularly in the "automatic differentiation" literature. To my knowledge, it was first introduced by H E Fettis, CACM 134, 1962. We'll see other applications of the method shortly.

Suppose

$$F(x) = 1 + \sum_{1}^{\infty} a_j x^j \quad \text{and} \quad H(x) = F(x)^p = 1 + \sum_{1}^{\infty} b_k x^k$$

Differentiate, multiply by F, and expand:

$$H' = pF^{p-1}F' \qquad FH' = pHF'$$

$$\left(1 + \sum a_j x^j\right)\left(\sum k b_k x^{k-1}\right) = p\left(\sum j a_j x^{j-1}\right)\left(1 + \sum b_k x^k\right)$$

Carry out the multiplications and equate like powers of x:

$$\sum_{j+k=r} (pj - k)a_j b_k = 0$$

This provides iterative formulas for $b_1, b_2, \cdots$, now realized. Note that the algorithm is $O(\text{max\_deg}^2)$.

```
procedure  real_power (const  F: power_series;
                              p: Real;
                       var    H: power_series);
   var   k, r: Word;
         sum: Real;

   begin
   if F[0] <> 1.0  then   Halt;
   H := One;
   if  p = 0.0  then   Exit;
   for  r := 1  to  max_deg  do
      begin
      sum := 0.0;
      for  k := 1  to  r - 1  do
         sum := sum +  (p*(r - k) - k)
                           *F[r - k]*H[k];
      H[r] := p*F[r] + sum/r;
      end;
   end;    { real_power }
```

If the exponent p is a Word in "real_power", then $F(0)$ can be 0. This special situation is implemented in "nth_power" below. But "nth_power" will be used in the implementation of "composition", whose idea is quite simple:

$$F(y) = \sum_{1}^{\infty} a_k y^k \qquad H(x) = F[G(x)] = \sum_{1}^{\infty} a_k G(x)^k$$

Only as needed, powers of G are computed by "nth_power", multiplied by the corresponding coefficient from F, and added.

```
procedure   composition(var   F, G, H: power_series);

       var   j, k: Integer;
             P: power_series;

       begin    { composition }
       if   (F[0] <> 0.0) or (G[0] <> 0.0) or
            (F[1] <> 1.0) or (G[1] <> 1.0)   then   Halt;
       for   k := 1   to   max_deg   do   H[k] := F[1]*G[k];
       for   k := 2   to   max_deg   do
          if   F[k] <> 0.0   then
             begin
             nth_power(G, k, P);
             for   j := k   to   max_deg   do
                H[j] := H[j] + F[k]*P[j];
             end;
       end;    { composition }
```

"Exponentiate" is a special case of "compose", but computed much faster directly by the differentiation device used in "real_power":

$$H(x) = e^{F(x)} \qquad H' = F'H$$

$$\sum rb_r x^{r-1} = \sum ja_j x^{j-1} \sum b_k x^k \qquad rb_r = \sum_{j+k=r} ja_j b_k$$

```
procedure   exponentiate(const   F: power_series;
                              var   H: power_series);

       var   j, r, deg_F: Word;
             sum: Real;

       begin
       H := One;
       deg_F := 0;
       while   (deg_F <= max_deg) and
               ( Abs(F[deg_F]) < eta )   do   Inc(deg_F);
       if   deg_F = 0   then   H[0] := Exp(F[0]);
       for   r := deg_F   to   max_deg   do
          begin
          sum := 0.0;
          for   j := deg_F   to   r   do
             sum := sum + j*F[j]*H[r - j];
```

```
        H[r]  := sum/r;
        end;
  end;    { exponentiate }
```

"Invert" uses another clever algorithm of H E Fettis. We seek the inverse power series $H(x)$, where

$$x = F(y) = y + \sum_2^\infty a_k y^k \qquad y = H(x) = \sum_2^\infty b_m x^m$$

We build up H "in place" as a sequence of polynomials of increasing degree, starting with $H_1(x) = x$. We propose to find $H_m(x)$ of degree m such that $H_m(x) \equiv y \pmod{x^{m+1}}$, which certainly holds for $m = 1$. If we have found H_{m-1}, then for $k \geq 2$,

$$H_{m-1}(x) = y + x^m L(x)$$
$$H_{m-1}(x)^k = y^k + ky^{k-1}x^m L(x) + \cdots \equiv y^k \pmod{x^{m+1}}$$

because $y \equiv 0 \pmod{x}$. We set

$$H_m(x) = x - \sum_{k=2}^m a_k H_{m-1}(x)^k$$

Then

$$H_m(x) \equiv x - \sum_{k=2}^m a_k y^k \equiv y \pmod{x^{m+1}}$$

as desired.

The main feature of "invert" is the computation of the polynomial in $H = H_{m-1}$ to get the next H. It is done by the Horner expansion starting with 0:

$$H := ((\cdots(((-a_m)H - a_{m-1})H - a_{m-2})H \cdots - a_2)H)H + x$$

Let P represent the accumulated polynomial up to some closing ")". Then at the next step, first P:=P-a, then P:=P*H (mod x^(m+1)). The subprocedure "truncated_product" upgrades P to PH. The outer **for** loop goes backwards, so no upgraded coefficient of P enters a subsequent computation.

```
    procedure  invert(const  F: power_series;
                      var  H: power_series);

    var      P: power_series;
         k, m: Word;
```

```
procedure  truncated_product
                (var   P: power_series;   m: Word);

    var   j, k: Word;
          sum: Real;

    begin
    for  j := m   downto   1  do
       begin
       sum := 0;
       for  k := 1  to  j  do
          sum := sum + P[j - k]*H[k];
       P[j]  := sum;
       end;
    end;    { truncated_product }

begin    { invert }
if   (F[0] <> 0.0) or (F[1] <> 1.0)   then   Halt;

H := Zero;
for  m := 2  to  max_deg  do
   begin
   P := Zero;
   for  k := m  downto  2  do
      begin
      P[0] := -F[k];
      truncated_product(P, m);
      end;
   truncated_product(P, m);
   P[1] := P[1] + 1.0;
   H := P;
   end;    { for  m }
end;    { invert }
```

The program for "compose_inverse" is an algorithm from A Nijenhuis and H S Wilf, Combinatorial Algorithms 2/e, Academic Press, 1978, p. 190. The algorithm is rather deep, and I shall only sketch its ideas. The problem is to find the expansion of $H(x)$, where

$$G(y) = \sum_{1}^{\infty} b_k y^k \qquad x = F(y) = y + \sum_{2}^{\infty} a_j y^j \qquad H(x) = \left(G \circ F^{-1}\right)(x)$$

Now $H = G \circ F^{-1}$ implies $G = H \circ F$. We shall construct two sequences of (truncated) series, F_k and G_k, satisfying $G_k = H \circ F_k$, $F_k \equiv x \pmod{x^{k+1}}$.

Initially, $F_1 = F$ and $G_1 = G$. If we have reached the k-th level, then the construction at the next level is

$$F_k(y) = y + ay^k + \cdots \qquad P(x) = x + ax^k \qquad G_{k+1} = G_k \circ P^{-1} \qquad F_{k+1} = F_k \circ P^{-1}$$

When we reach m = max_deg, then G_m will be the required H. We actually build the G_k in place in the memory occupied by the variable H. Note that F is a value parameter because we modify it in place (without altering its original).

```
procedure   compose_inverse(F: power_series;
                                const   G: power_series;
                                var    H: power_series);

   var   r, j, k: Word;
               t: Real;

   begin
   if   (F[0] <> 0.0) or (F[1] <> 1.0) or
        (G[0] <> 0.0)    then   Halt;
   H := G;
   for  k := 2  to  max_deg  do
        begin
        t := F[k];
        for  j := 0  to  max_deg - k  do
            for  r := j + k  to  max_deg  do
                begin
                F[r] := F[r] - t*F[r - k + 1];
                H[r] := H[r] - t*H[r - k + 1];
                end;
        end;    { for k }
   end;    { compose_inverse }
```

"Reversion of series" is an old problem in analysis. We are starting with

$$F(x) = \sum_1^\infty a_j x^j \qquad G(y) = y + \sum_2^\infty b_k y^k \qquad G(y) = F(x)$$

with solution

$$
\begin{bmatrix} y \\ y^2 \\ \cdot \\ \cdot \\ \cdot \\ y^m \end{bmatrix}
= C
\begin{bmatrix} x \\ x^2 \\ \cdot \\ \cdot \\ \cdot \\ x^m \end{bmatrix}
$$

The relations $y^{j+1} = y \cdot y^j$ and $G(y) = F(x)$ imply respectively

$$c_{j+1, i} = \sum_{k=1}^{i-j} c_{1k} c_{j, i-k} \qquad a_j = \sum_{k=1}^{m} b_k c_{kj}$$

These iterations are exactly what we need for an algorithm, due to H C Thatcher, Jr, CACM 273, 1965.

```
procedure   reversion1(F, G: power_series;
                    var   C: matrix);

   var   i, j, k: Word;   sum: Real;

   begin
   for   i := 1   to   max_deg   do
      begin
      for   j := i - 1   downto   1   do
         begin   sum := 0;
         for   k := 1   to   i - j   do
            sum := sum + C[1, k]*C[j, i - k];
         C[j + 1, i]   := sum;
         end;   { for   j }
      sum := F[i];
      for   j := 2   to   i   do
         sum := sum - G[j]*C[j, i];
      C[1, i]   := sum;
      end   { for   i }
   end;   { reversion1 }
```

The following algorithm for "reversion2", due to H E Fettis, CACM 193, 1962, is like the algorithm above for "compose_inverse", but easier. We define two sequences F_n and G_n of series, satisfying

$$G_n(y) = F_n(x) \qquad G_n(y) \equiv y \pmod{y^{n+1}}$$

The iterative definitions are simply

$$G_1 = G \quad F_1 = F \qquad G_{n+1} = G_n - b(G_n)^{n+1} \qquad F_{n+1} = F_n - b(F_n)^{n+1}$$

where $G_n = y + by^{n+1} + \cdots$. At the point where we truncate, $G_m = y$, so $F_m = H$.

```
procedure   reversion2(F, G: power_series;
                    var   H: power_series);
```

```
var   j, n: Integer;
         b: Real;
      D, E: power_series;

begin    { reversion2 }
for   n := 2  to   max_deg   do
     begin
     b := G[n];    { previous G }
     nth_power(F, n, D);
     nth_power(G, n, E);
     for  j :=   2  to   max_deg   do
        begin
        F[j] := F[j] - b*D[j];
        G[j] := G[j] - b*E[j];
        end;
     end;
H := F;
end;    { reversion2 }
```

The final procedure, for n-th powers of a series with no constant term is again based on the differentiation idea:

$$G = F^n \qquad G' = nF^{n-1}F' \qquad FG' = nF'G$$

The latter relation, expanded, gives the iteration. It is necessary to be careful with the ranges of the various indices.

```
procedure   nth_power(const   F: power_series;
                         n: Word;
                         var G: power_series);

var   deg_F, deg_G,
         j, k, r: Integer;
       p, sum, t: Real;

begin
if   F[0] <> 0.0   then   Exit;
G := Zero;
deg_F := 0;
repeat
   Inc(deg_F)
until   (Abs(F[deg_F]) >= eta) or
        (deg_F = max_deg);
deg_G := deg_F*n;
if   (F[deg_F] = 0.0) or
     (deg_G > max_deg)   then   Exit;
```

```
        p := 1.0;   t := F[deg_F];
        for  j := 1  to  n  do  p := p*t;     { p = t^n }
        G[deg_G] := p;

        for  r := deg_G + 1  to  max_deg  do
            begin
            sum := 0.0;   k := r;
            for  j := deg_F + 1
                    to  r + deg_F - deg_G  do
                begin
                Dec(k);
                sum := sum + (k - n*j)*F[j]*G[k];
                end;
            G[r] := - sum/( (r - n*deg_F)*F[deg_F] );
            end;    { for  k }
        end;     { nth_power }

begin
for   k := 0  to  max_deg  do  Zero[k] := 0.0;
One := Zero;
One[0] := 1.0;
end.    { unit  pow_ser }
```

We have completed a long unit with some remarkable algorithms. We complete this section with "realpoly", a unit of a rather different class of algorithms dealing with polynomials. These algorithms come from the book of Nijenhuis and Wilf cited above, pp. 171—177. We start as usual with the interface.

Program 8.5.3

```
unit  realpoly;    { 04/28/95 }

interface

const  max_deg = 100;

type         Real = Extended;
        polynomial = record
                            deg: Word;
                         coeff: array[0..max_deg]
                                 of Real;
                         end;
                nodes = array[1..max_deg] of Real;
```

```
function   value(const   F: polynomial;
                  const   a: nodes;   x: Real): Real;
procedure  replace(var   F:  polynomial;
                   var   a: nodes;   b: Real);

procedure  expand1(var   F:  polynomial;
                   var   a: nodes;   b: Real);
procedure  expand2(var   F:  polynomial;
                   var   a: nodes;   b: Real);
procedure  factorial(var   F:  polynomial);
procedure  inverse_factorial(var   F:  polynomial);
```

In this unit, "polynomial" will have several possibilities in addition to the standard one. Given a sequence a_1, a_2, $\cdots$ of *nodes*, we can form

$$F(x) = c_0 + c_1(x - a_1) + c_2(x - a_1)(x - a_2) + \cdots$$
$$+ c_n(x - a_1)(x - a_2) \cdots (x - a_n)$$

with the special case of expansion into *factorial polynomials*

$$F(x) = c_0 + c_1 x + c_2 x(x - 1) + \cdots + c_n x(x - 1) \cdots (x - n + 1)$$

I'll explain what each procedure does as we implement it.

The first is "value"; it is simply the Horner evaluation of the polynomial formed with nodes, as displayed above.

implementation

```
function   value(const   F: polynomial;
                  const   a: nodes;   x: Real): Real;

   var   t: Real;   j: Word;

   begin
   t := F.coeff[F.deg];
   for   j := F.deg   downto   1   do
      t := F.coeff[j - 1] + (x - a[j])*t;
   value := t;
   end;   { value }
```

The purpose of "replace" is to replace the nodes $a_1, \cdots, a_n$ by $b, a_1, \cdots, a_{n-1}$, where b is a new node, then compute $F(x)$ expanded into the new node set, in place.

```
procedure  replace(var   F:  polynomial;
                   var   a: nodes;   b: Real);
```

```
var   j: Word;

begin
for   j := F.deg - 1   downto   1   do
   begin
   F.coeff[j] := F.coeff[j]
                         + (b - a[j + 1])*F.coeff[j + 1];
   a[j + 1] := a[j];
   end;
F.coeff[0] := F.coeff[0] + (b - a[1])*F.coeff[1];
a[1] := b;
end;    { replace }
```

Both "expand1" and "expand2" start with a node expansion of a polynomial with the node set a, and express the polynomial in powers of $x - b$. The second is probably faster.

```
procedure   expand1(var   F:   polynomial;
                    var   a:  nodes;   b: Real);

   var   j: Word;

   begin
   for   j := 1   to   F.deg   do   replace(F, a, b);
   end;

procedure   expand2(var   F:   polynomial;
                    var   a:  nodes;   b: Real);

   var   j, k: Word;
         t: Real;

   begin
   for   j :=   F.deg - 1   downto   0   do
      begin
      t :=   b - a[j + 1];
      for   k := j   to   F.deg - 1   do
         F.coeff[k]   := F.coeff[k] + t*F.coeff[k + 1];
      a[j + 1] := b;
      end;
   end;    { expand2 }
```

"Factorial" starts with a polynomial in powers of x, and expresses it as a factorial polynomial, coefficients replaced in place. "Inverse_factorial" is the inverse operation; note how the code for each is the inverse of the code for the other. Having both makes tests easy.

```
procedure  factorial(var  F:  polynomial);

  var  j, k: Word;

  begin
  for  j := 1  to  F.deg - 1  do
      for  k := F.deg - 1  downto  j  do
          F.coeff[k] := F.coeff[k] + j*F.coeff[k + 1];
  end;    { factorial }

procedure  inverse_factorial(var  F:  polynomial);

  var  j, k: Word;

  begin
  for  j := F.deg - 1  downto  1  do
      for  k := j  to  F.deg - 1  do
          F.coeff[k] := F.coeff[k] - j*F.coeff[k + 1];
  end;    { inverse_factorial }

end.    { real_poly }
```

Exercises

1. Complete the implementation of the unit "orthpoly".
2. Write a program to test all the orthogonal polynomials in "orthpoly".
3. Write a program to test all the procedures of "pow_ser".
4. Write a program to test the procedures of the unit "realpoly".

The implementations of the various procedures in the unit "pow_ser" are hopefully clear, but they are not as efficient as they could be. Please pay particular attention to the inner loops, where small improvements greatly speed up execution. No computation should be performed twice; "inc" and "dec" are faster than + and −, etc. These things can be fixed by introducing a few local variables. The remaining exercises suggest some inner loops ripe for improvement.

5. "Quotient" has an inner loop subtraction.
6. The "real_power" inner loop computes $k - j$ twice. Can its multiplication by p be replaced by a (much faster) addition?
7. The inner loop in "composition" multiplies by F[k] many times with fixed k. Computing the offset of an array element takes needless time.
8. "Exponentiate" has an inner loop subtraction.
9. "Truncated_ product" (in "invert") has an inner loop subtraction.
10. The inner loop in "compose_inverse" computes $r - k + 1$ twice.
11. The inner loop of "reversion1" has a subtraction.

12. The integer expression computed in the inner loop of "nth_power" has a multiplication and a subtraction.

Section 6. Integer Matrices

In this section we deal with matrices of integers, and we cover two topics: reduction to row-echelon form, and reduction to invariant factor (Smith normal) form. Let us look at the interface section of the unit "int_matr":

Program 8.6.1

```
unit   int_matr;    { Integer Matrices   04/26/95 }

interface

const   max_m = 20;   max_n = max_m + 1;

type   matrix = array[1..max_m, 1..max_n] of Integer;

var   A, B: matrix;

procedure   get_matrix(var   A: matrix);
procedure   print_matrix(const   A: matrix;
                         name: Char);
procedure   echelon_form(const   A: matrix;
                         var   B: matrix);
procedure   invariant_factors(const   A: matrix;
                              var   rank: Word;
                              var   C: matrix);
```

Echelon Form

We'll omit listing "get_matrix" and "print_matrix" here as we have seen enough of such procedures in previous chapters. An integer matrix can be changed to row echelon form by a sequence of elementary row transformations. These are (a) change the signs of all elements of a row, (b) interchange two rows, and (c) add an integer multiple of one row to another row. It is shown in linear algebra texts that an $m \times n$ matrix B is obtained from a matrix A by a sequence of elementary row transformations if and only if $B = PA$, where P is an $m \times m$ integer matrix of determinant ± 1. The determinant condition is the same as P having an integer matrix inverse. A nonsingular square integer matrix whose inverse is an integer matrix is called a *unimodular matrix*.

An $m \times n$ matrix A is in *row echelon form* provided

$$m = 1 \quad \text{or} \quad A = \begin{bmatrix} B & C \\ 0 & D \end{bmatrix}$$

where the blocks B and D are in row-echelon form (inductive definition). For example,

$$A = \begin{bmatrix} 3 & -4 & 1 & 7 & -8 & 2 \\ 0 & 0 & 5 & -5 & 0 & 3 \\ 0 & 0 & 0 & -2 & 1 & -7 \\ 0 & 0 & 0 & 0 & 0 & 2 \end{bmatrix}$$

The reduction of a matrix to row echelon form is often called *Gaussian elimination* when it is applied to solving systems of equations. We'll go into that more in the next section. But with that application in mind, we shall allow still another elementary row transformation: divide out a common factor of the elements of a row, but keep track of the product "det" of these common factors.

Let us examine the implementation of "echelon_form". It is an iterative procedure; it assumes that the upper block $[B \; C]$ has been already processed, and it remains to process the lower right block D, whose first row and column indices are "row0" and "col0". We shall refer to this block as "the submatrix". Signs are changed as necessary, so all of the first column elements of the submatrix are nonnegative. Subprocedure "find_pivot" locates the smallest element of this column, and "transpose_rows" moves it to the first row. Then "subtract" subtracts multiples of this first row of the submatrix from its other rows, to reduce their first elements to numbers smaller than the "pivot" element of the upper left corner. Whenever a resulting first element is nonzero, the gcd of its row is factored out by "gcd". "Fix_column" orchestrates this sequence of events, repeating it until all of the first column of the submatrix is 0 after its first row. Then the process moves down a row, moves right, and starts over again.

implementation

```
var    det: Integer;
       m, n: Word;

procedure    echelon_form(const   A: matrix;
                          var    B: matrix);
    var    row0, col0,
           row, col,
               pivot: Word;

    procedure    find_pivot;

        var    row, min, h: Word;
```

```
begin
min := 0;   row := row0;   pivot := 0;
while  (min = 0) and (row <= m)   do
   begin
   h := B[row, col0];
   if  h  = 0   then   Inc(row)
   else
      begin   min := h;   pivot := row;   end;
   end;
if  min > 0   then
   begin
   for  row := pivot + 1   to   m   do
      begin
      h := B[row, col0];
      if  (h < min) and (h > 0)   then
         begin
         min := h;   pivot := row;
         end;
      end;   { for   row }
   end;   { if   min }
end;   { find_pivot }

procedure  transpose_rows(pivot: Integer);

var  col: Word;   h: Integer;

begin
det := -det;
for  col := col0   to   n   do
   begin
   h := B[row0, col];
   B[row0, col] := B[pivot, col];
   B[pivot, col] := h;
   end
end;

procedure  gcd(row: Word);

var       h, q,
      col, min: Word;
            v: array[1..max_n] of Integer;
```

```
begin
for  col := col0  to  n  do
   v[col] := Abs(B[row, col]);
min := v[col0];  q := col0;
for  col := col0 + 1  to  n  do
   begin
   h := v[col];
   if  (h > 0) and (h < min)  then
      begin  q := col;  min := h  end;
   end;
if  q > col0  then
   begin
   v[q] := v[col0];  v[col0] := min;
   end;
col := col0 + 1;
while  col <= n  do
   begin
   if  v[col] > 0  then
      begin
      v[col] := v[col] mod min;
      if  v[col] > 0  then
         begin
         min := v[col];  v[col] := v[col0];
         v[col0] := min;  col := col0;
         end;
      end;
   Inc(col);
   end;
h := v[col0];
det := det*h;
for  col := col0  to  n  do
   B[row, col] := B[row, col] div h;
end;    { gcd }

procedure  subtract(pivot: Integer);

   var  k:  Word;
        q: Integer;

   begin
   q := B[pivot, col0] div B[row0, col0];
   for  k := col0  to  n  do
      B[pivot, k] := B[pivot, k] - q*B[row0, k];
   end;
```

```
procedure  fix_column;

   var   row: Word;

   begin
   row := row0 + 1;
   while  row <= m  do
      begin
      if  B[row, col0] > 0   then
         begin
         subtract(row);
         if  B[row, col0] > 0   then
            begin
            transpose_rows(row);
            gcd(row0);   row := row0;
            end;
         end;    { if  B }
      Inc(row);
      end;     { while }
   end;     { fix_column }

begin    { echelon_form }
det := 1;   row0 := 1;   col0 := 1;
B := A;
repeat
   for  row := row0  to  m  do
      begin
      if  B[row, col0] < 0   then
         begin    { Change sign of row row }
         det := -det;
         for  col := col0  to  n  do
            B[row, col] := -B[row, col];
         end;
      if  B[row, col0] > 0   then  gcd(row);
      end;    { for  row }
   find_pivot;
   if  pivot = 0   then   Inc(col0)
   else
      begin
      if  pivot > row0   then
         transpose_rows(pivot);
      if  row0 < m   then   fix_column;
      Inc(row0);
      end
until  (row0 > m) or (col0 > n);
end;    { echelon_form }
```

Invariant Factors

An $m \times n$ integer matrix A is in *invariant factor form* or *Smith normal form* provided all elements off the main diagonal are 0 and each (positive) element on the main diagonal divides the next element, with zeros all at the end. For example

$$A = [B \; 0_{8,n}] \qquad B = \text{diag}\{1 \; 3 \; 15 \; 30 \; 300 \; 0 \; 0 \; 0\}$$

The main result in this subject is that if A is an $m \times n$ integer matrix, then there exist $m \times m$ and $n \times n$ unimodular matrices P and Q such that PAQ is in invariant factor normal form. The left multiplication by P can be effected by a sequence of elementary row transformations, and the right multiplication by Q can be effected by a sequence of elementary column transformations: (a) change the signs of all elements of a column, (b) interchange two columns, and (c) add an integer multiple of one column to another.

Procedure "invariant_factors" has two parts. First, A is changed to diagonal form by a sequence of elementary row and column transformations. This part is very similar to the reduction to row_echelon form. We assume that A has been reduced to the form

$$\begin{bmatrix} D & 0 \\ 0 & E \end{bmatrix} \qquad D = \text{diag}\{b_1 \cdots b_{d-1}\}$$

so that the upper left corner of E has position d, d. "Find_least_element" finds the least (in magnitude) element of E, moves it to the upper left corner by row and column transpositions, and changes signs of a row if necessary to make that element positive. However, if $E = 0$, then the boolean "zero_block" is set.

Then "decrease_elements" subtracts multiples of the first row of E from other rows to make all other elements of its first column less than the first element, and does the same for columns. These procedures are repeated until the first row and column of I are zero, except for the first element; then the diagonal matrix D gets another row, d is increased, and E is decreased in size. Eventually A is reduced to the diagonal form above, with $E = 0$.

```
procedure  invariant_factors(const  A: matrix;
                             var    rank: Word;
                             var    C: matrix);

       var     min_dim: Word;
               change,
            zero_block: Boolean;
```

```
procedure  find_least_element(d: Integer);

   var     min, v, t: Integer;
           row0, col0,
             row, col: Word;

   begin
   min := 0;   row0 := d;   col0 := d;
   for  row := d  to  m  do
      for  col := d  to  n  do
         begin
         v := Abs(C[row, col]);
         if  (v > 0) and
             ((v < min) or (min = 0))   then
            begin
            min := v;   row0 := row;   col0 := col;
            end;
         end;    { for }
   zero_block := (min = 0);
   change := false;
   if  not zero_block   then
      begin
      if  row0 > d   then
      { transpose rows d and row0 }
         for  col := d  to  n  do
            begin
            change := true;
            t := C[d, col];
            C[d, col] := C[row0, col];
            C[row0, col] := t;
            end;
      if  col0 > d   then
      { transpose cols d and col0 }
         for  row := d  to  m  do
            begin
            change := true;
            t := C[row, d];
            C[row, d] := C[row, col0];
            C[row, col0] := t;
            end;
      if  C[d, d] < 0   then
         for  row := d  to  m  do
            C[row, d] := -C[row, d];
      end;    { if  not zero_block }
   end;    { find_least_element }
```

```
procedure   decrease_elements (d: Integer);

          var   row, col: Word;
                quotient: Integer;

      begin   { decrease_elements }
        for   col := d + 1   to   n   do
          if   Abs (C[d, col]) >= C[d, d]   then
              begin
              change := true;
              quotient := C[d, col] div C[d, d];
              for   row := d   to   m   do
                 C[row, col] :=
                     C[row, col]
                          - quotient*C[row, d];
              end;
        for   row := d + 1   to   m   do
          if   Abs (C[row, d]) >= C[d, d]   then
              begin
              change := true;
              quotient := C[row, d] div C[d, d];
              for   col := d   to   n   do
                 C[row, col] :=
                     C[row, col]
                          - quotient*C[d, col];
              end;
      end;   { decrease_elements }
```

We go to work on the nonzero part of the diagonal, two elements at a time.
The key step is to show that any pair of diagonal elements a and b can be replaced
by their gcd and lcm. Recall that the gcd is a linear combination of a and b with
integer coefficients. Also, the gcd divides both a and b:

$$d = \gcd(a,\ b) = ax + by \qquad a = du \qquad b = dv$$
$$m = \operatorname{lcm}(a,\ b) = duv$$

We now easily check

$$P = \begin{bmatrix} 1 & 1 \\ -vy & ux \end{bmatrix} \qquad Q = \begin{bmatrix} x & -v \\ y & u \end{bmatrix} \qquad PAQ = P \begin{bmatrix} a & 0 \\ 0 & b \end{bmatrix} Q = \begin{bmatrix} d & 0 \\ 0 & m \end{bmatrix}$$

and P and Q are unimodular. "Process_diagonal" does this replacement process
with the first element of the diagonal and each of its following elements. The
result is that the first element now divides all the others. Then "process_diagonal"
starts over again with the second element of the diagonal, etc.

```
procedure  process_diagonal;

    var   a, b, j, k: Word;
                   d: Integer;

    function  gcd(a, b: Integer): Integer;

        var  t: Integer;

        begin
        while  b > 0  do
           begin
           t := b;  b := a mod b;  a := t;
           end;
        gcd := a;
        end;    { gcd }

    begin    { process_diagonal }
    for  j := 1  to  rank - 1  do
       for  k := j + 1  to  rank  do
          begin
          a := C[j, j];  b := C[k, k];
          d := gcd(a, b);
          C[k, k] := (a div d)*b;    { lcm(a, b) }
          C[j, j] := d;
          end;    { for k }
       end;    { process_diagonal }

begin    { invariant_factors }
if  m < n  then  min_dim := m
else  min_dim := n;
C := A;
rank := 1;
repeat
   repeat
      find_least_element(rank);
      if  not zero_block  then
         decrease_elements(rank);
   until  not change;
   Inc(rank);
until  (rank > min_dim) or zero_block;
Dec(rank);
process_diagonal;  WriteLn;
end;    { invariant_factors }

end.    { unit  int_matr }
```

Exercises

1. Write a program to test "echelon_form".
2. Write a program to test "invariant_factors".
3. In the next exercise you will need a declaration
   procedure gcd(a, b: integer;   var   d, x, y: integer);
   to produce d = (a, b) = ax + by. Write such a procedure and imbed it in a program to test it.
4. [Cont.] Modify "invariant_factors" so that its output is not only the Smith normal form C, but also the unimodular matrices P and Q such that $PAQ = C$.
5. [Cont.] Write a program to test the new "invariant_factors".

Section 7. Linear Algebra

The previous section dealt with integer matrices. Now we move on to real matrices. Our first item is the solution of linear systems by Gaussian elimination. There is an enormous literature on this subject and a great deal of software. The following unit "lin_sys" contains a fairly careful program for solving linear systems. Its data will be a matrix $A = [B \ C]$ of two blocks, where the left block is the square system matrix, and the right block represents several columns that serve as the right hand sides of systems of equations. The problem to be solved is $BX = C$. We allow any number of columns of C from 1 to the number of columns of B. An important special case is $C = I$ with solution $X = C^{-1}$.
   We start with the interface section of "lin_sys":

Program 8.7.1

```
unit  lin_sys;    { 05/22/95 }

interface

   const   max_no_rows = 20;    { 20 x 40 max }

   type    Real = Extended;
           matrix = array[1..max_no_rows,
                          1..2*max_no_rows] of Real;

   var     no_rows,
           no_cols: Word;
           singular: Boolean;
```

```
procedure   get_system_matrix(var   A: matrix);
procedure   get_rhs(var   A: matrix);
procedure   print_matrix(const   A: matrix;   ...);
procedure   triangularize(var   A: matrix);
procedure   back_subtract(var   A: matrix);
procedure   check_answer(const   A, B: matrix;
                      var   C: matrix);
```

We'll omit the I/O procedure listings (ditto in the other program of this section) and move to the first part of the Gaussian elimination algorithm, a sequence of elementary row transformations that reduces the system matrix B to upper triangular form. "Pivoting" is an important device for gaining numerical stability. After the first p rows have been processed, we find the largest (in magnitude) element in column $p+1$ from row $p+1$ to the final row. Then we transpose the row where the largest element is found with row $p+1$ and proceed to subtract multiples of (new) row $p+1$ from the rows below it, to make zeros. This is the job of "triangularize".

implementation

```
uses   CRT;

const   eps = 1.0e-10;

var        change: Boolean;
        no_vectors: Word;

procedure   subtract_rows(var   A: matrix;
                          p, q: Word;   x: Real);

    var   j: Word;

    begin
    for   j := p + 1   to   no_cols   do
        A[q, j] := A[q, j] - x*A[p, j];
    end;    { subtract_rows }

procedure   triangularize(var   A: matrix);

    var   p, q, k: Word;
```

```
function  partial_pivot(p: Word): Word;

   var  j, q: Word;
        x, y: Real;

   begin
   q := p;  x := Abs(A[p, p]);
   for  j := p + 1  to  no_rows  do
      begin
      y := Abs(A[j, p]);
      if  y > x  then
         begin  q := j;  x := y  end;
      end;
   singular := (Abs(x) < eps);
   partial_pivot := q;
   end;   { partial_pivot }

procedure  transpose_rows(p, q: Word);

   var  j: Word;
        x: Real;

   begin
   for  j := p  to  no_cols  do
      begin
      x := A[p, j];  A[p, j] := A[q, j];
      A[q, j] := x;
      end
   end;   { transpose_rows }

begin   { triangularize }
p := 1;
repeat
   k := partial_pivot(p);
   if  not singular  then
      begin
      if  k > p  then  transpose_rows(p, k);
      for  q := p + 1  to  no_rows  do
         subtract_rows(A, p, q,
                          A[q, p]/A[p, p]);
      end;
   Inc(p);
until  (p > no_rows)  or  singular;
end;   { triangularize }
```

The solution is completed by "back_subtract", which completes the reduction of the left block to the identity matrix by more elementary row transformations. The theory is pretty straightforward. Any sequence of elementary row transformations is equivalent to left multiplication by a nonsingular matrix:

$$P[B\ C] = [I\ X] \qquad PB = I \qquad P = B^{-1} \qquad X = PC = B^{-1}C$$

```
procedure   back_subtract(var   A: matrix);

   var   p, q:   Word;

   procedure   normalize_row(p: Word);

      var   k: Word;
            x: Real;

      begin
      x := A[p, p];   A[p, p]   := 1.0;
      for   k := p + 1   to   no_cols   do
         A[p, k]   := A[p, k]/x;
      end;   { normalize_row }

   begin
   for   p := 1   to   no_rows   do   normalize_row(p);
   for   p := no_rows   downto   2   do
      for   q := p - 1   downto   1   do
         subtract_rows(A, p, q, A[q, p]);
   end;   { back_subtract }

end.   { lin_sys }
```

Symmetric Matrices

Our next topic is characteristic roots and vectors of symmetric matrices. Let A be a real symmetric $n \times n$ matrix. By standard theory, all characteristic roots of A are real, and there is an orthonormal basis of corresponding characteristic vectors.

We begin with an algorithm "iterate" for finding the largest characteristic root in magnitude, provided that all other characteristic roots are smaller in magnitude. Then there is a unique (up to sign) characteristic unit vector corresponding to that largest root. The following unit "symm_mat" contains "iterate" (and another method discussed later).

Program 8.7.2

```
unit   symm_mat;     { 05/22/95 }

interface

const   max_dim = 50;

type                   Real = Extended;
        symmetric_matrix = array[1..max_dim,
                                  1..max_dim] of Real;
                  vector = array[0..max_dim] of Real;
                    Str4 = string[4];

var   dim: Word;

procedure   get_matrix(var   A: symmetric_matrix);
procedure   print_matrix(var   A: symmetric_matrix);
procedure   iterate(const   A: symmetric_matrix;
                    var   v: vector;
                    var   r: Real;
                    var   fail: Boolean);
procedure   check_char(const   A: symmetric_matrix;
                       const   v: vector;
                       const   r: Real);
procedure   print_char_pair(var   v: vector;
                            r: Real;
                            failed: Boolean);
procedure   QR_algorithm(var   A: symmetric_matrix;
                         var   char_root: vector;
                         var   W: symmetric_matrix);
```

The procedure "iterate" iterates the vector function $\phi(\mathbf{v}) = \pm A\mathbf{v}/|A\mathbf{v}|$. We start with a unit vector $\mathbf{e}_1$ and define $\mathbf{e}_{n+1} = \phi(\mathbf{e}_n)$. Except for rare poor choices of the initial $\mathbf{e}_1$, the sequence $\{\mathbf{e}_n\}$ converges geometrically to a characteristic vector corresponding to the largest characteristic root. The method is known as the *power method* and is discussed in most standard references on numerical linear algebra, e.g, Burden and Faires, Numerical Analysis 5/e, PWS-Kent, 1993, 506–523.

```
implementation

uses   CRT;

const   tol = 1.0e-30;
```

```
procedure   iterate(const   A: symmetric_matrix;
                     var   v: vector;
                     var   r: Real;
                     var   fail: Boolean);

    var               x: Real;
         count, j, k: Word;
                    w: vector;
                 test: Boolean;

    function   norm_squared(const   v: vector): Real;

        var  x: Real;
             j: Integer;

        begin
        x := 0.0;
        for  j := 1  to  dim  do  x := x + Sqr(v[j]);
        norm_squared := x;
        end;

    function  distance_squared
                    (const   v, w: vector): Real;

        var  x: Real;
             j: Integer;

        begin
        x := 0.0;
        for  j := 1  to  dim  do
           x := x + Sqr(v[j] - w[j]);
        distance_squared := x;
        end;

    begin    { iterate }
    count := 0;  fail := false;  test := false;
    repeat
       Inc(count);
       w := v;
       for  j := 1  to  dim  do
          begin    { v := Av }
          x := 0.0;
          for  k := 1  to  dim  do
             x := x + A[k, j]*w[k];
          v[j] := x;
          end;
```

```
    x := norm_squared(v);
    if   x < tol   then   fail := true
    else
       begin
       r := 0.0;
       for   j := 1   to   dim   do
           r := r + v[j]*w[j];
       x := 1.0/Sqrt(x);
       if   r < 0.0   then   x := -x;
       for   j := 1   to   dim   do   v[j] := x*v[j];
       test := (distance_squared(v, w) < tol);
       end;   { else }
 until   fail or test or (count = 200);
 fail := not test;
 end;   { iterate }
```

The QR Algorithm

We now look at a very remarkable algorithm for finding all of the characteristic roots and vectors of any real, nonsingular, symmetric matrix A. This *QR algorithm* was published independently in the early 60's by J. G. F. Francis and V. N. Kublanovskaya. See Burden and Faires, *loc cit*, pp. 523–541.

Very briefly, start with $A_0 = A$. If we have reached A_n, then by the Gram-Schmidt orthonormalization process, there is a unique factorization $A_n = U_n T_n$, where U_n is orthogonal and T_n is upper triangular with positive diagonal elements. Set $A_{n+1} = T_n U_n$. Then the sequence $\{A_n\}$ generally converges to a diagonal matrix, and its diagonal elements are the characteristic roots of A. Amazing! Multiple characteristic roots can wreck convergence. But convergence is assured if all characteristic roots are simple and have distinct absolute values.

To gain a slight feel for why the algorithm works, note that

$$A_{n+1} = T_n U_n = U_n^{-1} A_n U_n$$

so that all of the matrices A_n are congruent (orthogonally equivalent), hence have the same characteristic roots.

The factorization $A = UT$ (originally QR) of a nonsingular symmetric matrix is slow in general. It is much faster if A is a *tridiagonal matrix*, that is, if all elements of A are 0 except for those on the main diagonal, the superdiagonal, and (its reflection) the subdiagonal. It can be shown that if $A = UT$ starts tridiagonal, then TU is again tridiagonal, hence the whole sequence $\{A_n\}$ is tridiagonal.

The implementation of the QR algorithm begins with a preliminary step (due to A. Householder) that replaces A by a congruent tridiagonal matrix. That is done by "tri_diagonalize". The UT factorization of A is then done by a sequence of plane rotations, (like a conic section is rotated into standard form). Actually, a small modification of what I have described speeds convergence: If λ is near a characteristic root of A_n then we factor and iterate:

$$A_n - \lambda I = U_n T_n \qquad A_{n+1} = \lambda I + T_n U_n$$

See "translate" and "factor" below. The purpose of "factor" is to factor the diagonal block of $A_n - \lambda I$ from its m-th to its n-th row. Then "multiply" computes the corresponding block of A_{n+1}.

```
procedure  QR_algorithm(var  A: symmetric_matrix;
                        var  char_root: vector;
                        var  W: symmetric_matrix);

   var      diagonal,
        sine, cosine,
          super_diag,
         super_diag2: vector;
            norm, eps: Real;
               m, n: Word;
                 A0: symmetric_matrix;
      c, s, x, x1,
         y, z, u,
       lambda, mu: Real;

   procedure  initialize_QR;

      var  j, k: Word;

      procedure  tri_diagonalize
                     (var  A, W: symmetric_matrix;
                      var  diagonal,
                      super_diag: vector);

         var  c, t, x, y: Real;
                       v: vector;
                 i, j, k: Word;

         procedure  Householder_step(i: Word);

            var  p: vector;
                 j, k: Word;
                 z: Real;
```

```
begin
for  j := i + 1  to  dim  do
   begin    { p := 2Av }
   t := 0.0;
   for  k := i + 1  to  dim  do
      t := t + A[j, k]*v[k];
   p[j] := 2.0*t;
   end;    { for j }
z := 0.0;    { z := 4v'Av }
for  j := i + 1 to  dim  do
   z := z + p[j]*v[j];
z := 2.0*z;
for  j := i + 1  to  dim  do
   for  k := j  to  dim  do
      A[j, k] := A[j, k] - v[j]*p[k]
                      - (p[j] - z*v[j])*v[k];
end;    { Householder_step }

procedure  update_W(i: Word);

   var  p: vector;
        j, k: Word;

   begin
   for  j := 1  to  dim  do    { p := 2wv }
      begin
      t := 0.0;
      for  k := i + 1  to  dim  do
         t := t + W[j, k]*v[k];
      p[j] := 2.0*t;
      end;    { for j }
   for  j := 1  to  dim  do
      for  k := i + 1  to  dim  do
         W[j, k] := W[j, k] - p[j]*v[k];
   end;    { update_W }

begin    { tri_diagonalize }
super_diag[dim] := 0.0;
for  i := 1  to  dim - 2  do
   begin
   diagonal[i] := A[i, i];
   t := 0.0;
   for  j := i + 1  to  dim  do
      t := t + Sqr(A[i, j]);
   c := Sqrt(t);    x := A[i, i + 1];
   if  x > 0.0  then  c := -c;
```

```
            super_diag[i] := c;
            if   c <> 0.0   then
                begin
                y := Sqrt(0.5*(1.0 - x/c));
                v[i + 1] := y;
                y := -1.0/(2.0*c*y);
                for  j := i + 2  to   dim   do
                    v[j] := y*A[i, j];
                Householder_step(i);
                update_W(i);
                end;    { if c }
            end;    { for i }
        k := dim - 1;
        diagonal[k] := A[k, k];
        diagonal[dim] := A[dim, dim];
        super_diag[k] := A[k, dim];
        super_diag[dim] := 0.0;
        end;    { tri_diagonalize }

    begin    { initialize_QR }
    for  j := 1  to   dim   do
        begin
        for  k := 1  to   dim   do  W[j, k] := 0.0;
        W[j, j] := 1.0;
        end;
    tri_diagonalize(A, W, diagonal, super_diag);
    norm := 0.0;   y := 0.0;
    for  j := 1  to   dim   do
        begin
        x := y + Abs(diagonal[j]);
        y := Abs(super_diag[j]);
        x := x + y;
        if  x > norm   then   norm := x;
        end;    { for j }
    eps := norm*tol;
    for  j := 0  to   dim   do
        begin
        sine[j] := 0.0;   cosine[j] := 0.0;
        end;
    end;    { initialize_QR }

procedure   translate(a, b, c: Real;
                    var   lambda, mu: Real);
    { Translation speeds convergence. }

    var   trace, det, lambda0, b2, srt: Real;
```

```
begin
b2 := Sqr(b);   det := a*c - b2;   trace := a + c;
srt := Sqrt(Sqr(trace) - 4.0*det);
if  trace >= 0.0  then   lambda := trace + srt
else  lambda := trace - srt;
lambda := 0.5*lambda;    { larger root }
lambda0 := det/lambda;    { smaller root }
if  Abs(lambda0 - mu) < 0.5*Abs(lambda0)   then
   begin  lambda := lambda0;   mu := lambda0   end
else
   if  Abs(lambda - mu) < 0.5*Abs(lambda)   then
      mu := lambda
else  begin  mu := lambda0;   lambda := 0.0;   end;
end;   { translate }

procedure  factor(m, n: Word);

   var  j: Word;

   begin
   diagonal[m] := diagonal[m] - lambda;
   u := super_diag[m];
   for  j := m  to  n - 1  do
      begin
      x := diagonal[j];
      x1 := diagonal[j + 1] - lambda;
      y := super_diag[j];
      z := Sqrt(Sqr(x) + Sqr(u));
      c := x/z;   s := u/z;
      cosine[j] := c;   sine[j]:= s;
      diagonal[j] := z;
      diagonal[j + 1] := -s*y + c*x1;
      super_diag[j] :=    c*y + s*x1;
      u := super_diag[j + 1];
      super_diag[j + 1] := c*u;
      super_diag2[j] := s*u;
      end;    { for j }
   end;    { factor }

procedure  multiply(m, n: Word);

   var  j, k: Word;
```

```
begin
for  k := m  to  n - 1  do
   begin
   c := cosine[k];   s := sine[k];
   x := diagonal[k];   y := super_diag[k];
   super_diag[k - 1] := super_diag[k - 1]*c
                        + super_diag2[k - 1]*s;
   diagonal[k] :=    x*c + y*s + lambda;
   super_diag[k] := -x*s + y*c;
   diagonal[k + 1] := diagonal[k + 1]*c;
   for  j := 1  to  dim  do
      begin    { Now form  wu' }
      x := W[j, k];   y := W[j, k + 1];
      W[j, k] :=        x*c + y*s;
      W[j, k + 1] := -x*s + y*c;
      end;    { for j }
   end;    { for k }
diagonal[n] := diagonal[n] + lambda;
end;    { multiply }

procedure  report_result;
procedure  check_QR;

begin    { QR_algorithm }
A0 := A;
initialize_QR;
n := dim;
mu := 0;
while  n > 0  do
   begin
   if  Abs(super_diag[n - 1]) < eps   then
      begin
      char_root[n] := diagonal[n];
      Dec(n);
      end
   else
      begin
      m := n - 1;
      repeat
         Dec(m)
      until  Abs(super_diag[m]) < eps;
      Inc(m);
      translate(diagonal[n - 1], super_diag[n - 1],
               diagonal[n], lambda, mu);
      factor(m, n);
      super_diag2[m - 1] := 0.0;
```

```
      super_diag[m - 1] := 0.0;
      multiply(m, n);
    end;   { else }
  end;   { while  n > 0 }
  report_result;
  check_QR;
  end;   { QR_algorithm }

end.   { unit  symm_mat }
```

Exercises

1. Write a program to test "lin_sys" on systems that you input from the keyboard.
2. The pivoting method above is called *partial pivoting*. A more reliable method is *complete pivoting*, where the largest element in the remaining lower right block of the system matrix is chosen. This requires some extra bookkeeping. Modify "lin_sys" to include complete pivoting.
3. Another method for obtaining more accurate results is *scaling*. This is an operation preliminary to solving. First, each row is divided by the largest element in that row in the system matrix block. Second, each column in the system matrix is divided by its largest element. This amounts to replacing each unknown by a multiple, so the factors have to be saved for later restoration. Modify "lin_sys" to include scaling too.
4. The $n \times n$ *Hilbert matrix* $H_n = [1/(j + k - 1)]$ is nonsingular, but has small determinant, so its inverse is quite sensitive to small changes in the elements of H_n. Test this by inverting H_n and by inverting matrices obtained from H_n by rounding its elements to a fixed number of decimal places.
5. Write a program to test "iterate".
6. Write a program to test the QR algorithm.

Section 8. Objects

I have avoided OOP (Object Oriented Programming) mostly because I don't think it will be very useful in scientific programming until Borland adds overloading of operators and procedures to Pascal. However, in all fairness, I should give at least one example of its use to illustrate the main ideas.

Let us consider the product of matrices. If we could overload operators, we would be able to write

```
  C := A*B
```

for the assignment of the product of two matrices. Well, we can't do this, but there is a related problem. How can we program the matrix multiplication

$$C = AB \qquad c_{ik} = \sum_j a_{ij} b_{jk}$$

once and for all, without regard as to where the *coefficients of the matrices* are coming from? The unit "matrices" that follows does just this, also for the sum of two matrices and for printing a matrix. It introduces some of the main features of OOP, which I'll try to explain as we go along. The interface is already quite complicated, and we have to take it in pieces. We'll apply the (as yet undefined) matrix operations to three examples: matrices of reals, rationals, and complexes. The beginning of the interface defines the scalar types "rational" and "complex" and matrix types with these and real coefficients. Nothing unexpected here. It also defines a type "matrix" as an array of pointers. This is new to us, and it is by type casting the pointers that we can make them point to reals, rationals, complexes, or any other type of element that we wish to do matrix algebra with, for instance, rational polynomials.

Program 8.8.1

```
{$r+}
unit  matrices;    { 06/25/95 }

interface

   uses  CRT;

   const  max = 3;

   type
      Real = Extended;
      str10 = string[10];
      str30 = string[30];
      matrix = array[1..max, 1..max] of Pointer;

      rational = record  num, den: LongInt;  end;
      rational_ptr = ^rational;
      rational_matrix = array[1..max, 1..max]
                             of rational;

      real_ptr = ^Real;
      real_matrix = array[1..max, 1..max] of Real;

      complex = record  re, im: Real;  end;
      complex_ptr = ^complex;
      complex_matrix = array[1..max, 1..max] of complex
```

Now we come to our first definition of an object. The reserved word pair **object....end** generalizes **record....end**. Like a record, an object has variables, or data fields, simply called *fields*. It also contains built-in procedures and functions, called *methods* for manipulating those fields. This is the first difference between a record and an object. There are two special kinds of methods with their own reserved words. A **constructor** sets up memory for a variable of the object type and is usually used to initialize its fields. Each object type definition should contain a constructor, and it is usually named *init*. A **destructor** is used to reclaim the memory after an object type variable is no longer needed. Each object type definition should also contain a destructor, and it is usually named *done*. The object type "matrix_obj" that we now define contains exactly one field, a matrix M. Keep in mind that each instance of "matrix_object" contains M.

```
matrix_obj
   = object
         M: matrix;
         constructor   init;
         procedure   set_to_zero
            (var   p: Pointer);   virtual;
         procedure   scalar_sum
            (const   p, q: Pointer;
             var   r: Pointer);   virtual;
         procedure   scalar_product
            (const   p, q: Pointer;
             var   r: Pointer);   virtual;
         procedure   matrix_sum
            (const   A, B: matrix;   var   C: matrix);
         procedure   matrix_product
            (const   A, B: matrix;   var   C: matrix);
         procedure   convert_to_string
            (p: Pointer;   var   tt: str30);   virtual;
         procedure   print(name: str10);
         destructor   done;
      end;
```

The crucial reserved word **virtual** certainly needs explanation. Let us concentrate on two related procedures: the virtual "scalar_sum" and the (not virtual) "matrix_sum". The input of "scalar_sum" is two pointers, for the moment pointing to no type of data whatever, and the output is another pointer that somehow is to point to their sum. The input of "matrix_sum" is two matrices A and B, and the output is supposed to be their matrix sum C. The implementation of "matrix_sum" will call the method "scalar_sum". In time, "scalar_sum" will be specialized to the sum of (pointers to) reals, rationals, and complexes. Because "scalar_sum" is declared **virtual**, "matrix_sum" *will use these specializations*. This is precisely the mechanism that allows us to define the sum of two matrices without knowing in advance what their coefficients are.

Similarly, "matrix_product" will call the **virtual** method "scalar_product". It will also call the **virtual** method "set_to_zero" in order to initialize to 0 the sum of products in the definition of matrix multiplication.

The method "print" will print an array of *strings*. It will call the **virtual** method "convert_to_string", which will convert a scalar to a string. Its specializations will deal with reals, rationals and complexes.

Now we come to the object type "real_matrix_obj". Note carefully the beginning of its definition: `real_matrix_obj = object(matrix_obj)`. This means that "real_matrix_obj" is a *specialization* (descendent) of "matrix_obj", and as such, it *inherits* all of the fields and methods of its parent "matrix_obj". But we are free to modify those fields and methods as we see fit.

```
real_matrix_obj
    = object(matrix_obj)
            constructor   init(M0: real_matrix);
            procedure   set_to_zero
                (var  p: Pointer);   virtual;
            procedure   scalar_sum
                (const  p, q: Pointer;
                 var   r: Pointer);   virtual;
            procedure   scalar_product
                (const  p, q: Pointer;
                 var   r: Pointer);   virtual;
            procedure   matrix_sum
                (const   A, B: matrix;   var   C: matrix);
            procedure   matrix_product
                (const   A, B: matrix;   var   C: matrix);
            procedure   convert_to_string
                (p: Pointer;   var   tt: str30); virtual;
            procedure   print(name: str10);
            destructor   done;
        end;
```

We complete our type definitions with two more descendents of "matrix_obj":

```
rational_matrix_obj
    = object(matrix_obj)
            constructor   init(M0: rational_matrix);
            procedure   set_to_zero
                (var  p: Pointer);   virtual;
            procedure   scalar_sum
                (const  p, q: Pointer;
                 var   r: Pointer);   virtual;
```

```
       procedure   scalar_product
           (const   p, q: Pointer;
            var   r: Pointer);   virtual;
       procedure   matrix_sum
           (const   A, B: matrix;   var   C: matrix);
       procedure   matrix_product
           (const   A, B: matrix;   var   C: matrix);
       procedure   convert_to_string
           (p: Pointer;   var   tt: str30);   virtual;
       procedure   print(name:  str10);
       destructor   done;
     end;
complex_matrix_obj
   = object(matrix_obj)
       constructor   init(M0: complex_matrix);
       procedure   set_to_zero
           (var   p: Pointer);   virtual;
       procedure   scalar_sum
           (const   p, q: Pointer;
            var   r: Pointer);   virtual;
       procedure   scalar_product
           (const   p, q: Pointer;
            var   r: Pointer);   virtual;
       procedure   matrix_sum
           (const   A, B: matrix;   var   C: matrix);
       procedure   matrix_product
           (const   A, B: matrix;   var   C: matrix);
       procedure   convert_to_string
           (p: Pointer;   var   tt: str30);   virtual;
       procedure   print(name:  str10);
       destructor   done;
     end;
```

We are ready for the implementations of this slew of methods. You are probably thinking at this point that this is an enormous amount of baggage to carry around, and it must be a lot faster simply to define matrix product as many times as needed for different coefficient domains. Probably true, but read on if you want to understand something about OOP.

We start with the methods of "matrix_obj". The first thing to notice in the implementation is that methods with the same names are distinguished by dereferencing: "matrix_obj.init" for the method "init" of object type "matrix_obj", etc. Next, the constructor "init" and the destructor "done" seem to do nothing. This is not so; they do a lot of (hidden) memory management. The three **virtual** methods "set_to_zero", "scalar_sum", and "scalar_product" really do nothing.

They act as place holders for their later specializations.   Read carefully the implementations of "matrix_sum" and "matrix_product", which call these virtual methods. Similarly, "convert_to_string" does nothing for the moment, but its specializations will be the methods called when we actually use "print". I'll omit some of the screen boxing details of "print" in the listing here; they are on disk, or you can refer back to the unit "matrices" in Section 2.5 for similar material.

implementation

```
    constructor  matrix_obj.init;

        begin  end;

    procedure  matrix_obj.set_to_zero(var  p: Pointer);

        begin  end;

    procedure  matrix_obj.scalar_sum
                    (const  p, q: Pointer;
                     var  r: Pointer);

        begin  end;

    procedure  matrix_obj.scalar_product
                    (const  p, q: Pointer;
                     var  r: Pointer);

        begin  end;

    procedure  matrix_obj.matrix_sum
                    (const  A, B: matrix;   var  C: matrix);

        var  i, j: Word;

        begin
        for  i := 1  to  max  do
            for  j := 1  to  max  do
                scalar_sum(A[i, j], B[i, j], C[i, j]);
        end;

    procedure  matrix_obj.matrix_product
                    (const  A, B: matrix;   var  C: matrix);

        var    i, j, k: Word;
               prod, sum: Pointer;
```

```
     begin
     GetMem(prod, 100);
     for  i := 1  to  max  do
        for  k := 1  to  max  do
           begin
           sum := C[i, k];
           set_to_zero(sum);
           for  j := 1  to  max  do
              begin
              scalar_product(A[i, j], B[j, k], prod);
              scalar_sum(sum, prod, sum);
              end;
           end;
     FreeMem(prod, 100);
     end;    { matrix_obj.matrix_product }

procedure  matrix_obj.convert_to_string
              (p: Pointer;  var  tt: str30);

     begin  end;

procedure  matrix_obj.print(name: str10);

     var  i, j, k, hold_y: Word;
                    temp: str30;

     begin  ...........
     for  i := 1  to  max  do
        begin
        ............
        for  j := 1  to  max  do
           begin
           convert_to_string(M[i, j], temp);
           Write(temp);
           end; ........
        end; .........
     end;    { matrix_obj.print }

destructor  matrix_obj.done;

     begin  end;
```

It remains to implement the methods of "real_matrix_obj", "ratio-nal_matrix_obj", and "complex_matrix_obj". We'll only list and comment on one of these batches, the rational one, and leave the others as exercises. First look at the implementations of "init" and "done". Notice that "init" first calls its parent "matrix_obj.init", which allocates object memory in general. Then "init" allocates a lot of data memory by calls of "new". "Done" recovers this memory in precisely the reverse order. Next, "set_to_zero" now really does initialize a pointee to the rational number 0. Notice the necessary type casting "rational_ptr". I remind you that M was the one field of "matrix_obj", and it is inherited by "rational_matrix_obj" and the other descendents.

```
constructor   rational_matrix_obj.init
                  (M0: rational_matrix);

   var   i, j: Word;

   begin
   matrix_obj.init;
   for   i := 1  to   max   do
      for   j := 1  to   max   do
         begin
         New(rational_ptr(M[i, j]));
         rational_ptr(M[i, j])^ := M0[i, j];
         end;
   end;     { rational_matrix_obj.init }

procedure   rational_matrix_obj.set_to_zero
                  (var   p: Pointer);

   begin
   rational_ptr(p)^.num := 0;
   rational_ptr(p)^.den := 1;
   end;
```

The auxiliary procedure "lowest_terms" reduces a rational to lowest terms; we can safely omit the details of dividing out by the gcd. Read carefully the implementations of the virtual methods "scalar_sum" and "scalar_product", which now have some content, unlike those of their parents.

```
procedure   lowest_terms(var   x, y: LongInt);
   { Assumes   y > 0 }
   . . . . . . . . . . . . . . . . .
```

```
procedure   rational_matrix_obj.scalar_sum
                  (const   p, q: Pointer;
                   var   r: Pointer);

   var   x, y, z: rational;

   begin
   x := rational_ptr(p)^;
   y := rational_ptr(q)^;
   z.num := x.num*y.den + y.num*x.den;
   z.den := x.den*y.den;
   lowest_terms(z.num, z.den);
   if   r = nil   then   New(rational_ptr(r));
   rational_ptr(r)^ := z;
   end;   { rational_matrix_obj.scalar_sum }

procedure   rational_matrix_obj.scalar_product
                  (const   p, q: Pointer;
                   var   r: Pointer);

   var   x, y, z: rational;

   begin
   x := rational_ptr(p)^;
   y := rational_ptr(q)^;
   z.num := x.num*y.num;
   z.den := x.den*y.den;
   lowest_terms(z.num, z.den);
   if   r = nil   then   New(rational_ptr(r));
   rational_ptr(r)^ := z;
   end;   { rational_matrix_obj.scalar_product }
```

Now observe that "matrix_sum" and "matrix_product" merely call their parents. But in this context, the parents' calls to "scalar_sum" and "scalar_product" *refer to the descendents here*, because these called methods are **virtual**.

```
procedure   rational_matrix_obj.matrix_sum
                  (const   A, B: matrix;   var   C: matrix);

   begin
   matrix_obj.matrix_sum(A, B, C);
   end;
```

```
procedure   rational_matrix_obj.matrix_product
                (const   A, B: matrix;   var   C: matrix);

    begin
    matrix_obj.matrix_product(A, B, C);
    end;
```

Similar remarks apply to the virtual "convert_to_string", which is fleshed out here, and "print", which merely calls its parent.

```
procedure   rational_matrix_obj.convert_to_string
                (p: Pointer;   var   tt: str30);

    var   t1, t2: string[11];
              j: Word;

    begin
    Str(rational_ptr(p)^.num, t1);
    Str(rational_ptr(p)^.den, t2);
    tt := t1 + '/' + t2;
    for   j := Length(t1) + Length(t2)   to   12   do
        tt := ' ' + tt;
    end;

procedure   rational_matrix_obj.print(name: str10);

    begin
    matrix_obj.print(name);
    end;

destructor   rational_matrix_obj.done;

    var   i, j: Word;

    begin
    for   i := 1   to   max   do
        for   j := 1   to   max   do
            begin
            Dispose(rational_ptr(M[i, j]));
            end;
    matrix_obj.done;
    end;    { rational_matrix_obj.done }
end.    { unit   matrices }
```

I hope this gives a little taste of the OOP world out there.

Remark

Borland's new product *Delphi*, for developing Windows applications, is based on Pascal, but on a variation Borland calls "Object Pascal". It differs from Borland Pascal 7.0 mainly in its treatment of OOP. There are substantial changes, starting with **class** replacing **object** and continuing with a slew of new reserved words.

Exercises

1. Finish the implementations of the methods of "real_matrix_obj".
2. Finish the implementations of the methods of "complex_matrix_obj".
3. Write a program to test everything in "matrices".

Chapter 9 Applications to Mathematical Analysis

Section 1. Approximation

One of the limitations of floating point arithmetic is *computer roundoff error*. It is measured by a number called *unit roundoff error* (URE), which is defined as the least positive real number u such that the computer cannot distinguish between $1.0+u$ and 1.0. Actually, this number is software dependent, and as you will surely test, with Pascal it is different for different real types. The following program is one approach to measuring URE. Its first approximation to u is the first negative power of 10 that doesn't change 1.0 when added to it. Its next approximation to u is a multiple of that, so the result is the first significant decimal digit of the true u. Then it goes for two more decimal digits, which is really close enough.

Program 9.1.1

```
program  unit_roundoff_error;    { 05/03/95 }

    var   s, u: Real;   k: Word;

    begin
    u := 1.0;
    while   1.0 + u > 1.0   do   u := 0.1*u;
    s := u;
    for   k := 1   to   3   do
        begin
        while   1.0 + u <= 1.0   do   u := u + s;
        u := u - s;
        s := 0.1*s;
        end;
    WriteLn('Unit roundoff error:  ', u:14);
    Write('Press <Enter>:  ');
    ReadLn;
    end.    { unit_roundoff_error }
```

Our next small topic is "Richardson extrapolation", also called "extrapolation to infinity". Suppose we seek $\lim_{x \to 0+} F(x)$. Our first approximation is $s_0 = F(1)$. Our second approximation is $s_1 = F\left(\frac{1}{2}\right)$. If the change from s_0 to s_1 indicates a trend, then we should take the linear function $L(x)$ that interpolates the two points $(1, s_0)$ and $\left(\frac{1}{2}, s_1\right)$, then use $L(0)$ as the best guess for the limit based on all information at hand.

Continuing, we compute $s_2 = F\left(\frac{1}{4}\right)$, form the quadratic polynomial $Q(x)$ that interpolates the two points above and $\left(\frac{1}{4}, s_2\right)$, and take $Q(0)$ as the best guess for the limit. This is the essence of extrapolation to infinity, and we'll see an important use for it later in Romberg integration.

The method works very well if we can assume a Maclaurin expansion $F(x) = \sum_0^\infty a_j x^j$. In general our data will consist of a finite sequence

$$s_0, \; s_1, \; s_2, \; \cdots, \; s_n$$

where $s_k = F\left(2^{-k}\right)$ (or more generally $s_k = F\left(2^{-k} \cdot h\right)$). Our problem is to compute $P_n(0)$ where $P_n(x)$ is the unique polynomial of degree at most n that interpolates the $n+1$ points $\left(2^{-k}, s_k\right)$. It seems like a difficult computation, but actually there is an elegant solution to the problem by computing a difference table. The solution is based on the following result.

Theorem Suppose $Q(x)$ is the unique polynomial of degree at most $n - 1$ that interpolates the n points

$$(1, \; s_0), \; \left(\tfrac{1}{2}, \; s_1\right), \; \cdots, \; \left(\tfrac{1}{2^{n-1}}, \; s_{n-1}\right)$$

and suppose that $R(x)$ is the unique polynomial of degree at most $n - 1$ that interpolates the n points

$$\left(\tfrac{1}{2}, \; s_1\right), \; \left(\tfrac{1}{4}, \; s_2\right), \; \cdots, \; \left(\tfrac{1}{2^n}, \; s_n\right)$$

Then

$$P(x) = \frac{\left(x - 2^{-n}\right)}{\left(1 - 2^{-n}\right)} Q(x) + \frac{(x - 1)}{\left(2^{-n} - 1\right)} R(x) = \frac{(2^n x - 1)Q(x) - 2^n(x - 1)R(x)}{(2^n - 1)}$$

is the unique polynomial of degree at most n that interpolates the $n + 1$ points

$$(1, \; s_0), \; \left(\tfrac{1}{2}, \; s_1\right), \; \cdots, \; \left(\tfrac{1}{2^n}, \; s_n\right)$$

and

$$P(0) = \frac{2^n R(0) - Q(0)}{2^n - 1}$$

Note that $P(x)$ has the correct degree, and that it is a linear combination of $Q(x)$ and $R(x)$ with coefficients that sum to 1. This takes care of its values at $\frac{1}{2}$ to $\frac{1}{2^{n-1}}$. It is routine to check that $P(x)$ has the correct values at the remaining two points 1 and $\frac{1}{2^n}$, and then to check that $P(0)$ is the value claimed.

This theory is implemented in a difference table

$$
\begin{array}{cccccc}
d_{00} & d_{01} & d_{02} & \cdots & & d_{0n} \\
d_{10} & d_{11} & d_{12} & \cdots & & d_{1,n-1} \\
\cdots & \cdots & \cdots & \cdots & & \\
d_{n0} & & & & &
\end{array}
$$

constructed one row at a time as follows. Row 0 is simply $d_{0k} = s_k$ and the j-th row is constructed according to

$$d_{j,k} = \frac{2^j d_{j-1,\,k+1} - d_{j-1,\,k}}{2^j - 1}$$

By the theorem, the elements of row 1 are the values at 0 of the linear polynomials passing through successive pairs of elements of row 0. Then the elements of row 2 are the values at 0 of the quadratic polynomials passing through three consecutive terms of row 0, etc. Eventually it is the final element d_{n0} that is the solution to our extrapolation problem.

The extrapolation unit that follows is economical in space in that only one linear array is used, at first holding row 0 of the difference table. Then subsequent rows are computed "in place" in this space. Note that each row is computed backwards to avoid overwriting array elements needed for subsequent computations.

The unit is also economical in time, and you might find it a nice exercise to write for yourself an implementation of the function "limit" to see how fast you can make it before reading the program that follows. I have set the limit at 16 terms. Actually, you will verify in experiments that computer roundoff error begins to dominate the calculation, and accuracy falls off, after about 10 terms.

Program 9.1.2

```
unit   extrapol;    { 06/02/95 }

interface

   const   max = 16;

   type        Real = Extended;
            sequence = array[0..max] of Real;

   function   limit(var s: sequence;   n: Word): Real;

implementation

   var   denom: array[1..max] of Real;

            k: Word;
            p: Real;
            s: sequence;
```

```
function  limit(var s: sequence;  n: Word): Real;

    var       diff: sequence;
          j, k, k1: Word;
            dk1, d: Real;

    begin
    if  n > max   then   n := max;
    diff := s;
    for  j := n  downto  1  do
        begin   { Compute j-th row }
        d := denom[j];
        k1 := 1;    { k + 1 }
        for  k := 0  to  j - 1  do
            begin
            dk1 := diff[k1];
            diff[k] := dk1 + d*(dk1 - diff[k]);
            Inc(k1);
            end;    { for k }
        end;    { for j }
    limit := diff[0];
    end;    { limit }

begin   { extrapol }
p := 1.0;
for  k := 1  to  max  do
    begin   { denom[k] = 1/(2^k - 1) }
    p := 2.0*p;
    denom[k] := 1.0/(p - 1.0);
    end;
end.    { unit  extrapol }
```

The next item is a very interesting algorithm for computing the z in $z^2 = x^2 + y^2$ without possible overflow or underflow. The source is Moler, C B and Morrison, D, Replacing square roots by Pythagorean sums, IBM Jour of Research and Development 27:6, Nov, 1993, 577–581. We'll imbed the algorithm in a program for converting rectangular to polar coordinates, and test it where overflow or underflow would occur if we computed the hypotenuse in the usual way.

Program 9.1.3

```
program  test_rect_polar;    { 08/11/95 }

    type  Real = Extended;
    { + range [3.4e-4932, 1.1e4932] }
```

```pascal
var   x, y, r, theta: Real;

procedure   rect_polar(x, y: Real;
                          var   r, theta: Real);

   const   pi_2 = 0.5*Pi;

   function   hypotenuse(x, y: Real): Real;

      var   t: Real;
            k: Word;

      begin
      x := Abs(x);   y := Abs(y);
      if   x < y   then
         begin   t := x;   x := y;   y := t;   end;
      { 0.0 <= y <= x }
      if   y <> 0.0   then
         begin
         for   k := 1   to   6   do
            begin
            { x^2 + y^2 is invariant,
              t --> 0,   y --> 0 }
            t := Sqr(y/x);
            t := t / (4.0 + t);
            x := x + 2.0*t*x;   y := y*t;
            end;
         end;
      hypotenuse := x;
      end;   { hypotenuse }

   begin   { rect_polar }
   r := hypotenuse(x, y);
   if   Abs(y) <= Abs(x)   then
      begin
      theta := Arctan(y/x);
      if   x < 0.0   then
         if   y >= 0.0   then   theta := theta + Pi
         else   theta := theta - Pi;
      end
   else
      begin
      theta := pi_2 - Arctan(x/y);
      if   y < 0.0   then   theta := theta - Pi;
      end;
   end;   { rect_polar }
```

```
begin    { test_rect_polar }
x := -3.0e4931;   y := 4.0e4931;
rect_polar(x, y, r, theta);
WriteLn('(x, y) = {r, ', #233, '}:');
WriteLn('(', x:17, ',', y:17, ') = {',
        r:17, ', ', theta:8:8, '}');
x := -3.0e-4931;   y := 4.0e-4931;
rect_polar(x, y, r, theta);
WriteLn('(x, y) = {r, ', #233, '}:');
WriteLn('(', x:17, ',', y:17, ') = {',
        r:17, ', ', theta:8:8, '}');
ReadLn;
end.    { test_rect_polar }
```

This program reminds us of an old subject called *trigonometry*; so let's have a unit for solving triangles, starting with the interface,

Program 9.1.4

```
unit   trig;    { 05/31/95 }

interface

uses   CRT;

type   Real = Extended;

var      problem_no, no_sol: Word;
                  good_data: Boolean;
                     a, b, c,
                  a1, b1, c1,
                  a2, b2, c2,
         alpha, beta, gamma,
      alpha1, beta1, gamma1,
      alpha2, beta2, gamma2,
      area, cir_rad, in_rad: Real;

procedure   printout
               (a, b, c, alpha, beta, gamma: Real);
procedure   sas(a, gamma, b: Real;
                var   alpha, c, beta: Real;
                var   good_data: Boolean);
procedure   asa(alpha, c, beta: Real;
                var   a, gamma, b: Real;
                var   good_data: Boolean);
procedure   sss(a, b, c: Real;
                var   alpha, beta, gamma: Real;
```

```
                        var   good_data: Boolean);
procedure   ssa(a, c, gamma: Real;
                var   alpha1, beta1, b1,
                      alpha2, beta2, b2: Real;
                var   good_data: Boolean;
                var   no_sol: Word);
procedure   get_sas_data;
procedure   put_sas_solution;
{ a, gamma, b --> alpha, c, beta }
procedure   get_asa_data;
procedure   put_asa_solution;
{ alpha, c, beta --> a, gamma, b }
procedure   get_sss_data;
procedure   put_sss_solution;
{ a, b, c --> alpha, beta, gamma }
procedure   get_ssa_data;
procedure   put_ssa_solution;
{ a, c, gamma --> alpha, b, beta }
procedure   globals(a, b, c: Real;
                var   area, cir_rad, in_rad: Real);
```

For each of the standard trigonometry problems (side-side-side, side-angle-side, etc) there is a solution procedure, a procedure for getting input, and a procedure for output of the solution. Only part of the implementation will be listed; the rest is left for you. Note that "globals" computes area, inradius, and circumradius.

implementation

```
const   deg2rad = Pi/180.0;
        rad2deg = 180.0/Pi;

procedure   printout(a, b, c, alpha, beta, gamma: Real);

    begin
    end;

function   arcsin(x: Real): Real;     { degrees }

    begin
    if   x  = 1.0  then   arcsin := 90.0
    else   if   x = -1.0   then   arcsin := -90.0
    else   arcsin := rad2deg *
                        Arctan(x/Sqrt(1.0 - Sqr(x)));
    end;
```

```pascal
function arccos(x: Real): Real;    { degrees }

   var  y: Real;

   begin
   if  x = 0.0  then   arccos := 90.0
   else
      begin
      y := rad2deg * Arctan( Sqrt(1.0 - Sqr(x))/x );
      if x > 0.0  then   arccos := y
      else  arccos := y + 180.0;
      end;
   end;

procedure  ssa(a, c, gamma: Real;
               var  alpha1, beta1, b1,
                    alpha2, beta2, b2: Real;
               var  good_data: Boolean;
               var  no_sol: Word);

   var  h: Real;

   begin    { ssa }
   good_data := (a > 0.0) and (c > 0.0)
                             and (gamma > 0.0)
                             and (gamma < 180.0);
   if  not good_data  then   Exit;
   h := a*Sin(gamma*deg2rad);
   if  c < h  then   no_sol := 0
   else  if  c = h  then
      begin
      no_sol := 1;
      alpha1 := 90.0;
      beta1 := 90.0 - gamma;
      b1 := a * Cos(gamma*deg2rad);
      end
   else  if  c >= a   then
      begin
      no_sol := 1;
      alpha1 := arcsin(h/c);
      beta1 := 180.0 - gamma - alpha1;
      b1 := c*Cos(alpha1*deg2rad)
              + a*Cos(gamma*deg2rad);
      end
   else    { a > c > h }
```

```
  begin
    no_sol := 2;
    alpha1 := arcsin(h/c);
    beta1 := 180.0 - gamma - alpha1;
    b1 := c*Cos(alpha1*deg2rad)
             + a*Cos(gamma*deg2rad);
    alpha2 := 180.0 - alpha1;
    beta2 := alpha1 - gamma;
    b2 := c*Cos(alpha2*deg2rad)
             + a*Cos(gamma*deg2rad);
  end;
end;    { ssa }

procedure  report_bad_data;

  begin
    WriteLn('The data is geometrically impossible.');
    WriteLn('No such triangle exists.');
    WriteLn;
    WriteLn('Please try again.');
  end;

procedure  get_ssa_data;

  begin
    WriteLn('(Enter angles in degrees ',
             'and decimals, like 35.76)');
    WriteLn;
    WriteLn('Input: sides   a,   c,   angle gamma:');
    WriteLn;
    Write('a = ');   ReadLn(a);
    WriteLn;
    Write('c = ');   ReadLn(c);
    WriteLn;
    Write('gamma = ');   ReadLn(gamma);
  end;   {  get_ssa_data }

procedure  put_ssa_solution;
{ a, c, gamma --> alpha, b, beta }

    var   area2, in_rad2, cir_rad2: Real;
```

```pascal
begin
if  good_data   then
   begin
   WriteLn;
   case  no_sol   of
      0:  WriteLn('No solutions');
      1:  begin
          WriteLn;
          WriteLn('One solution:');
          WriteLn;

          Write('Given:');
          GotoXY(20, WhereY);
          WriteLn('Solution:');

          Write('a    = ',  a:6:4);
          GotoXY(20, WhereY);
          WriteLn('alpha = ', alpha1:4:2, ' deg');

          Write('c    = ', c:6:4);
          GotoXY(20, WhereY);
          WriteLn('b    = ', b1:6:4);

          Write('gamma = ', gamma:4:2, ' deg');
          GotoXY(20, WhereY);
          WriteLn('beta  = ', beta1:4:2, ' deg');

          globals(a, b1, c, area,
                  cir_rad, in_rad);
          GotoXY(20, WhereY);
          WriteLn('area = ', area:6:4,
                  ' sqr units');
          GotoXY(20, WhereY);
          WriteLn('inradius = ', in_rad:6:4);
          GotoXY(20, WhereY);
          WriteLn('circumradius = ', cir_rad:6:4);
          end;
      2:  begin
          WriteLn;
          WriteLn('Two solutions');
          .......................
          end;
      end;
   end
else  report_bad_data;
end;  { put_ssa_solution }
```

```
procedure   globals(a, b, c: Real;
                    var    area, cir_rad, in_rad: Real);

   var   s: Real;

   begin
   s :=   0.5*(a + b + c);
   area := Sqrt( s*(s - a)*(s - b)*(s -   c) );
   cir_rad := (a*b*c)/(4.0*area);
   in_rad := area / s;
   end;

end.    { unit  trig }
```

Exercises

1. The disk file "epsilon.pas" contains another program "ure" for computing unit roundoff error. Examine it, and compare it to "unit_roundoff_error" above.

2. Add variables as needed and replace the **for** j loop in the function "limit" of "extrapol" by

```
       for  j := 1  to  n  do
            begin   { Compute j-th row }
            k1 := j;   j_k := 1;
            for  k := j - 1  downto  0  do
                 begin
                 dk1 := diff[k1];
                 diff[k] := dk1 + denom[j_k]*(dk1 - diff[k]);
                 Dec(k1);   Inc(j_k);
                 end;    { for k }
            end;    { for j }
```
   Show that it yields the same result as the original. (Its advantage is that additional data points can be added one at a time.)

3. Test the function "limit" in the unit "extrapol" on $\lim_{x \to 0+} (1 - \cos x)/x^2$.

4. Test the function "limit" in the unit "extrapol" on $\lim_{x \to 0+} x^x$. Any observations?

5. The series $\sum_1^\infty n^{-2}$ converges to $\pi^2/6$. Let s_k be the sum of the first 2^k terms of this series, for k from 1 to 9. Apply "limit" in the unit "extrapol" to this sequence for an estimate of the sum of the series.

6. (Cont.) Apply the same idea to $\sum_1^\infty F(1/n)$, where $F(x) = (\arctan x) \ln(1 + x)$, only go further, maybe to 2^{12} terms. The series is known to converge to $(\pi/2) \ln 2$.

7. Complete the unit "trig" for solving triangles.

8. Write a user-friendly program for solving triangles, using "trig" of course.

9. In case calculators and computers drop out of fashion, you may again need trig tables. Write a program that produces a 4-place table of sine, cosine, and tangent. (You may have to look up an old one to remind yourself of the format.)

Section 2. Sequences and Sums

We begin with a program for computing e to 1000 decimal places, starting from its factorial expansion

$$e = 1 + \sum_{k=1}^{\infty} \frac{1}{k!} = 2 + \sum_{k=1}^{\infty} \frac{1}{(k+1)!}$$

In general, any number b satisfying $0 \le b < 1$ has a unique *factorial expansion*

$$b = \sum_{k=1}^{\infty} \frac{b_k}{(k+1)!} \qquad b_k \in \mathbf{Z} \qquad 0 \le b_k \le k$$

If we truncate the factorial expansion

$$b = \sum_{k=1}^{\infty} \frac{b_k}{(k+1)!} = \sum_{k=1}^{N} \frac{b_k}{(k+1)!} + \sum_{k=N+1}^{\infty} \frac{b_k}{(k+1)!}$$

then we can bound the error:

$$\sum_{k=N+1}^{\infty} \frac{b_k}{(k+1)!} \le \sum_{k=N+1}^{\infty} \frac{k}{(k+1)!} = \sum_{k=N+1}^{\infty} \left[\frac{1}{k!} - \frac{1}{(k+1)!} \right] = \frac{1}{(N+1)!}$$

(telescoping series). As $1/451! \approx 1.3 \times 10^{-1003}$, we obtain better than 1000 decimal place accuracy with $N = 450$ terms of the factorial expansion, and this applies to e.

If $b = 0.d_1 d_2 d_3 \cdots$, then

$$10b = d_1 + 0.d_2 d_3 d_4 \cdots \approx \sum_{k=1}^{N} \frac{10 b_k}{(k+1)!} = \sum_{k=1}^{N} \frac{c_k}{(k+1)!}$$

The c's are computed as follows. First all b's are multiplied by 10 to form new b's (maybe too large). Starting with b_N and working down, each c_k is set equal to $b_k \bmod k$ and the carry is added to b_{k-1}.

In the listing, note that the multiplications by 2 and by 8 (power of 2) are compiled as shifts left. Also, the computation of coeff[j] in the next line, using a multiplication and a subtraction, is faster than another long division.

Program 9.2.1

```
program  compute_e;    { 08/12/95 }



   const                 N = 451;
              no_digits = 1000;

   var   j, k, q, s, t: Word;
                   coeff: array[1..N] of Word;

   begin    { compute_e }
   { Initialize }
   coeff[1] := 0;
   for  j := 2  to  N  do  coeff[j] := 1;
   Write('e = 2.');
   { Compute decimal expansion }
   for  k := 1  to  no_digits  do
      begin
      coeff[N] := 10*coeff[N];
      for  j := N  downto  2  do
         begin
         s := coeff[j - 1];
         t := coeff[j];
         q := t div j;
         coeff[j - 1] := 8*s + 2*s + q;
         coeff[j] := t - j*q;      { t mod j }
         end;    { for j }
      Write(coeff[1]);
      coeff[1] := 0;
      if  k mod 50 = 0  then
         begin
         WriteLn;  Write(' ':6);
         end
      else  if  k mod 5 = 0  then  Write(' ');
      end;    { for k }
   WriteLn;  WriteLn;
   Write('Press <Enter>: ');  ReadLn;
   end.    { compute_e }
```

Our next project is to estimate $\sum_1^n k!$. This sum is obviously very large for large n, so we really want to estimate its logarithm. We define sequences

$$s_n = \sum_{k=1}^n k! \qquad t_n = (n+1)s_{n-1} \qquad x_n = \ln s_n \qquad y_n = \ln t_n$$

Then
$$s_n = s_{n-1} + n! = s_{n-1} + n \cdot (n-1)! = s_{n-1} + n[s_{n-1} - s_{n-2}]$$
$$= (n+1)s_{n-1} - n \cdot s_{n-2} = t_n - t_{n-1}$$

and we have iterative formulas for the logarithms:
$$x_n = \ln(t_n - t_{n-1}) = \ln t_n + \ln\left(1 - \frac{t_{n-1}}{t_n}\right) = y_n + \ln[1 - \exp(y_{n-1} - y_n)]$$

$$y_n = \ln(n+1) + \ln s_{n-1}$$

The following program is based on these formulas.

Program 9.2.2

```
program  factorial_sum;    { 12/24/93 }

   var             x, log_s,
                   old_y, y,
                   mantissa: Extended;
                        j, n,
                characteristic: Word;

   begin    { factorial_sum }
   repeat
      Write('Enter N (0 to quit): ');
      ReadLn(n);
      if  n = 0   then   Halt;
      if  n = 1   then   x := 0.0
      else
         begin
         x := Ln(3.0);   old_y := x;
         for  j := 3  to  n  do
            begin
            y := x + Ln(j + 1.0);
            x := y + Ln(1.0 - Exp(old_y - y));
            old_y := y
            end;    { for j }
         end;    { else }
      log_s := x/Ln(10.0);
      characteristic := Trunc(log_s);
      mantissa := Exp(
          (log_s - characteristic)*Ln(10.0) );
      WriteLn;
      WriteLn('s(', n,') = ', mantissa:12:10,
              ' e', characteristic);
      WriteLn;
   until  false;
   end.    { factorial_sum }
```

Our next program is a short unit, with two implementations, of the Euler transformation for summing alternating series. The problem is to estimate the sum $S = \sum_0^\infty (-1)^n a_n$, where $a_n > 0$ and where a_n decreases to 0, but not too rapidly, say, $a_n = O(1/n)$. The Euler transform changes the alternating series into one that often converges much faster. We start with the difference operators Δ and E defined by

$$\Delta a_n = a_{n+1} - a_n \qquad E a_n = a_{n+1}$$

Then formally, $E = 1 + \Delta$ and $a_n = E^n a_0$. We write

$$S = \sum_{n=0}^\infty (-1)^n a_n = \sum_{n=0}^\infty (-1)^n E^n a_0 = \frac{1}{1+E} a_0$$

But

$$\frac{1}{1+E} = \frac{1}{2+\Delta} = \frac{1}{2} \frac{1}{1 + \frac{1}{2}\Delta} = \frac{1}{2} \sum_{n=0}^\infty \frac{(-1)^n}{2^n} \Delta^n$$

We conclude that

$$S = \sum_{n=0}^\infty (-1)^n a_n = \frac{1}{2} \sum_{n=0}^\infty \frac{(-1)^n}{2^n} \Delta^n a_0$$

For example, consider the alternating harmonic series

$$S = \sum_{n=0}^\infty \frac{(-1)^n}{n+1} = \ln 2 \approx 0.693147$$

It takes about 10^6 terms of the series for 6-place accuracy. In this case the transformed series is

$$S = \sum_{n=0}^\infty \frac{1}{2^n(n+1)} \qquad \text{and} \qquad \sum_{n=0}^{20} \frac{1}{2^n(n+1)} \approx 0.693147$$

a considerable improvement! See the difference table below whose first column contains the entries for $\Delta^n a_0$

n	0	1	2	3	4	5
a_n	1	1/2	1/3	1/4	1/5	1/6
Δa_n	−1/2	−1/6	−1/12	−1/20	−1/30	
$\Delta^2 a_n$	1/3	1/12	1/30	1/60		
$\Delta^3 a_n$	−1/4	−1/20	−1/60			
$\Delta^4 a_n$	1/5	1/30				
$\Delta^5 a_n$	−1/6					

The unit "alt_ser" begins with "euler1", which implements directly the theory above. Instead of a rectangular array for the difference table, only a single row is used, and each subsequent row is computed on top of the current row, that is, "in place". To avoid fuss with signs, differences are computed backwards instead of forwards.

There is a refinement of the Euler transformation due to A van Wijngaarden. Its rather complicated derivation can be found in Goodwin, E T, ed, Modern Computing Methods, 2/e, Her Majesty's Stationery Office, 1961, 125-126. It is implemented in "euler2" below.

Program 9.2.3

```
{$N+}
unit   alt_ser;     { 06/01/95 }

interface

const   abs_max = 50;

type        Real = Extended;
        sequence = array[0..abs_max] of Real;

function   euler1(var   A: sequence;   max: Word): Real;
function   euler2(var   A: sequence;   max: Word): Real;

implementation

function   euler1(var   A: sequence;   max: Word): Real;

    var         k, n: Word;
                diff: sequence;
          pow2, sum: Real;

    begin    { euler1 }
    pow2 := 1.0;
    diff := A;
    sum := diff[0];
    for   n := 1   to    max   do
        begin
        pow2 := 0.5*pow2;
        for   k := 0   to    max - n    do
            diff[k]  := diff[k] - diff[k + 1];
        sum := sum + pow2*diff[0];
        end;
    euler1 := 0.5*sum;
    end;    { euler1 }
```

```
function   euler2(var   A: sequence;   max: Word): Real;

   const   max_count = 4;
                 eps = 1.0e-16;

   var             sign: Integer;
              diagonal: sequence;
         first, term,
                  sum: Real;
        first_eff_col,
          n, k, count: Word;

   begin    { euler2 }
   { Initialize }
   n := 0;
   first_eff_col := 0;
   count := 0;
   sign := 1;
   sum := 0.5*A[0];
   diagonal[0] := sum;
   { End initializations }
   repeat
      Inc(n);
      diagonal[n] := 0.5*A[n];
      for  k := n - 1  downto  first_eff_col  do
         diagonal[k]
             := 0.5*(diagonal[k] - diagonal[k + 1]);
      first := diagonal[first_eff_col];
      if  Abs(diagonal[first_eff_col + 1])
                    < Abs(first)   then
          begin
          term:= 2*sign*first;
          sign := -sign;
          Inc(first_eff_col);
          end
      else   term := sign*first;
      sum := sum + term;
      if  Abs(term) < eps   then   Inc(count)
      else   count := 0;
   until   (count >= max_count) or (n = max);
   euler2 := sum;
   end;    { euler2 }

end.   { unit  alt_ser }
```

We computed e above; now let's look at π. There is a long history of algorithms for computing π, and perhaps the state of the art is the algorithm of the Borwein brothers, Peter and Jonathan. Their book, Pi and the AGM, Wiley, 1987, contains a survey of the modern theory. A brief exposition of their algorithm is in my article, Computing pi, Coll. Math. J. 17, 1986, 230-235. The algorithm goes as follows: Three sequences $\{x_n\}$, $\{y_n\}$, and $\{p_n\}$ are defined by the initializations

$$x_0 = 2^{-1/2} \qquad y_0 = 0 \qquad p_0 = 2$$

and the iterations

$$\begin{cases} x_{n+1} = \frac{1}{2}\left(x_n^{1/2} + x_n^{-1/2}\right) \\ y_{n+1} = \left(y_n x_n^{1/2} + x_n^{-1/2}\right)/(y_n + 1) \qquad (n \ge 0) \\ p_{n+1} = [(x_n + 1)/(y_n + 1)]p_n \end{cases}$$

Then $\{p_n\}$ is a strictly decreasing sequence, $p_n \to \pi$ rapidly, in fact, and for $n \ge 4$,

$$0 < p_n - \pi \le 10^{-2^{n+1}}$$

The program "compute_pi" implements this algorithm. I remind you that Exercise 4 of Section 8.3 suggests a multi-precision integer implementation of the algorithm. You might wish to refer back to that now.

Program 9.2.4

```
program  compute_pi;    { 06/01/95 }
{ Algorithm of J. M. and P. B. Borwein }

    var            x, y, approx_pi,
            sqrt_x, recip_sqrt_x: Extended;
                           n: Word;
```

```
begin
{ Initialize }
x := Sqrt(0.5);          { x(0) }
y := 0.0;                { y(0) }
approx_pi := 2.0;        { pi(0) }
WriteLn('pi     = ', Pi:20:18);
WriteLn('pi(0) = ', approx_pi:20:18);
WriteLn('error =', approx_pi - Pi:10);
{ Iterate }
for  n := 1  to  5  do
    begin
    { The order of computing is important. }
    sqrt_x := Sqrt(x);
    recip_sqrt_x := 1.0/sqrt_x;
    approx_pi := (x + 1.0)*approx_pi/(y + 1.0);
    y := (y*sqrt_x + recip_sqrt_x)/(y + 1.0);
    x := 0.5*(sqrt_x + recip_sqrt_x);
    WriteLn;
    WriteLn('pi     = ', Pi:20:18);
    WriteLn('pi(', n, ') = ', approx_pi:20:18);
    WriteLn('error =', approx_pi - Pi:10);
    end;
ReadLn;
end.    { compute_pi }
```

Anyone interested in another approach to computing π to high accuracy will find a program "pi_spigot" on the disk. It implements an algorithm recently published in Rabinowitz, S and Wagon, S, A spigot algorithm for the digits of pi, Amer. Math. Monthly 102, 1995, 195-203. It is like the program "compute_e" above.

Finally, there is a program "fft" on disk for implementing the Fast Fourier Transform, which is an efficient way to handle numerical data in many situations. It works in the complex domain, so it uses the unit "complexs" of Section 3.4.

Exercises

1. Test the unit "alt_ser" on the alternating harmonic series.
2. Test it also in the case $a_n = 1/\ln(n+2)$.
3. Bracket MaxLongint between two successive Fibonacci numbers.

In the next three exercises, eps $= 10^{-16}$ and all variables are real. Predict what each does.

4. **function** unknown1: Real;
```
      begin
      x := 1.0;
      repeat
           y := x;
           x := 1.0/(1.0 + y);
      until  Abs(x - y) < eps;
      unknown1 := x;
      end;
```
5. **function** unknown2: Real;
```
      begin
      x := 0.0;
      repeat
           y := x;
           x := Sqrt(12.0 - y);
      until  Abs(x - y) < eps;
      unknown2 := x;
      end;
```
6. **procedure** unknown3(**var** x, y: Real);
```
      begin
      x := 0.0;   y := 0.0;
      repeat
           z := x;   w := y;
           x := Sqrt(7.0 - y);
           y := Sqrt(7.0 + x);
      until  (Abs(x - z) < eps) and
             (Abs(y - w) < eps);
      end;
```
7. Write a unit for estimating derivatives of a function by various finite difference approximations.

Section 3. Functions

The following unit contains some interesting algorithms for computing values of some standard real functions, some under conditions of limited computing facilities. Let's look at the interface and discuss what's in it.

Program 9.3.1

```
unit   funct_1;    { 08/12/95 }

interface

var   error_flag: Boolean;
```

```
function   reciprocal(x: Real): Real;
function   power(x: Real;   n: LongInt): Real;
function   square_root(x: Real): Real;
function   exponential_1(x: Real): Real;
function   nat_log(x: Real): Real;
function   loga(a, x: Real): Real;
procedure  quadratic_formula(a, b, c: Real;
                        var   r1, r2: Real);
```

"Reciprocal" computes $1/x$, where $0 < x < 2$, using only addition, subtraction, and multiplication. "Power" is the standard method of computing integer powers of a real number, using the binary expansion of the exponent. "Square_root", like "reciprocal", uses only $+$, $-$, $\times$, but produces $\sqrt{x}$.

"Exponential_1" is a crude numerical approximation to e^{-x}, reasonably accurate only for x near 0. "Nat_log" uses only the 4 arithmetic operations and $\sqrt{}$ for a very good approximation to $\ln x$. "Loga" is another very good approximation, for $\log_a x$, using only the 4 arithmetic operations. Its source is Knuth, D E, vol I, p. 24. Finally, "quadratic_formula" solves real quadratic equations with real roots. It uses some care to avoid roundoff errors that might result from division by a very small denominator. The implementation of "funct_1" follows.

implementation

```
const   eps = 1.0e-15;

function   reciprocal(x: Real): Real;
     { Assumes   0 < x < 2 }

   var   a, b: Real;

   begin
   a := 1.0;   b := 1.0 - x;
   repeat    { Loop invariant: ax = (1 - b) }
      a := a*(1.0 + b);   b := Sqr(b);
   until   b < eps;
   { b --> 0, so a --> 1/x }
   reciprocal := a;
   end;

function   power(x: Real;   n: LongInt): Real;

   var      m: LongInt;
         y, z: Real;
```

```
begin
if  n = 0  then  power := 1.0
else
    begin
    m := Abs(n);
    y := 1.0;   z := x;
    while  m > 0  do
        begin
        if  Odd(m)  then
            begin  y := y*z;   Dec(m);   end
        else
            begin  z := Sqr(z);   m := m shr 1;   end;
        end;
    if  n >= 0  then  power := y
    else
        begin
        if  y = 0.0  then  Halt
        else  power := 1.0/y;
        end;
    end;
end;   { power }

function  square_root(x: Real): Real;
    { Assumes  x >= 0 }

    var  a, b, m: Real;

    begin
    if  x = 0  then  square_root := 0
    else
        begin
        m := 1.0;
        { Normalize }
        while  x >= 2.0  do
            begin  x := 0.25*x;   m := 2.0*m   end;
        while  x < 0.5  do
            begin  x := 4.0*x;   m := 0.5*m   end;
        a := x;   b := 1.0 - x;
        repeat    { Loop invariant:   a^2 = x(1 - b) }
            a := a*(1 + 0.5*b);
            b := 0.25*(3.0 + b)*Sqr(b);
        until  b < eps;    { b --> 0, so a --> sqrt(x) }
        { Denormalize }
        square_root := a*m;
        end    { else }
    end;    { square_root }
```

```
function  exponential_1(x: Real): Real;
   { Estimates exp(-x) }

   const  a0 = 0.05;    a1 = 6.9;
          a2 = 205.8;   a3 = 42.0;

   var  y, z: Real;

   begin
   y := Sqr(x);   z := a2/(y + a3);
   z := a0*y + x + a1 - z;
   exponential_1 := 1.0 - 2.0*x/z;
   end;    { exponential_1 }

function  nat_log(x: Real): Real;
    { Assumes  x > 0 }

   var  a, b: Real;

   begin
   a := 0.5*(x - 1.0/x);   b := 0.5*(x + 1.0/x);
   repeat
      b := Sqrt(0.5*(1.0 + b));   a := a/b
   until  b < 1.0 + eps;
   nat_log := a;
   end;    { nat_log }

function  loga(a, x: Real): Real;
   { Assumes  a > 1.0,   x > 0.0 }

   var  logarithm,
        y, error: Real;
           expon: Integer;

   begin    { loga }
   { Normalize:   x = (a^expon)y,   1.0 <= y < a }
   y := x;
   expon := 0;
   if  y >= 1.0   then
       while  y >= a   do
          begin  Inc(expon);   y := y/a   end
   else
      while  y < 1.0   do
          begin  Dec(expon);   y := a*y   end;
   { end normalize }
   logarithm := expon;   error := 1.0;
```

```
   repeat    { Binary expansion }
      y := Sqr(y);     { y < a }
      error := 0.5*error;
      if  y >= a   then
          begin
          y := y/a;   logarithm := logarithm + error;
          end
      { Now y < a }
   until   error < eps;
   loga := logarithm;
   end;    { loga }

procedure   quadratic_formula(a, b, c: Real;
                              var   r1, r2: Real);
   { Soluton of   ax + 2bx + c = 0,    a <> 0,
     nonnegative discriminant }

   var   root_delta, delta, den: Real;

   begin
   error_flag := False;
   delta := b*b - a*c;
   if  delta < 0.0   then
       begin
       r1 := 0.0;   r2 := 0.0;
       error_flag := True;   Exit;
       end
   else  if  delta = 0.0   then
       begin  r1 := -b/a;   r2 := r1;   end
   else
       begin
       root_delta := Sqrt(delta);
       if  b >= 0.0   then
           begin
           den := b + root_delta;
           r1 := -c/den;   r2 := -den/a;
           end
       else
           begin
           den := root_delta - b;
           r1 := den/a;   r2 := c/den;
           end;
       end;
   end;    { quadratic_formula }

end.    { funct_1 }
```

Our next unit "funct_2" presents algorithms for some standard transcendental functions, based on their continued fraction approximations. The source is More-lock, J C, CACM 129, 1964. Information on continued fraction approximations in general can be found in Ralston, A and Rabinowitz, P, First Course in Numerical Analysis, 2/e, McGraw-Hill, 1978. An interesting point is that the algorithms work in the complex domain as well as the real domain, so could be combined with the unit "complexs" of Section 3.4 to handle the standard transcendental functions in the complex domain.

We start with the continued fraction

$$F(z) = 2 + \cfrac{z}{6} + \cfrac{z}{10} + \cfrac{z}{14} + \cdots + \cfrac{z}{4n+2} + \cdots$$

which means

$$F(z) = 2 + z/F_1(z) \qquad F_1(z) = 6 + z/F_2(z) \qquad \cdots$$
$$F_n(z) = (4n+2) + z/F_{n+1}(z)$$

The continued fraction converges rapidly, and the basic formulas are

$$F(z^2) = z\frac{e^z + 1}{e^z - 1} = \frac{z}{\tanh \frac{1}{2}z} \qquad F(-z^2) = \frac{z}{\tan \frac{1}{2}z}$$

The second follows from the first by substituting iz for z. From these formulas follow

$$\sin z = \frac{2zF(-z^2)}{z^2 + F^2(-z^2)} \qquad \cos z = \frac{F(-z^2) - z^2}{F(-z^2) + z^2} \qquad \tan z = \frac{2zF(-z^2)}{F^2(-z^2) - z^2}$$

and

$$\sinh z = \frac{2zF(z^2)}{F^2(z^2) - z^2} \qquad \cosh z = \frac{F(z^2) + z^2}{F^2(z^2) - z^2}$$

In the program we truncate the continued fraction expansion after n terms. For instance, if we cut at $n = 2$, then

$$z\frac{e^z + 1}{e^z - 1} = F(z^2) \approx 2 + \frac{z}{6 + z/10} = \frac{12z + 120}{z + 60}$$

This approximation, for $z = 0.5$, is accurate to 6 decimal places. The interface of "funct_2" follows.

Program 9.3.2

```pascal
unit  funct_2;    { 08/12/95 }

interface

   const  eps = 1.0e-16;

   type  Real = Extended;

   var  error_flag: Boolean;

   function  _Exp(x: Real;  n: Word): Real;
   function  _Sin(x: Real;  n: Word): Real;
   function  _Cos(x: Real;  n: Word): Real;
   function  _tan(x: Real;  n: Word): Real;
   function  sinh(x: Real;  n: Word): Real;
   function  cosh(x: Real;  n: Word): Real;
   function  tanh(x: Real;  n: Word): Real;
   function  arcsin(x: Real): Real;
   function  x_over_tanh_x(x: Real;  n: Word): Real;
```

The two final functions are in a different class. "Arcsin" shows how to use the built-in "arctan" function and a case analysis to define the arcsin function. The function "x_over_tanh_x" returns $x/\tanh x$ and is based on a different continued fraction, the *Lambert continued fraction*. It is like the continued fraction above for $F(z^2)$ except the sequence of integers is 1, 3, 5, 7, ... instead of 2, 6, 10, ... We pass to the implementation, which is quite straightforward.

```pascal
implementation

   function  cont_fract(z: Real;  n: Word): Real;

      var  i: Word;
           t: Real;

      begin
      t := 4.0*n + 2.0;
      for  i := n  downto  1  do
         t := 4.0*i - 2.0 + z/t;
      cont_fract := t;
      end;

   function  _Exp(x: Real;  n: Word): Real;

      var  s: Real;
```

```pascal
    begin
    s := cont_fract(x*x, n);
    _Exp := (s + x)/(s - x);
    end;    { _exp }

function _Sin(x: Real;   n: Word): Real;

    var  r, s: Real;

    begin
    r := x*x;
    s := cont_fract(-r, n);
    _Sin := 2.0*x*s/(r + s*s);
    end;    { _sin }

function  _Cos(x: Real;   n: Word): Real;

    var  r, s: Real;

    begin
    r := x*x;
    s := cont_fract(-r, n);
    s := s*s;
    _Cos := (s - r)/(s + r);
    end;    { _cos }

function  _tan(x: Real;   n: Word): Real;

    var  r, s: Real;

    begin
    r := x*x;
    s := cont_fract(-r, n);
    _tan := 2.0*x*s/(s*s - r);
    end;    { _tan }

function  sinh(x: Real;   n: Word): Real;

    var  r, s: Real;

    begin
    r := x*x;
    s := cont_fract(r, n);
    sinh := 2.0*x*s/(s*s - r);
    end;    { _sinh }
```

```
function   cosh(x: Real;   n: Word): Real;

   var   r, s: Real;

   begin
   r := x*x;
   s := cont_fract(r, n);
   s := s*s;
   cosh := (s + r)/(s - r);
   end;    { _cosh }

function   tanh(x: Real;   n: Word): Real;

   var   r, s: Real;

   begin
   r := x*x;
   s := cont_fract(r, n);
   tanh := 2.0*x*s/(s*s + r);
   end;    { _tanh }

function   x_over_tanh_x(x: Real;   n: Word): Real;

   var   j: Integer;   y, z: Real;

   begin
   j := 2*n - 1;   y := Sqr(x);   z := 0.0;
   repeat
      z := y/(j + z);   Dec(j, 2);
   until   j = 1;
   x_over_tanh_x := 1.0 + z;
   end;

function   arcsin(x: Real): Real;

   const   half_pi = 0.5*Pi;

   var   t: Real;

   begin
   error_flag := False;
   if   Abs(x) <= 0.707107   then
      begin
      t := Sqrt(1.0 - Sqr(x));
      arcsin := Arctan(x/t);
      end
```

```
        else
           begin
           t := 1.0 - Sqr(x);
           if  t < 0.0  then
               begin
               arcsin := 0.0;   error_flag := True;
               end
           else
               begin
               t := Sqrt(t);
               if  x > 0.0  then
                   arcsin := half_pi - Arctan(t/x)
               else   arcsin := -half_pi - Arctan(t/x);
               end;
           end;
        end;    { arcsin }

end.    { unit  funct_2 }
```

Exercises

1. Write a program to test everything in "funct_1".
2. Write a program to test everything in "funct_2".

Section 4. Zeros of Functions

We have used Newton's method for solving real equations; let us look at a complex version of the method for solving polynomial equations, based on the unit "complexs" of Section 3.4. (We omit listing the I/O procedures here.) The Horner evaluation of a polynomial and its first derivative is adopted from the unit "real_poly" of Section 3.9. As you know, the success of Newton in converging to a simple zero depends on the choice of an initial guess. Always keep in mind the example $F(z) = z^2 + 1$ where any initial real z must fail.

Program 9.4.1

```
program  complex_newton;    { 05/01/95 }

   uses  CRT, complexs;    { ch3\s4 }

   const  eps = 1.0e-10;
          eta = 1.0e-4;
          max = 50;
```

```
type   poly = ^term;
       term = record
                    deg: Word;
                    coef: complex;
                    next: poly;
                end;

var   z, w, w1, t: complex;
                F: poly;
                j: Word;
               ch: Char;

(* procedure  get_poly(var F: poly);
   procedure  get_initial_z;
   procedure  print_poly(F: poly;  n: word); *)

procedure  evaluate(F: poly;  c: complex;
                       var  d0, d1: complex);
     { Returns  d0 = F(c)  and  d1 = F'(c) }

     var  i, k: Word;
          run: poly;

     begin
     d0 := F^.coef;
     d1.re := 0.0;
     d1.im := 0.0;
     run := F;
     while  run <> nil  do
        begin
        if  run^.next = nil  then  k := run^.deg
        else  k := run^.deg - run^.next^.deg;
        for  i := k  downto  1  do
           begin
           product(d1, c, d1);
           sum(d1, d0, d1);
           product(d0, c, d0);
           end;
        run := run^.next;
        if  run <> nil  then  sum(d0, run^.coef, d0);
        end;    { while }
     end;    { evaluate }
```

```
begin    { complex_newton }
ch := 'Y';
repeat
    if  UpCase(ch) = 'Y'  then  get_poly(F);
    get_initial_z;
    j := 0;
    evaluate(F, z, w, w1);
    while  (modulus(w1) >= eta) and (j < max) and
           (modulus(w) >= eps)  do
        begin    { Newton-Raphson iteration }
        quotient(w, w1, t);
        difference(z, t, z);
        evaluate(F, z, w, w1);
        Inc(j);
        end;    { while }
    if  modulus(w1) < eta  then
        begin
        Write('F'' too small at  z = ');
        print_complex(z, 10);  WriteLn;
        end
    else if  modulus(w) < eps  then
        begin    { Zero found }
        end
    else
        begin    { Report max iterations }
        end;
    { etc }
until  UpCase(ch) = 'Q';
end.    { complex_newton }
```

Newton's method can be extended to vector functions of several real variables. Let $\mathbf{R}^n$ denote the space of n-dimensional real *column* vectors. Let F be a smooth function

$$F: \mathbf{D} \longrightarrow \mathbf{R}^n \qquad \mathbf{D} \subseteq \mathbf{R}^n$$

The ordinary first derivative used in Newton steps for one variable is replaced by the Jacobian matrix $J = \|\partial F_i / \partial x_j\|$. Given a point $\mathbf{x}$ in $\mathbf{D}$, the Newton step is

$$\mathbf{x} \longleftarrow \mathbf{x} - J^{-1}\Big|_{\mathbf{x}} F(\mathbf{x})$$

If the original $\mathbf{x}$ is a good approximation to a zero of F, then the new $\mathbf{x}$ should be a better one.

A program based on this algorithm must include a method of computing all first partial derivatives. The most direct is simply to write code for the derivative

functions. Better is to have a procedure built in for symbolic differentiation. Best
of all is "automatic differentiation", but that is another subject altogether.

Our example is for the modest $n = 2$. We try to find extrema of a real-valued
function $F(x, y)$ by the first derivative test: finding common zeros of the two first
partials. Their Jacobian matrix is the (Hessian) matrix of second derivatives of F.

Program 9.4.2

```
program  nonlinear_system;    { 05/01/95 }

   const   max = 20;
           eps = 1.0e-10;
           eta = 1.0e-3;

   type   vector_fun = function(j: Word;
                                 x, y: Real): Real;

   var   x, y: Real;
         i, n: Word;
            ch: Char;

   function   F(x, y: Real): Real;

      begin
      F := Cos(x)*Cos(y) - 0.1*x - 0.2*y + 0.15*x*y;
      end;

   {$F+}
   function  G(j: Word;   x, y: Real): Real;

      begin
      if  j = 1   then
         G := -Sin(x)*Cos(y) - 0.1 + 0.15*y
      else
         G := -Cos(x)*Sin(y) - 0.2 + 0.15*x;
      end;     { F }

   function  dG(j, k: Word;   x, y: Real):   Real;

     begin
     if  j = 1   then
         begin
         if  k = 1   then   dG := -Cos(x)*Cos(y)
         else   dG := Sin(x)*Sin(y) + 0.15;
         end
```

```pascal
     else
        begin
        if  k = 1   then   dG := Sin(x)*Sin(y) + 0.15
        else   dG := -Cos(x)*Cos(y);
        end;
     end;    { dF }
   {$F-}

   procedure   solve(G:vector_fun;   var   x, y: Real);

      var       j, k: Integer;
             r, s, t: Real;
            singular: Boolean;
               u, v: Real;
                 ch: Char;

      procedure   Newton_step(var   x, y: Real);

         var    det: Real;
               j, k: Word;

         begin
         u := G(1, x, y);   v := G(2, x, y);
         r := dG(1, 1, x, y);
         s := dG(1, 2, x, y);
         t := dG(2, 2, x, y);
         det := r*t - s*s;
         if  Abs(det) < eta   then   singular := true
         else
            begin
            singular := false;
            x := x - (t*u - s*v)/det;
            y := y - (-s*u + r*v)/det;
            end
         end;    { Newton_step }

      begin    { solve }
      repeat
         Newton_step(x, y);
         Dec(i);
      until   singular or (i = 0) or
              ( Abs(u) + Abs(v) < eps );
      if  singular   then   WriteLn('System Crash.')
      else  if  i = 0   then
         begin
         WriteLn('After   ', n,'  iterations,');
```

```
      WriteLn('F(', x, ', ', y, ') = (',
                     u, ', ', v, ')')
      end
   else   WriteLn('Solution: (x, y) = (',
                     x, ', ', y, ')');
   end;    { solve }

begin      { nonlinear_system }
x := 0.0;   y := 0.0;
i := max;
solve(G, x, y);
WriteLn('F(x, y)                    =', F(x, y));
WriteLn;
Write('Press <Emter>: ');   ReadLn;
end.
```

The rate of convergence of Newton's method is quadratic. There are methods that converge faster; Halley's method is the best known; its convergence is cubic. It also has a nice geometric interpretation. Recall that in Newton's method you construct the tangent to $y = F(x)$ at $(x, F(x))$. Then the new x is the intersection of this tangent line with the x-axis. In Halley's method, you construct the rectangular hyperbola that has second order contact with $y = F(x)$ at $(x, F(x))$. The new x is where this hyperbola meets the x-axis.

Halley's method is due to Edmund Halley (of Halley's comet), and the original source is Halley, E, A New and general method of finding the roots of equations, Philos Trans Royal Soc London 18 (1694), 136. As far as I know, the geometric interpretation is due to Salikov, G S, On the convergence of the process of tangent hyperbolas (Russian), Doklady Akad. Nauk CCCP 82 (1952), 525-528. Please see also Richmond, H W, On certain formulae for numerical approximation, J London Math Soc 19 (1944), 31-38, and Traub, J F, Iterative Methods for the Solution of Equations, Prentice-Hall, 1964.

The computation details of the following program are included with the disk program in a note at the end of the listing. The note also includes material on several other methods of third—fifth order. The disk also contains a program "laguerre" with a rather obscure third order method of Edmund Laguerre for solving polynomial equations.

Program 9.4.3

```
program   halley;    { 06/03/95 }

   type   Real = Extended;

   var   c0, c1: Real;
```

```
procedure   F(t: Real;   var   f0,   f1, f2: Real);
    { Outputs a function,  first derivative,
        and second derivative values of a function }

    begin
    f0 := Cos(t) - t;
    f1 := -Sin(t) - 1;
    f2 := -Cos(t);
    end;

function   iterate(t: Real): Real;

    var   f0, f1, f2,
                u, v: Real;

    begin
    F(t, f0, f1, f2);
    if  Abs(f1) < 0.001  then  Halt;
    u := f0/f1;   v := f2/f1;
    iterate := t - u / (1.0 - u*v);
    end;

begin
WriteLn('Halley method');
c1 := 5.0;
repeat
    WriteLn('    c = ', c1:18:18);
    c0 := c1;
    c1 := iterate(c0);
until  Abs(c1 - c0) < 1.0e-15;
WriteLn('zero = ', c0:18:18);
ReadLn;
end.
```

We conclude this section with a program "all_zeros" to find all the zeros of a complex polynomial, a constructive proof of Gauss's famous "Fundamental Theorem of Algebra". Note the **uses** clause reference to the unit "complexs" of Chapter 3, Section 4. You should review that unit at this time.

The Fundamental Theorem asserts that each non-constant complex polynomial $F(z)$ has a complete factorization

$$F(z) = a_n(z - r_1)(z - r_2) \cdots (z - r_n)$$

where a_n is the leading coefficient and $r_1, \cdots, r_n$ are the zeros of the polynomial, with multiplicities.

Suppose that $z = a$ is a point of the complex plane where $F(a) \neq 0$. We want to move from a to another point z such that $|F(z)| < |F(a)|$. We may suppose

$$F(z) = b + c(z - a)^k + O\left(|z - a|^{k+1}\right) \qquad b = F(a) \neq 0, \quad c \neq 0, \quad k \geq 1$$

We choose

$$z = a + re^{i\theta} \qquad \left(re^{i\theta}\right)^k = -b/c$$

Program 9.4.4

```
program  all_zeros;    { 08/12/95 }
{ Input: F(z);   Output: all zeros of F(z) }

   uses   DOS, CRT, complexs;    { ch3\s4 }

   const   max_deg = 100;
               eps = 1.0e-16;
               eta = 1.0e-4;

   type    polynomial = record
                            deg: Word;
                            coeff: array[0..max_deg]
                                     of complex;
                        end;

   var            a: complex;
                  j: Word;
               F, G: polynomial;
               root: array[1..max_deg] of complex;
         v, r, theta: Real;

   procedure   translate(const  F: polynomial;
                                 a: complex;
                         var  G: polynomial);
   { The elements of  G  are the coefficients of
       F(z) expressed as a polynomial in  z - a. }

      var   j, k: Word;
            w: complex;
```

```
      begin
      G.deg := F.deg;
      G.coeff[0] := F.coeff[F.deg];
      for  j := 1  to  F.deg  do  G.coeff[j] := zero;
      for  j := 1  to  F.deg  do
         for  k := j  downto  0  do
            begin
            product(G.coeff[k], a, w);
            if k > 0  then
               sum(w, G.coeff[k - 1], G.coeff[k])
            else
               sum(w, F.coeff[F.deg - j], G.coeff[0]);
            end;    { for  k }
      end;    { translate }

   procedure  evaluate(const  F: polynomial;
                              z: complex;
                         var  w: complex);
   { Output: w = F(z) }

      var  j:  Word;

      begin
      w := F.coeff[F.deg];
      for  j := F.deg - 1  downto  0  do
         begin
         product(w, z, w);
         sum(w, F.coeff[j], w);
         end;
      end;    { evaluate }

   procedure  find_next_zero(const  F: polynomial;
                              var  a: complex);
   { Finds a single zero; initial guess: a. }

      var  done: Boolean;
           w: complex;
           k: Word;

      procedure  minimize(const  F: polynomial;
                             r, theta: Real;
                          var  v: Real;
                          var  a: complex);
         { Minimize |F| on the ray from a with
           angle theta; initial step length r. }
```

```
const   one_third = 1.0/3.0;

var     s, t: complex;
        v1, v2: Real;

begin
s.re := r*Cos(theta);   s.im := r*Sin(theta);
evaluate(F, a, t);   v1 := modulus(t);
while   r >= eps   do
   begin
   repeat
      { Step until a decrease in |F|. }
      if   KeyPressed   then   Halt;
      v2 := v1;
      sum(a, s, a);   evaluate(F, a, t);
      v1 := modulus(t);
   until   v1 >= v2*(1.0 - eps);
   { Retreat a step and decrease step size. }
   difference(a, s, a);   v1 := v2;
   r := one_third*r;
   s.re := one_third*s.re;
   s.im := one_third*s.im;
   end;   { while }
v := v2;
end;   { minimize }

begin   { find_next_zero }
repeat
   translate(F, a, G);
   done := (modulus(G.coeff[0]) < eps);
   if   not done   then
      begin
      if   KeyPressed   then   Halt;
      k := 1;
      while   modulus(G.coeff[k])
           < eta*modulus(G.coeff[0])   do   Inc(k);
      quotient(G.coeff[0], G.coeff[k], w);
      negate(w);
      polar_form(w, r, theta);
      theta := theta/k;   r := Exp(Ln(r)/k);
      minimize(F, r, theta, v, a);
      done := (v < eps);
      end;   { if not done }
until   done;
root[F.deg] := a;
end;   { find_next_zero }
```

```
procedure   factor_out_zero(a: complex;
                                var   F: polynomial);
{ Input:  F and a zero  a  of F.
  Output:   F := F(z)/(z - a) }

   var   j: Word;
          t, w: complex;

   begin
   t := F.coeff[F.deg];
   for  j := F.deg - 1  downto  1  do
       begin
       w := F.coeff[j];  F.coeff[j] := t;
       product(a, t, t);
       sum(w, t, t);
       end;
   F.coeff[0] := t;
   Dec(F.deg);
   end;       { factor_out_zero }

begin   { allzeros }
{ Input }
for  j := 1  to  F.deg  do
    begin
    find_next_zero(F, a);
    factor_out_zero(a, F);
    end;
  { Output }
end.    { allzeros }
```

Section 5.   Integrals

We now look at numerical integration. Besides standard textbooks on numerical analysis, a good source is Davis, P J and Rabinowitz, P, Numerical Integration, Blaisdell, 1967. Our unit "integral" includes several procedures (functions actually) for numerical integration, both for simple and multiple integrals. Let us examine its interface.

Program 9.5.1

```
unit   integral;    { 06/23/95 }

interface

const   max_dim = 10;    { for gauss_n }
        max_deg = 96;    { for Gaussian quadrature }
```

```
type          Real = Extended;
          real_fun = function (x: Real): Real;
         real_fun2 = function (x, y: Real): Real;
          real_vec = array [1..max_dim + 1] of Real;
             index = array [1..max_dim + 1] of Word;
           vec_fun = function (j: Word;
                               x: real_vec): Real;

var   no_evaluations,           { for adaptive_simpson }
         highest_level: Word;   { for adaptive_simpson }

function   simpson (F: real_fun;   x0, x1: Real;
                     div_no: Word): Real;
function   double_simpson (F: real_fun2;
                           x0, x1, y0, y1: Real;
                           x_div, y_div: Word): Real;
function   adaptive_simpson
               (F: real_fun; x0, x1, eps, eta: Real): Real;
function   romberg (f: real_fun;
                    x0, x1, eps, eta: Real;
                    min, max: Word): Real;
function   gauss3 (F: real_fun; x0, x1: Real;
                   n: Word): Real;
procedure  compute_gauss_coeffs (n: Word);
function   gauss (F: real_fun; x0, x1: Real;
                  deg: Word): Real;
function   gauss_n (F, low_bd, upp_bd: vec_fun;
                    dim, p: Word;   no_divs: index): Real;
```

The first function "simpson" is the standard Sinpson's rule integrator for the integral of $F(x)$ over $[x_0, x_1]$, where the number of divisions of the interval is decided in advance. "Double_simpson" is a direct extension of Simpson's rule to the integration of $F(x, y)$ over a rectangle.

"Adaptive_simpson" is a procedure for simple integrals which adjusts the number and size of the subdivisions of the interval to meet error bounds given in advance. This is an important technique called "automatic integration" or "adaptive integration", and all modern integration programs are adaptive in one way or another. In this case the source is McKeeman, W M, CACM 145, 1962. The interfaced variables "no_evaluations" and "highest_level" are used by "adaptive_simpson" to report how the procedure operated. The function "romberg" is another adaptive integrator, probably the one currently used the most.

The family of Gaussian quadrature procedures depends on vectors of zeros and weights, held in the interfaced variables "zero" and "weight". The zeros are zeros of Legendre polynomials of given degree "deg", and the procedure "compute_gauss_coeffs" computes these zeros and their corresponding weights—details below. Function "gauss" is the one-dimensional version of Gaussian integration, without adaptive error correction. The final procedure "Gauss_n" is an integrator for iterated integrals of many variables.

Let us now pass on to the implementation, starting with "simpson". Recall the single interval idea, sometimes referred to as "3-point Simpson". Let m be the midpoint of $[x_0, x_1]$ and let $Q(x)$ be the unique quadratic polynomial that interpolates $F(x)$ at x_0, m, and x_1. The integral of $F(x)$ over $[x_0, x_1]$ is estimated by the integral of $Q(x)$:

$$\int_{x_0}^{x_1} F(x)\, dx \approx \int_{x_0}^{x_1} Q(x)\, dx = \frac{x_1 - x_0}{6}[F(x_0) + 4F(m) + F(x_1)]$$

The estimate is exact if $F(x)$ is a polynomial of degree 3. In "simpson", the interval is divided into div_no equal parts, and 3–point Simpson is applied to each. When the midpoints are added, the original interval is really divided by 2*div_no equally spaced points, which results in the well-known sequence of weights 1, 4, 2, 4, 2, $\cdots$, 4, 1. Please keep in mind for later reference that this general Simpson's rule is based on a piecewise quadratic approximation to the integrand.

implementation

uses   CRT;

```
function   simpson(F: real_fun;   x0, x1: Real;
                   div_no: Word) : Real;

   var    x, dx, sum: Real;
                 j: Word;

   begin
   dx := (x1 - x0)/(2.0*div_no);
   sum := F(x0) + F(x1);
   x := x0;
   for  j := 1  to  2*div_no - 1  do
       begin
       x := x + dx;
       if  Odd(j)  then  sum := sum + 4.0*F(x)
       else  sum := sum + 2.0*F(x);
       end;
   simpson := dx*sum/3.0;
   end;    { simpson }
```

The next function "double_simpson" is essentially the tensor product of two simple Simpson rules, and its code is pretty transparent:

```
function   double_simpson (F: real_fun2;
                           x0, x1, y0, y1: Real;
                           x_div, y_div: Word) : Real;

   var   dx, dy,
         x, sum: Real;
             i: Word;

   function   simple_simpson (x: Real) : Real;
       { Like "simpson" above }

   begin   { double_simpson }
   dx := (x1 - x0)/(2.0*x_div);
   dy := (y1 - y0)/(2.0*y_div);
   x := x0;
   sum := simple_simpson(x0) + simple_simpson(x1);
   for  i := 1  to   2*x_div - 1   do
      begin
      x := x + dx;
      if   Odd(i)    then
          sum := sum + 4.0*simple_simpson(x)
      else   sum := sum + 2.0*simple_simpson(x);
      end;
   double_simpson := dx*dy*sum/9.0;
   end;    { double_simpson }
```

Our first integrator with adaptive step size "adaptive_simpson" needs much explanation. First note its parameters "eps" and "eta", which measure absolute error and relative error respectively. Their roles are best explained by an inequality:

$$\left| \int_a^b F(x)\, dx - estimate \right| < eps + eta \int_a^b |F(x)|\, dx$$

To see the point of the term involving eta, delete it, and suppose the I is the estimate of the integral on the left, with absolute error eps. Then $10I$ should be just as good an estimate of the integral of $10F(x)$. Is it so? Of course not if we ignore relative error. Well, you can see from this that we will be estimating both the integral of $F(x)$ and of $|F(x)|$ simultaneously.

"Adaptive_simpson" is a driver for the recursive procedure "simpson3point", which computes integrals over an interval $[x_0, x_0 + \delta x]$, where x_0 is not necessarily the original left endpoint. If the Simpson 3-point rule is not good enough over an interval, then that interval is divided into 3 equal parts, and the rule is applied over each part. This makes 7 division points, but we need to evaluate F only at 4 of these, as the values at the other three are passed as the last three parameters of "simpson3point". There is one peculiar technical point here. When we pass to the subintervals of one-third the length, you would expect to divide the *eps* and *eta* by 3. A lot of experimental work has indicated that this is excessive (it assumes that the error will be about equally distributed in the three subdivisions) and that $\sqrt{3}$ is a more practical divisor.

```
function   adaptive_simpson
             (F: real_fun; x0, x1, eps, eta: Real): Real;

  const   max_level = 35;

  var   k, nest_level: Word;
          integral_abs: Real;

  function   simpson3point
             (x0, delta_x, estimate, integral_abs, eps,
              eta, left, middle, right: Real): Real;

    var           dx3, sum,
                eps3, eta3,
                      factor,
                  left_integ,
              middle_integ,
                right_integ,
           F1, F2, F4, F5: Real;

    begin    { simpson3point }
    { Initializations }
    Inc(nest_level);
    dx3  := delta_x/3.0;
    F1  := F(x0 + 0.5*dx3);
    F2  := F(x0 + dx3);
    F4  := F(x0 + 2.0*dx3);
    F5  := F(x0 + 2.5*dx3);
    Inc(no_evaluations, 4);
    factor  := dx3/6.0;
    left_integ := factor*(left + 4.0*F1 + F2);
    middle_integ := factor*(F2 + 4.0*middle + F4);
    right_integ := factor*(F4 + 4.0*F5 + right);
    sum := left_integ + middle_integ + right_integ;
```

```
integral_abs
    := integral_abs - Abs(estimate)
        + Abs(left_integ) + Abs(middle_integ)
        + Abs(right_integ);
    { End initializations }
    if  (nest_level > 1 ) and
        ( (nest_level = max_level) or
        (Abs(sum - estimate)
            <= eps + eta*integral_abs) )    then
        { guarantee at least one recursive call }
        simpson3point := sum
    else
        begin   { subdivide into thirds }
        if  nest_level > highest_level   then
            Inc(highest_level);
        eps3 := 0.577*eps;    { 1/sqrt(3) ~ 0.577 }
        eta3 := 0.577*eta;
        { Avoid 80x87 stack overflow by separating
            the three recursive calls. }
        left_integ   := simpson3point(x0, dx3,
                            left_integ, integral_abs,
                            eps3, eta3, left, F1, F2);
        middle_integ := simpson3point(x0 + dx3, dx3,
                            middle_integ, integral_abs,
                            eps3, eta3, F2, middle, F4);
        right_integ  := simpson3point
                            (x0 + 2.0*dx3, dx3,
                            right_integ, integral_abs,
                            eps3, eta3, F4, F5, right);
        simpson3point := left_integ + middle_integ
                                    + right_integ;

        end;    { else }
    Dec(nest_level);
    end;    { simpson3point }

begin    { adaptive_simpson }
nest_level := 1;
highest_level := 1;
no_evaluations := 3;
adaptive_simpson
    := simpson3point
            (x0, x1 - x0, 0.0, 0.0, eps, eta,
            F(x0), F(x0 + 0.5*(x1 - x0)), F(x1));
end;    { adaptive_simpson }
```

A recursive version of "adaptive_simpson" (ALGOL) is in McKeeman, W M and Tesler, L, CACM 182, 1963 and (FORTRAN) in Davis and Rabinowitz, *loc cit*, 198.

Romberg integration is based on the trapezoidal rule which comes from a piecewise linear approximation to the integrand. (Remember that Simpson's rule was based on a piecewise quadratic approximation.) The basic fact about the error in the trapezoidal rule is this.

Theorem Let $F(x)$ be a smooth function on $[a, b]$ and divide this interval into n equal parts, each of length $h = (b - a)/n$. Let $I(h)$ denote the corresponding trapezoidal approximation:

$$I(h) = \tfrac{1}{2}h[f_0 + 2f_1 + 2f_2 + \cdots + 2f_{n-1} + f_n] \qquad f_j = F(a + jh)$$

Then

$$\int_a^b F(x)\,dx = I(h) - \sum_1^\infty a_k h^{2k}$$

where the a_k are certain constants.

The main point is that the error in trapezoidal approximation is expressible as a power series in *even* powers of h. For details see Hildebrand, F B, Introduction to Numerical Analysis, McGraw-Hill, 1956, 149-153 and 177-179.

We now use the "Richardson extrapolation" of Section 1, but with a difference: because the error is in powers of h^2, we use powers of 4 rather than of 2 in forming the Romberg difference table:

$$
\begin{array}{cccccc}
T_{00} & T_{01} & T_{02} & \cdots & T_{0n} \\
T_{10} & T_{11} & T_{12} & \cdots & T_{1,\,n-1} \\
T_{20} & T_{21} & \cdots & & \\
& & & & \\
\multicolumn{5}{c}{\cdots\cdots\cdots\cdots\cdots\cdots\cdots\cdots\cdots\cdots\cdots} \\
T_{n0} & & & &
\end{array}
$$

In the 0-th row, $T_{0k} = I\big((b - a)/2^k\big)$, so that T_{00}, T_{01}, $\cdots$ are successive trapezoidal approximations to the integral, each with twice as many intervals as the previous one. By the theorem,

$$T_{0k} = \int_a^b F(x)\,dx + a_1 h^2 + O\big(h^4\big) \qquad h = (b - a)/2^k$$

It follows that

$$T_{0,\,k+1} = \int_a^b F(x)\,dx + \tfrac{1}{4}a_1 h^2 + O\big(h^4\big)$$

so we set

$$T_{1k} = \frac{4T_{0,k+1} - T_{0k}}{4 - 1} = \int_a^b F(x)\,dx + a_2 t^4 + O(t^6)$$

In general, we form the j-th row of the Romberg table by

$$T_{jk} = \frac{4^j T_{j-1,k+1} - T_{j-1,k}}{4^j - 1}$$

and have the error estimate

$$T_{jk} = \int_a^b F(x)\,dx + O(h^{2j+2}) \qquad h = (b-a)/2^k$$

Clearly, we do not need the whole table, only the last row computed, which we update "in place" to form the next row.

We double the number of sample points at each step. Note that the function must only be evaluated at the new points; they are the midpoints of the previous divisions. The following equation shows this better than 1000 words:

$$F_0 + 2F_1 + 2F_2 + \cdots 2F_{2n-1} + F_{2n}$$
$$= (F_0 + 2F_2 + 2F_4 + \cdots 2F_{2n-2} + F_{2n}) + 2(F_1 + F_3 + \cdots + F_{2n-1})$$

Thus to update the previous trapezoidal approximation, we must compute the sum of the function values at the new midpoints. This is done in the **for** k loop below. There is an added complication in this loop, which we'll pause to explain.

```
function   romberg (F: real_fun;
                    x0, x1, eps, eta: Real;
                    min, max: Word): Real;

const   abs_max = 30;

var           p, dx, error,
          F_of_x0, F_of_x1,
                   F_of_xk,
              roundoff_error,
              integral_abs,
                   tolerance,
          previous_estimate,
           current_estimate,
          mid_sum, temp_sum,
              mid_sum_abs: Real;
                    table: array[0..abs_max] of Real;
                    j, n: Word;
                    k, r: LongInt;
                    done: Boolean;
                   denom: array[1..abs_max] of Real;
```

```
begin    { romberg }
p := 1.0;
for  k := 1  to   abs_max   do
    begin
    p := 4.0*p;   denom[k] := 1.0/(p - 1.0);
    end;
dx := x1 - x0;
F_of_x0 := F(x0);   F_of_x1 := F(x1);
current_estimate := 0.0;
previous_estimate := 0.0;
done := false;
table[0] := 0.5*dx*(F_of_x0 + F_of_x1);
integral_abs := 0.5*Abs(dx)*
                        (Abs(F_of_x0) + Abs(F_of_x1));
n := 1;   r := 1;
repeat
    dx := 0.5*dx;
    mid_sum := 0.0;   mid_sum_abs := 0.0;
    roundoff_error := 0.0;
```

We interrupt the **repeat** loop to explain the following **for** sub-loop, whose main job is to sum the 2^{n-1} function values $F(x_k)$ at the new midpoints. It is possible at any point that the accumulated sum is very large compared to the next summand, and roundoff error will wipe out all, or most of the contribution of that summand. This might happen many times, so that actually the final sum is inaccurate.

Heinz Rutishauser introduced the device of keeping track of the accumulated roundoff error, trying to put it back at each step, and then updating this roundoff error to the portion that was not accepted. This code should be studied carefully as it implements an important idea. (Only a rough estimate is needed for the absolute value sum, so the device is not applied to it.) We complete "romberg":

```
for  k := 1  to  r  do
    begin    { Compute sum of midpoint values }
    F_of_xk := F(x0 + (2*k - 1)*dx);
    mid_sum_abs := mid_sum_abs + Abs(F_of_xk);
    { Add in the accumulated roundoff error }
    F_of_xk := F_of_xk + roundoff_error;
    temp_sum := mid_sum + F_of_xk;
    { Keep the part of the roundoff error
      that has not been absorbed. }
    roundoff_error
        := (mid_sum - temp_sum) + F_of_xk;
    mid_sum := temp_sum;
```

```
          if   KeyPressed   then   Halt;
          end;
       table[n]  :=  0.5*table[n - 1] + dx*mid_sum;
       integral_abs  :=  0.5*integral_abs
                            + Abs(dx)*mid_sum_abs;
       for  j := n - 1  downto   0   do
          table[j]  :=  table[j + 1]
                           + denom[n - j]
                              *(table[j + 1] - table[j]);
       if   n >= min   then
          begin
          tolerance  :=  eta*integral_abs + eps;
          error  :=  Abs(table[0]  - current_estimate)
                        + Abs(current_estimate
                                 - previous_estimate);
          done  :=  (error < tolerance);
          end;
       Inc(n);
       done  :=  done  or  (n > max);
       previous_estimate  :=  current_estimate;
       current_estimate  :=  table[0];
       r  :=  2*r;     { Prepare for next pass }
    until   done;
    romberg  :=  current_estimate;
    end;     { romberg }
```

This brings us to Gaussian quadrature, a family of integration rules based on *unequal* divisions of the basic interval. Simpson's rule is one of a family of "Newton-Cotes rules", all of which use equally-spaced division points. The basic 3-point Simpson's rule has 3 function evaluations and is exact for cubic polynomials. Surprisingly, the basic Gauss 2-point rule is also exact for cubics. In general, the Gauss n-point rule is exact for polynomials of degree $2n - 1$. Our first example "gauss3" applies the basic Gauss 3-point rule to each of the n equal part divisions of the interval. For the interval $[-1, 1]$, the function is evaluated at the zeros of the third Legendre polynomial $P_3 = \frac{1}{2}(5x^3 - 3x)$ and these values are added with special weights to form the approximation to the integral. (You might try to check directly that with the division points and weights in the code, the approximation is exact for polynomials of degree 5.) For any other interval, an affine transformation to $[-1, 1]$ is used.

```
function   gauss3(F: real_fun; x0, x1: Real;
                  n: Word): Real;
```

```
var    t, sum, x, z, dx: Real;
                  i, k: Word;
            zero, weight: array[1..3] of Real;

procedure   initialize_constants;

    var   s, t: Real;
              j: Word;

    begin    { initialize_constants }
    zero[1]  := -Sqrt(0.6);
    zero[2]  := 0.0;
    zero[3]  := Sqrt(0.6);
    weight[1]  := 5.0/9.0;
    weight[2]  := 8.0/9.0;
    weight[3]  := 5.0/9.0;
    for  j := 1  to  3  do
        begin    { Affine map [-1, 1] to [0, 1] }
        zero[j]  := 0.5*(1.0 + zero[j]);
        weight[j]  := 0.5*weight[j];
        end;
    end;    { initialize_constants }

begin    { gauss3 }
initialize_constants;
dx := (x1 - x0)/n;
x := x0;
sum := 0.0;
for  i := 0  to  n - 1  do
    begin
    t := 0.0;
    for  k := 1  to  3  do
        begin
        z := x + dx*zero[k];
        t := t + weight[k]*F(z)
        end;
    sum := sum + dx*t;
    x := x + dx;
    end;    { for i }
gauss3 := sum;
end;    { gauss3 }
```

For Gaussian quadrature in general, we need a digression on Legendre polynomials, programmed in the unit "orthopoly" in Section 8.5. Let's review some basic facts about this most useful family of polynomials.   References are Abramowitz, M and Stegun, L A, Handbook of Mathematical Functions, National Bureau of Standards, 1964, 773-785, 887-888, 916-919, Hildebrand, F B, Introduction to Numerical Analysis, McGraw-Hill, 1956, 272-274, 319-315, 386-388, Stoer, J and Bulirsch, R, Introduction to Numerical Analysis, Springer-Verlag, 1980, 142-148, Cheney, W and Kincaid, D, Numerical Mathematics and Computing 2/e, Brooks/Cole, 1985, 192-197, and Burden, R L and Faires, J D, Numerical Analysis 5/e, PWS-Kent, 1993, 205-208, 457-458. Be warned that these and other sources may have slight differences in normalization of the Legendre polynomials, and do contain a few misleading typos. I hope I can get it right here.

The Legendre polynomial $P_n(x)$ has degree exactly n. The recursive definition of the Legendre polynomials is

$$P_0(x) = 1 \qquad P_1(x) = 1 \qquad P_n(x) = \frac{1}{n}[(2n-1)xP_{n-1}(x) - (n-1)P_{n-2}(x)]$$

They form an orthogonal family on $[-1, 1]$ (not quite orthonormal):

$$\int_{-1}^{1} P_m(x)P_n(x)\, dx = 0 \quad (m \neq n) \qquad \int_{-1}^{1} P_n(x)^2\, dx = \frac{2}{2n+1}$$

The degree n condition and the second integral normalizes the family. (Instead of the integral of the square, the condition $P_n(1) = 1$ serves as well.) The leading coefficient of $P_n(x)$ is $(2n)!/\left[2^n(n!)^2\right]$. More general is the expansion

$$P_n(x) = \frac{1}{2^n} \sum_{k=0}^{n\ \text{div}\ 2} (-1)^k \binom{n}{k} \binom{2n-2k}{n} x^{n-2k}$$

Another explicit formula is the "Rodrigues' formula"

$$P_n(x) = \frac{1}{2^n n!} \frac{d^n}{dx^n} (x^2 - 1)^n$$

The first few Legendre polynomials are

$$P_0 = 1 \qquad P_1 = x \qquad P_2 = \tfrac{1}{2}(3x^2 - 1) \qquad P_3 = \tfrac{1}{2}(5x^3 - 3x)$$
$$P_4 = \tfrac{1}{8}(35x^4 - 30x^2 + 3) \qquad P_5 = \tfrac{1}{8}(63x^5 - 70x^3 + 15x)$$

It should be apparent that in general the odd degree Legendre polynomials are odd functions and the even degree ones are even functions.

The generating function for the Legendre polynomials is

$$(1 - 2xt + t^2)^{-1/2} = \sum_{n=0}^{\infty} P_n(x)t^n \qquad (|x| < 1 \quad |t| < 1)$$

Polynomials $P_n(x)$ satisfy the differential and differential-difference equations

$$\left(1 - x^2\right) P_n''(x) - 2x P_n'(x) + n(n+1) P_n(x) = 0$$
$$\left(1 - x^2\right) P_n'(x) = -nx P_n(x) + n P_{n-1}(x)$$

Let us fix n and look at the zeros of $P_n(x)$. The basic fact is that they are simple and in the open interval $(-1, 1)$. Thus

$$-1 < x_1 < x_2 < \cdots < x_n < 1 \qquad P_n(x_j) = 0$$

The corresponding Gaussian integration formula (with error) is

$$\int_{-1}^1 F(x)\, dx = \sum_{j=1}^n w_j F(x_j) + R_n \qquad R_n = \frac{2^{2n+1}(n!)^4}{(2n+1)[(2n)!]^3} F^{(2n)}(\xi)$$

where $-1 < \xi < 1$. The weights are given by several equivalent formulas:

$$w_j = \frac{2}{n P_{n-1}(x_j) P_n'(x_j)} = \frac{2}{\left(1 - x_j^2\right)\left[P_j'(x_j)\right]} = \frac{2\left(1 - x_j^2\right)}{(n+1) P_{n+1}(x_j)^2}$$

Our first procedure is "compute_gauss_coeffs" to compute the zeros and weights. Its subprocedure "legendre_poly" returns $P_n(x)$, $P_{n-1}(x)$, and $P_n'(x)$. The last is obtained by differentiating the basic recursion formula for $P_n(x)$:

$$P_n' = \frac{1}{n}\left[(2n - 1)\left(P_{n-1} + x P_{n-1}'\right) - (n - 1) P_{n-2}'\right]$$

The zeros are found by preliminary subdivision and secant steps, followed by Newton steps, which use the derivative values. Then the first formula for the weights—also using derivative values—is applied. (Two disk programs z-legend.pas and get_gauss.pas save/load these numbers.)

```
var    zero, weight: array[1..max_deg] of Real;

procedure  compute_gauss_coeffs(n: Word);

    const   eps = 6.0e-20;

    var                 i, index: Word;
              P0k, P0k_1, D0k,
              P1k, P1k_1, D1k,
          x0, x1, y, z, dx,
                    x, u: Real;
```

```
procedure  legendre_poly(n: Word;   x: Real;
                          var   Pk, Pk_1, Dk: Real);

    var   Pk_2, Dk_1, Dk_2: Real;
               i, j, k: Word;

    begin    { legendre_poly }
    if   n = 0   then
        begin   Pk := 1.0;   Dk := 0.0;    end
    else
        begin
        Pk_1 := 1.0;   Pk := x;
        Dk_1 := 0.0;   Dk := 1.0;
        i := 3;   j := 1;      { i = 2k - 1,   j = k - 1 }
        for   k := 2   to   n   do
            begin
            Pk_2 := Pk_1;   Pk_1 := Pk;
            Dk_2 := Dk_1;   Dk_1 := Dk;
            Pk := ( i*x*Pk_1 - j*Pk_2 )/k;
            Dk := ( i*(Pk_1 + x*Dk_1) - j*Dk_2 )/k;
            Inc(i, 2);   Inc(j);
            end;
        end
    end;    { legendre_poly }

begin    { compute_gauss_coeffs }
index := (n + 1) div 2;
dx := 1.0/(10.0*n);
x0 := 0.0;   x1 := x0 + dx;
if   Odd(n)    then
    begin
    zero[index] := 0.0;
    legendre_poly(n, x0, P0k, P0k_1, D0k);
    weight[index] := 2.0/(P0k_1*D0k*n);
    end;
for  i := 0   to   10*n - 1   do
    begin
    x0 := x1;
    x1 := x1 + dx;
    legendre_poly(n, x0, P0k, P0k_1, D0k);
    legendre_poly(n, x1, P1k, P1k_1, D1k);
    if   P0k*P1k <= 0.0   then
        begin
        { One secant step }
        x := x0 - P0k*dx /(P1k - P0k);
        { Newton steps }
```

```
        legendre_poly(n, x, P0k, P0k_1, D0k);
        u := P0k/D0k;
        y := x - u;
        while  Abs(x - y) >= eps  do
            begin
            x := y;
            legendre_poly(n, x, P0k, P0k_1, D0k);
            u := P0k/D0k;
            y := x - u;
            end;
        Inc(index);
        legendre_poly(n, y, P0k, P0k_1, D0k);
        zero[index] := y;
        weight[index] := 2.0/(P0k_1*D0k*n);
        if  index = n  then  Break;
        end;    { if  P0k*P1k <= 0.0 }
    end;    { for  i }
{ Apply symmetry for [-1, 0] }
for  i := 1  to  n  div 2  do
    begin
    zero[i] := - zero[n - i + 1];
    weight[i] := weight[n - i + 1];
    end;
end;    { compute_gauss_coeffs }
```

With the zeros and weights under control, we are ready for "gauss" itself. This is a single Gauss step on the interval. Of course, a serious application will divide the interval into smaller intervals and apply this procedure on each, possibly in adaptive fashion.

```
function  gauss(F: real_fun; x0, x1: Real;
                deg: Word): Real;

    var       index: Word;
        a, b, sum: Real;

    begin    { gauss }
    a := 0.5*(x1 - x0);
    b := 0.5*(x1 + x0);
    sum := 0.0;
    for  index := 1  to  deg  do
        sum := sum + F(a*zero[index] + b)*weight[index];
    gauss := a*sum;
    end;    { gauss }
```

We now consider iterated integrals such as

$$\int_1^2 x^3\,dx \int_0^{x^2} y^2\,dy \int_0^{y/x} z\,dz \quad \left(=\frac{273}{8}\right)$$

In general, we consider integrals of the form

$$\int_{L_1}^{U_1} F_1(x_1)\,dx_1 \int_{L_2(x_1)}^{U_2(x_1)} F_2(x_1,\,x_2)\,dx_2 \cdots$$

$$\int_{L_n(x_1,x_2,\cdots,x_{n-1})}^{U_n(x_1,x_2,\cdots,x_{n-1})} F_n(x_1,\,x_2,\cdots,\,x_n)\,dx_n$$

The integrand is split into factors as indicated to cut down the cost of (many) function evaluations. Think of the word parameter j as a subscript. Then each factor must be of type vec_fun, which is a function of a word variable and a vector of (up to) 10 real components. (10 is an upper bound for the dimension.) Indeed, each F_j can be a function of the first j variables only. Similarly, the various upper and lower bound functions are of type vec_fun, and each U_j and L_j can be a function of the first $j - 1$ variables only. Gaussian quadrature for one variable is used in each coordinate, and the trick is to organize the work to avoid a great deal of storage.

```
function   gauss_n(F, low_bd, upp_bd: vec_fun;
                   dim, p: Word;   no_divs: index): Real;

    var   c, a, sum, x, dx: real_vec;
                div_no, x_no: index;
                        j: Word;

    procedure   initialize(i: Word);
        { x = at + c:
          [-1, 1] --> [L, L + dx] }

        var   k: Word;
```

```
begin
for  k := i  to  dim  do
    begin
    dx[k]  := (upp_bd(k, x)
                    - low_bd(k, x))/no_divs[k];
    c[k]  := low_bd(k, x) + 0.5*dx[k];
    a[k]  := 0.5*dx[k];
    { Initial x in dimension k }
    x[k]  := a[k]*zero[1] + c[k];
    sum[k]  := 0.0;
    div_no[k]  := 1;
    x_no[k]  := 1;
    end;
j := dim;
end;    { initialize }

begin    { gauss_n }
sum[dim + 1]  := 1.0;   { Initializatons explain }
a[dim + 1]  := 1.0;     { the extra dimension }
initialize(1);
repeat
    sum[j]  := sum[j] + (a[j + 1]*sum[j + 1])
                        *(weight[x_no[j]]*F(j, x));
    if  x_no[j] < p  then
        begin    { Advance to next x. }
        Inc(x_no[j]);
        x[j]  := a[j]*zero[x_no[j]] + c[j];
        if  j < dim  then  initialize(j + 1);
        end
    else  if  div_no[j] < no_divs[j]  then
        begin    { Advance into next divison. }
        c[j]  := c[j] + dx[j];
        Inc(div_no[j]);
        x_no[j]  := 1;
        x[j]  := a[j]*zero[x_no[j]] + c[j];
        if  j < dim  then  initialize(j + 1);
        end
    else  Dec(j);
until  j = 0;
gauss_n := a[1]*sum[1];
end;    { gauss_n }

end.    { unit integral }
```

Exercises

In the following exercises, remember that functions used as parameters in the various integration rules *must* be compiled in the far state: {$F+}.

1. Write a program to test "simpson".
2. Write a program to test "double_simpson".
3. Write a program to test "adaptive_simpson".
4. Write a program to test "romberg".
5. Write a program that compares a simple summation with/without the Rutishauser device of absorbing roundoff error. The program should prove conclusively the advantage of the device.
6. Write a program to test Romberg integration with/without the Rutishauser device of absorbing roundoff error.
7. Write a program to test "gauss3".
8. Write a program to test "gauss". (Check the answer given with degree 96 on [-20, 20].)
9. Write a program to test "gauss_n" on the text example of a triple integral.

Section 6.   Differential Equations

We shall concentrate on the initial value problem (IVP) for systems of ordinary differential equations (ODE), with particular emphasis on Runge-Kutta approximation methods.

IVP for ODE's

We work on an interval $[t_0, t_1]$. The IVP we consider is

$$\frac{dx}{dt} = F(t, x) \qquad x(t_0) = x_0 \qquad \text{Find} \quad x(t_1)$$

The dependent variable x can be a scalar function or, more generally, a vector-valued function. We explore numerical methods for estimating $x(t_1)$, always assuming that $F(t, x)$ is "sufficiently smooth".

Runge-Kutta Methods

Most primitive is the *Euler Method*:

$$x(t_1) \approx x_0 + F(t_0, x_0)\, dt \qquad dt = t_1 - t_0$$

Geometrically, it approximates the solution curve by its closest linear approximation at (t_0, x_0).

The *Modified Euler Method* is a two step method. Briefly, it uses the Euler method for a first approximation at t_1, computes the slope there, then starts again at t_0 with the average of the computed slopes to get the final estimate at t_1; two evaluations of F are required:

$$s_1 = F(t_0, x_0) \qquad z = x_0 + s_1 dt$$
$$s_2 = F(t_1, z) \qquad x_1 = x_0 + \tfrac{1}{2}(s_1 + s_2)dt$$

In general, a Runge-Kutta method is an *n*-step method. We have a sequence of *n* points (abscissas) on $[t_0, t_1]$, and we compute a sequence of slopes. Having computed several slopes, we use a weighted average of them for a line segment from the initial point to compute an estimate of the solution at the next abscissa, then compute the slope at that point for the next slope in the sequence. Finally, a weighted average of all the computed slopes is used to go from the initial point to the estimated solution $x(t_1)$.

The following unit RK is a collection of Runge-Kutta methods of increasing complexity. It starts with the Modified Euler Method, and continues with methods we call *Heun_1, Shampine-Bogacki_23, Kutta-Merson_34, Runge-Kutta-Norsett_34, Runge-Kutta_4, Runge-Kutta-Fehlberg_45, Shampine-Bogacki_45, Runge-Kutta-Verner_56, Prince-Dormund_78*

The error in a method of order *n* is $O(dt^n)$. Each method in RK is set up to compute two approximations. The methods with a single suffix compute two approximations of the same order, e.g, Runge-Kutta_4 computes two approximations of order 4. The two-digit suffix methods compute one estimate of each order, e.g, Kutta-Merson_34 computes approximations of orders 3 and 4. Each method also computes the difference (error) between its two approximations. I'll explain details as we move along. Only part of the unit RK will be printed here, just enough to explain the ideas; the complete unit is on your disk. As usual we begin with the interface section:

Program 9.6.1

```
unit   RK;    { 05/10/95 }

interface

const  max_dim = 10;

type          Real = Extended;
            vector = array[1..max_dim] of Real;
        derivative = procedure(const   t: Real;
                               const   x: vector;
                               var   w: vector);

function  max(a, b: Real): Real;
function  sup_norm(const   dim: Word;
                   const   z: vector): Real;
function  x2y(x, y: Real): Real;
```

```
procedure   Heun_1(dim: Word;
                   F: derivative;
                   const   t0, dt: Real;
                   const   x0: vector;
                   var   x1, y1, error: vector);

procedure   RKF_45(dim: Word;
                   F: derivative;
                   const   t0, dt: Real;
                   const   x0: vector;
                   var   x1, y1, error: vector);
```

As we are not allowed *vector* output of a Pascal function, we use the procedure **type** derivative instead of what we would really like to write: $\mathbf{w} = F(t, \mathbf{x})$. The functions max, sup_norm, and x2y are routine:

implementation

```
function   max(a, b: Real): Real;

   begin
   if   a >= b   then   max := a   else   max   := b;
   end;

function   sup_norm(const   dim: Word;
                    const   z: vector): Real;

   var   k: Word;
         w: Real;

   begin
   w := 0.0;
   for   k := 1   to   dim   do   w := max(w, Abs(z[k]));
   sup_norm := w;
   end;   { sup_norm }

function   x2y(x, y: Real): Real;

   begin
   x2y := Exp(y*Ln(x));
   end;
```

The procedure Heun_1 is a slight improvement on the Modified Euler Method. There are two parts to its implementation. Here is the first part:

```
procedure   Heun_1(dim: Word;
                   F: derivative;
                   const  t0, dt: Real;
                   const  x0: vector;
                   var    x1, y1, error: vector);

   const   a = 2.0/3.0;

   var   slope0, slope1, z, w: vector;
                             j: Word;

   begin   { Heun_1 }
   F(t0, x0, slope0);
   for  j := 1  to  dim  do
       z[j] := x0[j] + a*dt*slope0[j];
   F(t0 + a*dt, z, slope1);
   for  j := 1  to  dim  do
        x1[j] := x0[j] + dt*(0.25*slope0[j]
                           + 0.75*slope1[j]);
```

As in Euler, we start with the slope at (t_0, x_0). We follow the line with this slope 2/3 of the way and compute the next slope. The two slopes are averaged with weights 1/4 and 3/4 to get the slope taking us from (t_0, x_0) to $(t_0 + dt, x_1)$. This calculation is carried out in each component.

The second approximation y_1 is computed by first cutting the interval in half. The process just given is applied to the left half, yielding an intermediate solution w. (Note that the slope at the initial point need not be recomputed.) Then w is used as the initial value for a second application of the method over the right half, to yield y_1, and finally the error, the difference between the two approximations, is computed:

```
   for  j := 1  to  dim  do
       z[j] := x0[j] + (0.5*a)*dt*slope0[j];
   F(t0 + (0.5*a)*dt, z, slope1);
   for  j := 1  to  dim  do
       w[j] := x0[j] + dt*(0.125*slope0[j]
                         + 0.375*slope1[j]);
   F(t0 + 0.5*dt, w, slope0);
   for  j := 1  to  dim  do
       z[j] := w[j] + (0.5*a)*dt*slope0[j];
   F(t0 + (0.5 + a)*dt, z, slope1);
   for  j := 1  to  dim  do
       begin
       y1[j] := w[j] + dt*(0.125*slope0[j]
                         + 0.375*slope1[j]);
```

```
      error[j] := x1[j] - y1[j];
      end;
 end;    { Heun_1 }
```

Heun_1 is typical of the methods where the second approximation is obtained by two steps over half intervals.

RKF_45 is typical of another class, of which I believe Kutta-Merson was the first. The source of Runge-Kutta-Fehlberg is Fehlberg, E, Klassische Runge-Kutta-Formeln vierter und niedrigerer Ordnung mit Schrittweiten-Kontrolle und ihre Anwendung auf Waermeleitungsprobleme, Computing 6 (1970), 61-71. It was the state of the art in this business for several years and was the basis for the FORTRAN codes RKF45 and DERKF. Let us first look at the declarations:

```
procedure   RKF_45(dim: Word;
                   F: derivative;
                   const   t0, dt: Real;
                   const   x0: vector;
                   var   x1, y1, error: vector);

 const        bound = 5;
      abscissa: array[1..bound] of Real
                   = (1./4., 3./8., 12./13.,
                      1., 1./2. );
      weight: array[1..bound, 0..bound - 1] of Real
      { 1 }      = ( (1./4., 0., 0., 0., 0. ),
      { 2 }          (3./32., 9./32., 0., 0., 0. ),
      { 3 }          (1932./2197., -7200./2197.,
                      7296./2197., 0., 0. ),
      { 4 }          (439./216., -8., 3680./513.,
                      -845./4104., 0. ),
      { 5 }          (-8./27., 2., -3544./2565.,
                      1859./4104., -11./40. ) );
      mult1: array[0..bound] of Real
                   = (25./216., 0., 1408./2565.,
                      2197./4104., -1./5., 0 );
      mult2: array[0..bound] of Real
                   = (16./135., 0., 6656./12825.,
                      28561./56430., -9./50., 2./55. );
 var      slope: array[0..bound] of vector;
              z: array[1..bound] of vector;
      s, s1, s2: Real;
       i, j, k: Word;
```

The array abscissa gives t (scaled to a unit interval) where slopes will be computed. Each row of the array weight sums to the corresponding abscissa, and gives the weights used to average slopes up to the current one in order to compute the next one. Five slopes are averaged with the weights in the vector mult1 to compute the first approximation x_1, an approximation of order 4.

Finding the various rational numbers involved in the constants above is a really difficult task. The most ingenious part of this algorithm is that by only one more function (slope) evaluation a *fifth* order approximation can be obtained; that is where mult2 comes in. The body of RKF_45 follows. (Keep in mind that "slope" is an array of vectors, so "slope[i]" is a valid third parameter of F, and "slope[k, j]" is a real, the j-th component of the vector "slope[k]".)

```
begin    { RKF_45 }
{ Compute initial slope }
F(t0, x0, slope[0]);
{ Compute other slopes }
for  i := 1  to  bound  do
   begin
   for  j := 1  to  dim  do
      begin
      s := 0.0;
      for  k := 0  to  i - 1  do
         s := s + weight[i, k]*slope[k, j];
      z[i, j] := x0[j] + dt*s;
      end;    { for j }
   F(t0 + abcissa[i]*dt, z[i], slope[i]);
   end;    { for i }
{ Compute estimates and error }
for  j := 1  to  dim  do
   begin
   s1 := 0.0;   s2 := 0.0;
   for  k := 0  to  bound  do
      begin
      s1 := s1 + mult1[k]*slope[k, j];
      s2 := s2 + mult2[k]*slope[k, j];
      end;
   x1[j] := x0[j] + dt*s1;    { 4th order approx }
   y1[j] := x0[j] + dt*s2;    { 5th order approx }
   error[j] := x1[j] - y1[j];
   end;    { for j }
end;    { RKF_45 }
```

Tests of RK

We shall use the following nonlinear IVP for our next program and for a later one:

$$\begin{cases} dx_1/dt = -x_2^3 \\ dx_2/dt = x_1^3 \end{cases} \qquad \begin{array}{l} x_1(0) = 1 \\ x_2(0) = 0 \end{array}$$

The solution is periodic; indeed, the graphs of $x_1(t)$ and $x_2(t)$ are like those of $\cos t$ and $\sin t$ except they are flatter at the peaks. Our first program, runge_kutta_driver, is an adaptive solver using RK_4 as its underlying engine. Its purpose is to tabulate the solution at 20 points over a full period. (Soon we shall see how to determine the period.) The procedure "rk_approx" should be studied carefully, as it shows how the step size is adjusted until the error requirements are met. Note that "eps" is a bound on absolute error and "eta" is a bound on relative error, that is a scale factor to make error relative to the size of the solution, in case the solution is very large. Because both solution functions in this example are bounded by 1.0, bounding relative error is not necessary in this particular case, but it is important in general.

(The program is a simplified version of P. Maur, CACM 9, 1960, and is a model for adaptive integrators over $[t,\ t+dt]$, first in a single step, then in two steps from t to $t+\frac{1}{2}dt$ to $t+dt$, then compare the results. The basic integrator can be Euler_1, Heun_1, or RK_4 from our collection. It is interesting to compare their speeds and accuracies.)

Program 9.6.2

```
program  runge_kutta_driver;    { 05/10/95 }

   uses  CRT, rk;

   const  eps = 1.0e-10;   eta = 1.0e-10;

   var  t_init, t_end, t, dt: Real;
                 x, z, x0, x1: vector;
                    dim, j, n: Word;

   {$F+}
   procedure  F(const   t: Real;
                const   x: vector;
                var   w: vector);

      begin
      w[1]  := -x[2]*x[2]*x[2];
      w[2]  := x[1]*x[1]*x[1];
      end;
   {$F-}

   procedure  rk_approx(dim: Word;
                        const   t, dt: Real;
                        const   x: vector;
                        var   z: vector);
   { Approximates  x  at  t + dt  given  x  at  t }
```

```
var            step_OK: Boolean;
         dt1, t1, t2: Real;
      x1, x2, y2: vector;
                  k: Word;
            error: vector;

begin   { rk_approx }
t1 := t;   x1 := x;
dt1 := dt;
repeat
   repeat   { Compare one step to two }
      RK_4(dim, F, t1, dt1, x1, x2, y2, error);
      dt1 := 0.5*dt1;
   until  sup_norm(dim, error)
            < eps + eta*sup_norm(dim, y2);
   { Prepare for next comparison }
   t1 := t1 + 2.0*dt1;
   dt1 := (t + dt) - t1;
   x1 := y2;
   until  t1 >= t + dt;
   { Done current subinterval }
   z := y2;
   end;   { rk_approx }

begin   { runge_kutta_driver }
dim := 2;
t_init :=  0.0;   t_end := 7.41629871;
x0[1] := 1.0;   x0[2] := 0.0;
n := 20;
WriteLn;
WriteLn('t':10, 'x':18, 'y':18);
WriteLn;
WriteLn(t_init:10:8, x0[1]:18:10, x0[2]:18:10);
t := t_init;   x := x0;
dt := (t_end - t_init)/n;
for  j := 1  to  n  do
   begin
   rk_approx(dim, t, dt, x, z);
   t := t + dt;   x := z;
   WriteLn(t:10:8, z[1]:18:10, z[2]:18:10);
   end;
WriteLn;
Write('Press <Enter>: '); ReadLn;
end.   { runge_kutta_driver }
```

We next consider the boundary value problem

$$\frac{d^2x}{dt^2} = -1 - (t^2 + 1)x \qquad x(-1) = 0, \quad x(1) = 0$$

The problem is to find a choice of $x'(-1)$ that leads to the final value of x at 1. First we change the second order equation to a system by setting $x_1 = x$ and $x_2 = x'$. Some preliminary tests show that the initial values 1 and 2 for $x_2(-1)$ lead to opposite signs for $x_1(1)$. The method of "shooting" varies the initial slope $x'(-1)$ until the shell hits its target. We use the KM_34 integrator. (To shorten the listing here, all output code is omitted; check the disk file for it.)

Several points should be noted. First, there is a special routine to choose an initial step size. Second, in case t is very large, adding a very small increment to it will have no effect, so the computer unit roundoff error (URE) is accounted for. Experience shows that about 26 times URE is the minimum meaningful increment. Third, if the error test is not satisfied, the step size is decreased by a factor that takes into account that the better of the two approximations computed by KM_34 is of *fourth* order.

Program 9.6.3

```
program  shoot;     { 05/10/95 }

    uses   CRT, rk;

    const  u26 = 26.0*5.5e-20;
           eps = 5.0e-15;
           eta = 5.0e-15;

    var    delta_t: Real;
           x0, x1: vector;
        fun_count: Word;
              k: Word;
{$F+}
    procedure  F(const  t: Real;   const  x: vector;
                  var   w: vector);

        begin
        w[1]  := x[2];
        w[2]  := -1 - (1 + Sqr(t))*x[1];
        end;
{$F-}
    procedure  solve_ode(dim: Word;
                         const   t0, t1: Real;
                         const   x0: vector;
                         var x1: vector);
```

```
const  max_calls = 20000;

var   dt, min_dt,
      t, tol, temp,
        norm_error: Real;
        current_x,
          next_x,
            error,
        y1, der_x: vector;
      output_count: Word;
          advanced,
              done: Boolean;

procedure  initialize_dt;

    var   norm_der_x: Real;

    begin    { initialize_dt }
    { Compute one derivative and starting  dt. }
    fun_count := 0;
    output_count := 0;
    t := t0;
    delta_t := t1 - t0;
    current_x := x0;
    F(t, current_x, der_x);
    tol := eta*sup_norm(dim, current_x) + eps;
    dt := Abs(delta_t);
    norm_der_x := sup_norm(dim, der_x);
    if  tol < norm_der_x*x2y(dt, 4.0)   then
        begin
        dt := u26*max(Abs(t), dt);
        dt := max(dt, x2y(tol/norm_der_x, 0.25));
        end;
    if  delta_t < 0  then   dt := -dt;
    end;    { initialize_dt }

begin    { solve_ode }
initialize_dt;
repeat
    if  Abs(dt) > 2.0*Abs(delta_t)   then
        Inc(output_count);
    if  output_count = 100  then  Halt;
```

```
if  Abs(delta_t) <= u26*Abs(t)    then
   begin   { t very close to t1; use Euler. }
   for  k := 1  to  dim  do
      x1[k] := current_x[k]
               + delta_t*der_x[k];
   Exit;
   end;
done := false;
advanced := true;
min_dt := u26*Abs(t);
delta_t := t1 - t;
if  Abs(delta_t) < 2.0*Abs(dt)   then
   begin
   if  0.9*Abs(delta_t) > Abs(dt)   then
      dt := 0.5*delta_t
   else
      begin
      done := true;   dt := delta_t;
      end;
   end;
repeat
   { Compute and test the solution;
     decrease |dt| if test fails. }
   if  fun_count > max_calls   then   Halt;
   KM_34(dim, F, t, dt, current_x,
         next_x, y1, error);
   tol := sup_norm(dim, current_x)
          + sup_norm(dim, next_x);
   tol := 0.5*tol*eta + eps;
   norm_error := sup_norm(dim, error);
   if  norm_error >= tol   then
      begin   { Failure; decrease dt. }
      advanced := false;
      done := false;
      if norm_error >= 6561.0*tol   then
         dt := 0.1*dt   { 9^4 = 6561 }
      else
         dt := 0.9*dt
                  /x2y(norm_error/tol, 0.25);
      if  Abs(dt) <= min_dt   then   Halt;
      end;   { if  norm_error >= tol }
until  norm_error < tol;
t := t + dt;
current_x := next_x;
F(t, current_x, der_x);
Inc(fun_count);
```

```
if   advanced   then
      begin     { (0.9/4)^4 = 2.56e-3 }
      if   norm_error <= (2.56e-3)*tol     then
         temp := 4.0
      else   temp := x2y(tol/norm_error, 0.25);
      temp := temp*Abs(dt);
      if   min_dt > temp    then    temp := min_dt;
      if   dt >= 0    then    dt := temp
      else    dt := -temp;
      end;     { if advanced }
   until   done;
   x1 := next_x;
   end;     { solve_ode }

var   dx02: Real;

begin     { shoot }
x0[1]  := 0.0;    x0[2]  := 1.0;
dx02 := 0.1;
for   k := 1   to   9   do
   begin
   repeat
      solve_ode(2, -1.0, 1.0, x0, x1);
      x0[2]  := x0[2] + dx02;
   until   x1[1] >= 0.0;
   x0[2]  := x0[2]  - 2.0*dx02;
   dx02 := 0.1*dx02;
   end;     { for   k }
end.     { shoot }
```

(The answer is about 1.736. It is interesting that this problem was first run in 1963 on an LGP-30 mainframe in about 90 minutes!)

The next program finds the period for the nonlinear IVP in program "rkdriver". It uses RKF_45 as its engine, and is more typical of adaptive integration methods than the previous program. These methods stem mostly from Lawrence Shampine, and can be read in detail in his book with M K Gordan, Computer Solutions of Ordinary Differential Equations, WH Freeman, 1975.

This is another "shooting" situation. The endpoint of the t-interval is varied until the solution there matches the initial conditions. Preliminary tests indicate that the period is between 7 and 8, so we start there. As with the previous program, we omit output code in the printed listing. Adjusting for errors is really quite sophisticated here.

Program 9.6.4

```
program  RKF45_driver;     { 05/10/95 }

   uses  CRT, rk;

   const  u26 = 26.0*5.5e-20;
          eps = 5.0e-15;
          eta = 5.0e-15;

   var      t0, t1,
          delta_t: Real;
            x0, x1: vector;
        fun_count: Word;
           dim, k: Word;
{$F+}
   procedure  F(const  t: Real;   const  x: vector;
                   var  w: vector);

       var  y: Real;

       begin
       w[1]  := -x[2]*x[2]*x[2];
       w[2]  := x[1]*x[1]*x[1];
       Inc(fun_count);
       end;
{$F-}

   procedure  solve_ode(dim: Word;
                        const  t0, t1: Real;
                        const  x0: vector;
                        var x1: vector);

       const  max_calls = 20000;

       var    dt, min_dt,
            t, tol, temp,
             norm_error: Real;
             current_x,
                 next_x,
                  error,
             y1, der_x: vector;
         output_count: Word;
              advanced,
                  done: Boolean;
```

```
procedure  initialize_dt;

   var  norm_der_x: Real;

   begin    { initialize_dt }
   { Compute one derivative and starting  dt. }
   fun_count := 0;
   output_count := 0;
   t := t0;
   delta_t := t1 - t0;
   current_x := x0;
   F(t, current_x, der_x);
   tol := eta*sup_norm(dim, current_x) + eps;
   dt := Abs(delta_t);
   norm_der_x := sup_norm(dim, der_x);
   if  tol < norm_der_x*x2y(dt, 5.0)    then
      begin
      dt := u26*max(Abs(t), dt);
      dt := max(dt, x2y(tol/norm_der_x, 0.2));
      end;
   if  delta_t < 0  then   dt := -dt;
   end;    { initialize_dt }

begin    { solve_ode }
initialize_dt;
repeat
   if  Abs(dt) > 2.0*Abs(delta_t)   then
      Inc(output_count);
   if  output_count = 100  then   Halt;
   if  Abs(delta_t) <= u26*Abs(t)   then
      begin   { t very close to t1; use Euler. }
      for  k := 1  to  dim  do
         x1[k] := current_x[k]
                   + delta_t*der_x[k];
      Exit;
      end;
   done := false;
   advanced := true;
   min_dt := u26*Abs(t);
   delta_t := t1 - t;
   if  Abs(delta_t) < 2.0*Abs(dt)   then
      begin
      if  0.9*Abs(delta_t) > Abs(dt)   then
         dt := 0.5*delta_t
```

```
      else
         begin
         done := true;   dt := delta_t;
         end;
      end;
   repeat
      { Compute and test the solution;
        decrease |dt| if test fails. }
      if   fun_count > max_calls   then   Halt;
      RKF_45(dim, F, t, dt, current_x,
             next_x, y1, error);
      tol := sup_norm(dim, current_x)
              + sup_norm(dim, next_x);
      tol := 0.5*tol*eta + eps;
      norm_error := sup_norm(dim, error);
      if   norm_error >= tol   then
         begin   { Failure; decrease dt. }
         advanced := false;
         done := false;
         if norm_error >= 5.9e4*tol   then
            dt := 0.1*dt     { 9^5 = 59049 }
         else
            dt := 0.9*dt
                      /x2y(norm_error/tol, 0.2);
         if   Abs(dt) <= min_dt   then   Halt;
         end;   { if   norm_error >= tol }
   until   norm_error < tol;
   t := t + dt;
   current_x := next_x;
   F(t, current_x, der_x);
   Inc(fun_count);
   if   advanced   then
      begin   { 0.18^5 = 1.89-4 }
      if   norm_error <= (1.89e-4)*tol   then
         temp := 5.0
      else   temp := x2y(tol/norm_error, 0.2);
      temp := temp*Abs(dt);
      if   min_dt > temp   then   temp := min_dt;
      if   dt >= 0   then   dt := temp
      else   dt := -temp;
      end;   { if advanced }
until   done;
x1 := next_x;
end;   { solve_ode }
```

```
var    delta_t1: Real;

begin    { RKF45_driver }
dim := 2;
t0 := 0.0;    t1 := 6.9;
x0[1] := 1.0;   x0[2] := 0.0;
delta_t1 := 0.1;
for  k := 1  to  9  do
    begin
    repeat
        t1 := t1 + delta_t1;
        solve_ode(dim, t0, t1, x0, x1);
    until   x1[2] >= 0.0;
    t1 := t1 - delta_t1;
    delta_t1 := 0.1*delta_t1;
    end;    { for  k }
end.    { RKF45_driver }
```

Predictor-Corrector Methods

There is another family of ODE solvers based on rather different principles. The best known goes under the name Adams-Bashforth-Moulton *predictor-corrector method.* It would take us too long to discuss here, but you will find two programs (A_B_M and test_ABM) on your disk covering the method and an application.

Exercises

1. Write a program that tests all the methods of RK over one step for various sample problems.
2. The harmonic difference equation, related to the Laplace equation, starts with initial values on the boundary vertices of a rectangular grid and arbitrary values on the interior vertices, perhaps all equal to the average of the boundary values. Each step replaces the value at each interior vertex by the average of its neighboring four values. The process converges. Write a program to illustrate this method.

Appendix 1 The Borland Pascal Package

Section 1.   Installation

When you open your carton of *Borland Pascal*, you may be frightened by the tens of kilos of books and mountain of disks. This Appendix will get you started on installation and use of the system to write your Pascal programs. Even if you have already installed *Borland Pascal* and are using it, you may find some useful tips here, so please thumb through these pages. (There is much useful information in the *Borland Pascal User's Guide*, which is in your package.)

For starters, your computer should have a goodly chunk of free disk space in one hard disk partition. If you install the complete *Borland Pascal* system, that will take about 30M. The programs with this book will fill about 2M, and when you start writing programs, who knows? Before you start installation, decide which partition to use, and note how much free disk space is available. Don't push a partition to its absolute limit.

Start the "Install" program by inserting Disk 1 into the A: drive and typing

```
A:INSTALL <Enter>
```

(By the time this book appears you will probably be able to acquire BP on CD-ROM and install it that way.) The "Install" program gives you lots of options, and explains what it is doing as it runs. If you have adequate disk space, the easiest course is to install everything. If you want to install only what you need for using this book, when the "Install" program prompts you for what to install/omit, you can omit the Windows version, the Assembler, the Profiler, the Debugger, the Turbo Vision package, and the On-line compilers. You can also eliminate either the real mode version or the protected mode version, preferably the former.

You should have at least 8M core memory, and it should be configured for XMS (EMS) memory. If your computer runs MS-Windows, then you are OK. Otherwise your DOS manual will tell you how to add a device driver like EMM386.SYS or HIMEM.SYS to your CONFIG.SYS file to configure the memory region beyond 1M. If the extra memory is not available, you can use the real mode program TURBO.EXE, but I strongly recommend using the protected mode program BP.EXE. Using BP means that you do not have the dreadful DOS 640K barrier. Your programs can be almost as large as you please, and your data can fill a lot of extended memory.

It is most likely that some previously installed program has modified your CONFIG.SYS file to have a lot of files and buffers. Anyhow, the lines

```
files=20
buffers=30
```

or higher values should be included. I also strongly recommend working from a RAMDisk if you have the memory for it. It should have at least 2.5M. This requires a line in the CONFIG.SYS file like

```
DEVICEHIGH=E:\NWDOS\VDISK.SYS 5120 512 128 /E
```

which sets up a 5M RAMDisk

The protected mode version of the *Borland Pascal* is started by typing "BP <Enter>". The real mode version is started by typing "TURBO <Enter>". Both of these assume that the directory "BP\BIN" is in your path, or that you have changed to that directory. There are configuration files that set a lot of options for how you work.

One particular point to note is that the Editor is configured to respond to *Word Star* commands (Crtl–K–X, Ctrl–Q–X, ...). These can mostly be changed to make use of the F-keys, which were not used in ancient *Word Star* days of yore. To save you the trouble, I have included on disk the configuration file BP.TP that I use and the source code for building it, which you can modify to your own preferences. I also include TURBO.TP in case you have to work in real mode.

My source code file is TURBO7.TEM. It is compiled with the program TEMC.EXE, which explains itself. However, the commands for modifying existing configuration files with it are

```
TEMC xxx[.tem]  bp.tp <Enter>
TEMC xxx[.tem]  turbo.tp <Enter>
```

Henceforth I'll assume that you are going with the protected mode IDE (Integrated Development Environment).

Gradually I'll tell you my own working conditions. For now, note that I work with a Northgate *Omni* keyboard that has F-keys on the left (on the top too). I can hit any left F-key blindfolded, and that is important to me. (I must look to find F-keys on top.) The *Omni* has both left and right Alt- and Ctrl- keys, so I can hit any of the 40 combinations Shift–, Alt–, Ctrl–, F1 — F10 easily.

Rename the official BP.TP to (say) BP.$TP; then copy my BP.TP into BP\BIN. My preferences for key strokes are in the ASCII file TURBO7.KEY, which you should print. Many of the F-keys are used by the IDE as "hot keys" (like F2 – Save, F3 – Load, F9 – Compile, etc), so there are not as many free as I would like.

Here is the most useful part of TURBO7.KEY:

```
F1   Help                        F2    Save file
F3   Open file
F5   Zoom window                 F6    Next window
F9   Make (compile/link)         F10   Menu


Shift-F1   Help index
Shift-F3   Move to block begin
                            Shift-F4    Move to block end
                            Shift-F6    Cycle windows
Shift-F7   Copy block       Shift-F8    Move block
Shift-F9   Read block       Shift-F10   Write block


Shift-Ins   Paste from clipbd
                            Shift-Del   Cut to clipboard


                            Alt-F2    Find
Alt-F3   Close window       Alt-F4    Replace
Alt-F5   User screen        Alt-F6    Repeat find/replace
Alt-F7   Set block beginning
                            Alt-F8    Set block end


Alt-I   Indent block;       Alt-U     Outdent block


                            Alt-X     exit


                            Ctrl-F6      Delete line
                            Ctrl-F10     Delete block


Ctrl-PgUp   Top of file     Ctrl-PgDn    End of file
Ctrl-Home   Top of screen   Ctrl-End     Bottom of screen
Ctrl-Left   Left one word   Ctrl-Right   Right one word


Ctrl-E      Delete To end of line
Ctrl-W      Delete word
Ctrl-Del    Delete block    Ctrl-Ins   Copy block
```

You should have a mouse connected. It provides a fast way to execute commands bound to keys. In general, it is used in the usual way. Note that a "block" is a portion of "selected" text in mouse parlance. The usual left-click, drag, shift-left-click methods of selecting text work in the IDE.

You use the TEMC method only to change key bindings. I hope you like my choices of keys, and never use TEMC to change them. Changing other options, like screen colors, can be done on the fly, in the IDE Options choices. I'll give examples later.

Note the keypress "Alt–F5": *User screen.* This means the ordinary DOS screen. That keypress will change from the editing window to the DOS screen, and then any keypress will return to editing. This is particularly useful when you are testing a program from the IDE. It may run so fast that it returns to editing before you see what has happened, or it may crash with a "Runtime Error", and you want to see the error message. Trust me; this happens often.

Section 2. Programming in the IDE

Change to the directory BP\BIN and type BP <Enter>. Look at the screen that comes up. There will probably be a file named NONAME00.PAS open. If not, press F3 and type "xxx" or anything in the prompt for a file name, and press <Enter>. In the big edit window, type

```
program  my_first;

   var  k: word;

   begin
   for  k := 1  to  10  do
      begin
      writeln('Heil dir, Sonne!');
      writeln('pi/4 =', pi/4);
      end;
   write('Press <Enter>: ');
   readln;
   end.
```

Press F9 to compile this masterpiece. Fix its errors if any are noted. Then run the program: Press F10 R R. Alternatively, click on Run on the (top) menu bar, then click on Run in the dialogue box that opens. The *Borland Pascal User's Guide* contains numerous nifty ways to fool around with menus, and I will not repeat much of this, but rather concentrate on programming.

Press F3 to open another program. Give it the name "yyy". Press F6 to change windows, back to xxx.pas. Press Ctrl–PgUp and Home to move the cursor to the very beginning of the file. Press Alt–F7 to start marking (selecting) a block. Press Ctrl–PgDn and End to move to the very end of the file. Press Alt–F8 to finish marking the block. The whole program should now be highlighted.

Press Ctrl–Insert to copy the marked block into the Clipboard. Press F6 to return to yyy.pas. Press Shift–Ins to copy the clipboard into yyy.pas. Now you have an identical program to xxx.pas. Eliminate the highlighting by pressing

Alt–F7 and Alt–F8. (An alternative to this round about method is to save xxx.pas, shell out to DOS, copy xxx,pas to yyy.pas, etc.)

Move the cursor to the H in Heil, and mark (select) the whole phrase in the single quotes. Press Ctrl–F10 to delete the selected text. The line should now read

```
writeln('');
```

Type between the single quotes your translation into French of the erased phrase. Run the program again.

Finally, press Alt–X to exit the IDE. You will be prompted to save or not to save each of the two open files–your choice. Then you are out and free.

Section 3. My Work Methods

I always work from RAMDisk. It makes everything go faster, and saves hard disk wear and tear. Also when you quit, all files *.BAK, *.EXE, *.TPP, etc. your work session has created disappear, a good thing. The only disadvantage of working from RAMDisk is that you lose all of your work in a power failure, unless you are careful about shelling out to DOS and saving frequently. (I use a UPS, so I don't worry much about lightning, etc.)

The batch file T7.BAT on disk copies the minimal necessary files to RAMDisk; you should modify it according to your partitions and RAMDisk drive letter. Actually I keep several batch files, one for each specific project I am working on. A sample is SCIPAS.BAT, whose purpose is to load the necessary BP files on RAMDisk and all the program files for this book, plus a few support files, a configuration file BP.TP specific to the project, a batch file S.BAT that saves any file that has been changed back to the proper directory on the hard disk. Let's look at SCIPAS.BAT:

```
rem    SCIPAS.BAT    12/07/93
CALL   C:\BATCH\T7
E:
CD   \SCIPAS
XCOPY   *.*   H:  /S
H:
ATTRIB   -A *.*   /S
BP
CALL   S
```

As you can see, I store batch files in directory C:\BATCH, which is on my DOS path. SCIPAS.BAT's first action is to call the batch file T7.BAT, which loads what I want from *Borland Pascal* onto my RAMDisk H:. The DOS "CALL" command executes another batch file, then returns control to the calling batch file.

Next SCIPAS.BAT logs onto E:\SCIPAS and XCOPY's everything there, including subdirectories (option /S) onto H:. Then it logs onto H: and resets the attribute bit of every single file there. That way I can later save only files that have been altered. Then it starts the IDE (line BP). When I quit the IDE, the file S.BAT is executed.

I include S.BAT on disk as a model for you. Let's look at it:

```
rem   file   SCIPAS\S.BAT      12/09/93
XDEL  *.BAK  /S  /N
rem   /S   Subdirectories
rem   /N   No permission needed
XCOPY   *.PAS   E:\SCIPAS /S/M
XCOPY   *.BAT   E:\SCIPAS /S/M
XCOPY   BP.TP    E:\SCIPAS /M
XCOPY   BP.DSK   E:\SCIPAS /M
rem   /M   Copy only files with archive bit set
rem        and reset the archive bit
EXIT
```

First S.BAT deletes any backup files that have been created in the editing process, to free up some RAMDisk space. (I use S.BAT frequently to make permanent copies of what I am working on, maybe every half hour, not just at the end of work sessions. Lightning *may* strike.) Next it saves Pascal source codes *that have been altered* plus some other files that have been altered. The /M parameter to XCOPY means copy files with the attribute set and then reset the bit. (Run DOS's ATTRIB <Enter> on your C:\ directory if you don't know what file attributes are.)

S.BAT saves the editor configuration file BP.TP because I sometimes do make small changes to my editing configuration. The file BP.DSK is a copy of the desktop, which then allows me to continue my next work session exactly where I left off. (Note: Delete BP.DSK before starting work if you want to clean up your history lists; that is the only way to do so.)

The final EXIT in S.BAT is there because if I have shelled to DOS from the IDE to run S, then EXIT returns me to the IDE. In case you forgot – "shelling to DOS" means temporarily leaving the IDE to use DOS, and then returning. You do this by choosing from the menu bar: File - DOS Shell.

To make sure that changes to the editing environment and to the desktop are saved, you must choose Options – Save Options from the menu bar. While it only says that BP.TP will be saved, BP.DSK will also be saved. If you fail to save in this way, changes to your current editing configuration and current desktop will be lost.

We should look at T7.BAT:

```
ECHO   OFF
Rem   T7.bat        06/25/94
Rem   Copies the files from F:\BP needed to run BP
Rem   This assumes Borland Pascal is in F:\BP
Rem   and the RAMDisk is H:
Rem
XCOPY   F:\BP\BIN\BP.EXE   H:
XCOPY   F:\BP\BIN\BP.TP   H:
XCOPY   F:\BP\BIN\TPP.TPL   H:
Rem     Turbo.Tpl is the library
Rem     for real mode applications
Rem     Only load it if you intend to compile
Rem     for real mode.
Rem     XCOPY   F:\BP\BIN\TURBO.TPL   H:
Rem     PRNTFLTR.EXE is required to print your workfile
Rem     from the IDE.
Rem     XCOPY   F:\BP\BIN\PRNTFLTR.EXE
XCOPY   F:\BP\BIN\TURBO.TPH   H:
XCOPY   F:\BP\BIN\RTM.EXE   H:
XCOPY   F:\BP\BIN\DPMI16BI.OVL   H:
XCOPY   F:\BP\BIN\GREP.COM   H:
XCOPY   F:\BP\BIN\G.BAT   H:
XCOPY   F:\BP\BGI\EGAVGA.BGI   H:
```

You see that very few of the thousands of the *Borland Pascal* files are actually needed. The IDE, editor, compiler, linker, and more are in BP.EXE. The file TPP.TPL is the library of standard units for protected mode programs: System, Graph, DOS, CRT, etc. The optional (Remarked out) file TURBO.TPL is the library of standard units for real mode applications. The compiler in the protected mode program BP.EXE has options allowing you to choose as your target for compiled programs either protected mode or real mode (or windows). The default set by my configuration file BP.TP is protected mode. If you write programs for others to use in real mode, you will need the real mode library TURBO.TPL.

To change target, in the IDE choose from the menu bar Compiler – Target – DOS real mode.

The file TURBO.TPH contains on-line help for the IDE. The editor configuration file BP.TP, you know all about. The two files RTM.EXE and DPI16BI.OVL make up Borland's protected mode manager. Both must be available for any protected mode program to run, and may be included free of royalties in any applications you develop and distribute. The file EGAVGA.BGI (BGI: *Borland Graphic Interface*) contains necessary data for running graphics programs on VGA or EGA monitors. (There is a way to attach this file to executable programs; see Programs 5.8.1 and 5.8.2.)

The utility GREP.COM is very useful for finding particular words in many files. You will want to use it to search through your source code files *.pas. As its parameters and usage are hard to remember, I wrote the little batch file G.BAT that does the job for you:

```
grep  -indw  %1  *.pas
```

This says search for the string given as its (only) parameter (%1), ignore (i) case, give line numbers (n), search whole words only (w), and search all files *.pas in the directory (d) tree, starting from where g.bat was called. ("grep" is a Unix term for string searching.) You use g.bat as follows:

```
g writeln
```

This finds all occurrences of "writeln" in files *.pas, and writes the line and its line number of each. As the search result pours out rapidly, have a finger poised over, and ready to strike the Pause key (or include the | MORE pipe). Another option is to save g.bat's output to a file:

```
g writeln > ww.pas
```

Then you can inspect the result in the IDE editor.

To search for a string involving characters other than letters, digits and underscores, put the search string in quotes:

```
g "writeln("
g "for  x"
```

The (Unix) rules for regular strings are very complicated, and sometimes searches will fail. Documentation for GREP is in BP\DOCS\UTILS.DOC.

I have written several utilities to help me manage and change files. The listings are included on your disk. "Destroy" is for wiping out a directory tree. It has safeguards in the form of a list of directories protected from destruction. "Size" determines the total size of a directory tree, hence whether it will fit on a diskette. The program "Se-re" uses a list of search/replace words prepared in advance, and then does the replacements in every file in a directory tree. For example it can replace every instance of "write" and "writeln" with "Write" and "WriteLn" in all files *.pas in directory "D:\scipas". "Se-re" does not change the originals, but slightly changes their names for the corresponding altered files, in case you want to undo its actions. "Clobber" does an ax job on the original files and restores the original file names.

Section 4.   The Menu Bar

This section contains tips on using the menu items. Let us start on the Menu Bar with "File". Again, I remind you that you can access this by simply (left) clicking on "File" in the Menu Bar, or by "F10 F". The choices you will use most are Open, Save, Exit, and DOS Shell. The first three have "hot keys": F3, F2, Alt-X, so you need not go to "file" at all for these. For some reason, "Close" is under "Window", not "File". You can close a file by "Alt-F3", or by clicking on the little box in the upper left hand corner of its window. If you have made any editing changes since the previous "Save", you will be prompted to save again.

When File–Open prompts you for a file name, you ordinarily only enter a name like "my_file", and "my_file.pas" is opened–the suffix ".pas" added free. If you want to edit other files than "*.pas", you enter "name.", or "name.*", or "*." or "*.*", etc. depending what you want opened. The name "*.bat" for instance will give you a choice of all files meeting that wild card description.

Note the history list at the bottom of the "File" dialogue box. It has all files you previously opened, and you can click on any file in this list to open it again. Also, as soon as you press "Open" (or get there via the "hot" key F3) there is a history list available, accessed by the down arrow key.

The "Edit" menu contains two items you will use, the most important of which is "Undo". For each open file, the IDE keeps a long record of changes you have made, and by hitting "F10 E U" (or "Alt-BackSpace") many times, you can undo many changes, in case you are unhappy with some editing action. Needless to say, its most frequent use is to restore an inadvertently deleted block.

The other useful "Edit" item (sometimes) is "Show Clipboard". The clipboard is a temporary file, containing all blocks that you have either copied (Ctrl–Ins) or cut (Shift–Del). The last such block is selected (highlighted), however, you can select another block if you wish to retrieve it. Close the clipboard by clicking its upper left box; Then Shift–Ins will insert the selected block into your edit file at the cursor.

I never use "Search" from the menu bar, as I have bound keys Alt–F2, Alt–F4, Alt–F6 to Find, Replace, Repeat Find/Replace respectively. You may prefer to choose these actions from the "Search" dialogue box.

There are a few points to note concerning searching. First, if the cursor is over a word in the edit window, that word is the default find text when you invoke either find or find/replace. It disappears if you start typing a new word. But if you want to modify the default find text slightly, just move over it a little with the left/right arrow keys; the find text changes color so you know that you can delete from or add letters to it.

Second, there are history lists for the find text and for the replace text. They are accessed by the DownArrow key when you are prompted for text. Using these lists can save a lot of typing. Next, you can continue a search or replace across different files. For instance, suppose you want to replace all instances of "write" with "Write" in xxx.pas and yyy.pas. With both files open, do the replacement in one of the files, bring up the window for the other file, and "Repeat Find/Replace".

An important point: <Tab> and <Shift-Tab> cycle you through dialogue boxes forward and backward. When, for instance, the "Replace" dialogue box is up, you first enter "Text to find", either by typing or choosing from the history list. Then do not press <Enter>, which would close the "replace" dialogue box and start replacing, but press <Tab> (or use the mouse) to move to the "New text" prompt.

Again, don't press <Enter>, but rather move into the "Options" area and pick your options carefully. <Tab> moves you through the options; <Space> on an option toggles it on/off, as does typing its highlighted letter. A few experiments with Find/Replace will do much more for you than I can explain. Here is one tip: You can search for symbols, but symbols are not delimited by letters and digits. For instance, if "Whole words only" is checked, then the ; in "end;" will not be found. But it will be found if "Whole words only" is not checked.

The "Run" box contains some useful stuff for integrated debugging, that you will want to postpone until much later. There is one choice in its dialogue box worth noting: "Parameters". In developing a program, you may test it many times from the IDE, before you are satisfied and willing to run it as a stand-alone program. But some programs take parameters on their command lines. You know many programs like this. For instance DOS's "COPY" takes (up to) two parameters, the source and the destination. If your Pascal program takes parameters, you can enter them in an input box, just for testing, right here.

The "Compile" dialogue box contains one item, not bound to a hot key, that is useful: "Build". This will recompile all of the units that a program uses before compiling the program. "Make" (F9) only recompiles units that have been changed. You may not see how this can ever be useful, but it is. Suppose that you have a large program that is split into many units, and you want two versions, one using the 80x87, one not using it. My program *MicroCalc* is exactly like this (because some schools still lack coprocessors in their lab computers[1]). Each unit

---

[1]   I suppose I should mention that it is possible to "emulate" a math coprocessor by means of the E+ compiler directive used in combination with the N+ directive. This costs a lot in execution speed.

and the main program contain the conditional compiler directive

```
{$ifopt n+}
type  Real = Extended;
{$endif}
```

One simple change in the "Options Compiler" dialogue box then takes care of the whole program, about 25 units. But I have to recompile all of these units to make the change, and that is what "Build" does for me.

The "Debug" box can be useful, but there is one big caveat: The integrated Debugger only works on programs compiled for Real Mode target! To use this IDE debugger, you must have options set for the compiler to generate debugging information.

That brings us to "Options" on the Menu Bar. ("Tools" are not for us now.) You should take a good look through the various possibilities here. If you split a large program into units kept in subdirectories, you will have to go into "Directories" to tell the compiler where to look. Under "Options—Environment—Colors" check out "Editor" and "Syntax". Under "Options—Compiler" note the possibilities. If you have a mathematical coprocessor, make sure that "8087/80287" is checked. You surely have an 80286 or higher, so check "286 instructions". Click on "OK"; then enter Options again and click on "Save".

One of the really great features of the IDE is having different elements of program code in different colors. This was a new feature with *Borland Pascal 7.0*, and was not in any previous version of *Turbo Pascal*. Having reserved words a distinct color from identifiers catches spelling typos immediately; having strings another color prevents forgetting to end them; having comments still another color makes their starts and finishes obvious, without looking for the matching } or *). I consider this Borland's finest effort in saving my time! If you don't like the colors I chose (after many experiments) be free to change them. Remember to save your changes by going back into "Options" and choosing "Save options".

"Window" offers "Zoom" and "Next window" (cycle windows) for which we have hot keys F5 and F6. It also has "Close" for which I find Alt-F3 more convenient. You should try its other possibilities.

"Help" is context sensitive, and most easily invoked by F1. It will bring up help, wherever you are, pertaining to the word under the cursor. If you need help on the Pascal language and the numerous procedures in its libraries, press <Shift-F1>. It will take a few moments the first time to build the *Help Index*, but subsequent uses are immediate. When the index is up, just start typing the word you want to know about and you will move to it. If the word has several choices under it, usually you should choose the first option (*Borland Pascal*) before you click or press <Enter>.

The box that comes up with the item you have chosen may be too small. Then drag it to the left by its top bar, and drag its lower right corner to enlarge it.

Three final points:

<Esc> gets you out of many situations. Do not hesitate to press <Esc> if you can't think of anything else.

The mouse right click brings up a useful short menu in the edit window.

Sometimes when you run a graphics program from the IDE, then return to the IDE, the screen appears broken up, and possibly the mouse is disconnected. You can repair most of the damage either by "F3 Esc", starting to open a new file, but then not doing so. Another possibility: "F10 W R". Here F10 moves to the menu bar (which you might not even see!), W chooses Window, and R chooses Refresh screen. Neither of these methods will restore the mouse; for that simply shell out to DOS and type exit <Enter> to return to the IDE.

Section 5.   The Program Disk

The diskette accompanying this book contains two files: READ.ME, a text file, and SCIPAS.EXE, a self-extracting archive file created with the public domain program LHArc. Make a directory, say D:\SCIPAS and change to it:

```
D:  <Enter>
CD \  <Enter>
MD SCIPAS <Enter>
CD SCIPAS <Enter>
```

Copy the whole disk by

```
XCOPY A:*.* /V <Enter>
```

To extract the archive:

```
SCIPAS <Enter>
```

SCIPAS.EXE will extract itself, and build many subdirectories as it does so. After the extraction process ends, you may delete the archive SCIPAS.EXE. The file "READ.ME" on your disk contains important information, and you should print it.

Up front, there is a file INDEX.PAS that contains a complete index of all the Pascal and data files. Each subdirectory corresponds to a chapter (or appendix) and contains an index file itself. These chapter subdirectories contain second level subdirectories for each section. You will find all the units and programs printed (or referenced) in this book and almost all program code answers to the exercises. You may also find some other programs that interest you.

Appendix 2 On Programming

Section 1.   Tips on Programming

The following tips are to make your programming easier and the program code files easier to read. You would be surprised how often a program written just a few days earlier can be impossible to understand.

1. Use a lot of extra white space in your programs, horizontal and vertical. Every book editor knows the value of white space for clarity; don't forget that your programs have to be *read*, even if only by you.
2. Keep identifier names short, particularly temporary variables that merely control loops or hold a value temporarily. Don't use "current_value_x" when "x" will do.
3. Don't be stingy with memory by trying to have the same variable serve several purposes, just to save a few bytes. Such a variable may have an unexpected value some time when it is referenced, causing a subtle error, or even causing a horrible crash.
4. Don't be afraid to insert extra semicolons if they add to clarity; they can't hurt anything.
5. Don't be afraid to insert extra **begin   end** pairs if they will make your program easier to read. They often help show what is subordinate to what, and they cost zilch.
6. In the same spirit, *indent* generously. I recommend 3 space indents.
7. Avoid long lines. (In your program code, not at the post office.)
8. Document your programs, i.e, insert comments that explain what is going on. Comments should be informative and succinct.
9. One of the simplest ways to look around a long program file and then return to where you were: Insert any character (outside of a comment) where you want to return to. Then press F9 (compile) when ready to return!
10. When you type **begin**, immediately type **end** directly under it; then fill in the statements between. (Ditto **repeat...until** and **case...end**.) This helps avoid some subtle errors and also keeps your indentation uniform.

You will find on your disk two "model" programs that you can use as templates for programs you write. One is for programs; let's look at the other, for units:

Program 2.1.1

```
{$N+,R+,S+,I+,O+}
unit  model_unit;    { 07/08/94 }


interface
```

```
const   rows = 5;   cols = 8;

type    Real = Extended;
        matrix = array[1..rows, 1..cols] of Real;

var  A, B: matrix;

procedure  proc(n: Word;   ch:  Char;
                 var A: matrix);
function   func(n: Word;   ch:  Char;
                 var A: matrix):  Real;

implementation

uses   CRT, DOS;

procedure  proc(n: Word;   ch:  Char;
                 var A: matrix);

   var     t: Real;
         j, k:  Word;

   begin
   for  j := 1 to 5  do
      begin
      end;    { for  }

   while  true  do
      begin
      end;    { while  }

   repeat

   until  false;

   case  k  of
      1:    ;
      2:    ;
   end;    { case }

   end;   { proc }

function  func(n: Word;   ch:  Char;
                var A: matrix):  Real;

   var  t: Real;
```

```
    begin
    end;    { func }

  end.    { unit  model_unit }
```

By marking and copying blocks from this template, you can save a lot of typing. Don't forget that Alt–I indents a marked block and Alt–U undents it (in my editing key setup).

Section 2. Bug Sources

The Pascal compiler is so fast by now that typos and syntax errors in your program code will be fixed up in split-seconds, so it is no longer necessary to consider mistakes detected by the compiler as bugs. When you run a program and it hangs the computer—causing you to reboot (worst scenario), or it merely crashes, or it produces incorrect output, then you have a true bug.

Protected mode is supposed to prevent hanging by protecting the operating system from being overwritten by a running program. This is about 90% true; it certainly is the case that programs running in protected mode will hang far less than programs running in real mode. So even if you are writing a program for real mode, you may save a lot of trouble by testing it in protected mode.

Infinite looping can occur in **repeat**...**until** and **while**...**do** loops if the terminating condition never occurs. For instance, the following is typical of a common oversight:

```
  k := 1;
  repeat
      <statements>
      inc(k, 2);
      <statements>
  until k = 10000;
```

The control should have been k >= 10000;

More frequent is

```
k := 1;
repeat
   <statements>
until k = 10000;
```

where you just forget to increment k. To avoid such extremely annoying situations (often requiring *rebooting*), simply insert the line

```
if  KeyPressed  then  Halt;
```

in every single **repeat** and **while** loop. ("KeyPressed" requires **uses** CRT.) After the program is thoroughly shaken out, you can remove these safeguards. Of course, if you have a persistent error that requires your pressing a key repeatedly to get out of trouble, then you should replace "Halt" by a statement reporting the states of several variables before halting.

Sometimes a program run from the IDE (BP) will stop an infinite loop by itself, while the same program run from the DOS command line will go on forever. The computer's general environment is different when running a program from within the IDE than from DOS. Sometimes you can stop an infinite loop by pressing the <Pause> key, followed by <Ctrl-Break> a couple of times and then <Enter> a couple of times. I said "sometimes", but anything short of <Alt-Ctrl-Del> is worth trying! Sometimes the screen goes completely blank. Again, don't panic before you try everything: <Esc>, <Ctrl-Break>, <Enter>, <Alt-X> (to quit the IDE), CLS <Enter>, each a couple of times.

Deeply nested recursive calls that never stop nesting until a crash is another source of difficult situations. The "KeyPressed" idea above may work if you are fast enough on the trigger. A more reliable method is to use a global Word variable "nest_level" initialized to 0 and incremented on each entry to the procedure, decremented on each exit from the procedure, and with a statement like

```
if  nest_level = 100  then  Halt;
```

A minimal example of such a program:

```
program  crash;
   procedure  a;
      begin  a;  end;
   begin  a;  end.
```

Run this under varying conditions: protected/real mode, compiler option S+/-, from the IDE (Integrated Development Environment), from the DOS prompt. Do try <Ctrl-Break> before rebooting if it hangs, but expect the worst!

The single most frequent source of bugs in my programs involves memory management. Suppose ptr is a pointer variable declared by

```
var ptr: ^my_type;
```

Somewhere down the line memory is allocated by either

```
New(ptr);    or    GetMem(ptr, n);
```

Later in the program there is an assignment of ptr, directly or subtly via a procedure call—the assignment done before the memory originally allocated to ptr^ has been released. This will usually cause a crash eventually, or some totally unpredictable behavior.

So be very careful when you use "New" and "GetMem". When you write either in a program statement, you should immediately write the corresponding "Dispose" or "FreeMem". In the case of "FreeMem" you should free exactly as much memory as allocated originally. If the "n" in "GetMem" displayed above is a variable rather than a constant, be absolutely sure that the value of n does not change between the allocation and the release

```
FreeMem(ptr, n);
```

Each introduction of "New" or "GetMem" is a good place for a { comment } alerting you to possible danger.

Disconnecting a pointer from the memory that it points to can set a time bomb that may cause the worst imaginable kind of crash. I have not yet had a hard disk clobbered by this, although I worry that it will happen some day. However I have had my CMOS totally destroyed by such a bug. CMOS (complementary metal oxide semiconductor) is a battery powered chip that holds the computer's basic setup: memory size, disk drives, type of hard disk, video adapter type, etc. This information is usually changed by a "setup" program that is on a chip itself. In grossly simplified form, the mistake is as follows:

```
New(p);
{ p^ is dealt with }
New(p);
```

The memory allocated by the first call of "New" is disconnected by the second call, without first calling "Dispose", and there is no way to access this memory, which tends to confuse the program greatly, and lead to its downfall. (An assignment p := q; would have the same effect.) To coin an awful phrase: *Beware of dangling pointers.*

There is less serious, but far more subtle error involving pointers used as value parameters of procedures. As you know, a value parameter (not a **var** parameter) is copied. *But...* what it points to is *not* copied. So if the procedure modifies what the pointer variable points to, that will modify permanently those variables pointed to. Simple example:

```
type   integer_ptr: ^integer;

var   P: integer_ptr;

procedure  screw_up(Q: integer_ptr);

   begin
   Q^ := 8;
   end;

begin
New(P);   P^ := 7;
screw_up(P);   WriteLn(P^);
end.
```

You might think that because Q is a copy of P, the assignment of Q^ has no effect on P^. True, Q is a copy of P, but the pointer variables Q and P point to the same thing, i.e, P^ = Q^. Believe me, this is a hard error to track down, so you can't be too careful.

Carelessness with the scope of variables can cause problems that are hard to correct. If a variable seems to change value in unexpected ways, check its scope. Remember that variables local to a procedure are *not* initialized automatically when that procedure is called. Just because variable x local to procedure P has been given some value by P, it cannot be assumed to have that value the next time P is called! This fragment typifies scope errors:

```
var   t: Real;

procedure   A;

   begin
   . . . . . .
   t := . . .;
   . . . . . . . .
   A;
   x := t;
```

```
      . . . . . . .
      end;
```

In dealing with structures like linked lists and binary trees, be most careful with nested **with** statements. The variable (for instance) "tag" which you may think is p^.tag actually is p^.next^.tag because you forget that you are within a "**with** next **do**" clause.

In the same setup, suppose you have a recursive "destroy" that unloads a whole tree from its point of entry. The sequence

```
with   p^   do

      begin
      place_holder := next;
      destroy(p);
      p := place_holder;
      end;
```

looks plausible at first glance, but it's a disaster. Think about it—it's an easy mistake to make if there are enough other statements around to confuse the issue.

Section 3. Runtime Checking

The first step is to test your programs with all the possible runtime checks activated. Each of these increases the length of your compiled executable .exe file, and will make your program run slower. Therefore it may be desirable to remove these options when your code is thoroughly tested.

'First is {$R+} for "Range Checking". When Range Checking is activated, the program has code to trap assignments to arrays and strings that go outside their bounds and to trap assignments to simple variables that are out of range.

Next is {$S+} for "Stack Checking". The stack is a portion of memory reserved for storing addresses and local variables whenever a procedure (function) is called. Overflowing the stack is particularly possible when procedures are called recursively.

The amount of memory allocated to the stack is up to you. In the IDE you use Options—Memory or you put a compiler option {$M...} at the very beginning of your main program (not in units). For protected mode applications you are allowed only to specify the stack size. If you have deeply nested recursive calls in a program you will want to increase the default of 16,384 bytes to something near the maximum 65,520 bytes. But keep in mind that space is not free; increasing stack size decreases space available for ordinary declared variables.

The format of the compiler M option depends on the mode. For protected mode applications, it is {$M stacksize }. But for real mode, there are three parameters:

```
($M stacksize, heapmin, heapmax}
```

In real mode applications, a portion of memory is specified for the "heap". This is where variables created dynamically are stored. The default values for M in real mode are heapmin = 0, the minimum possible, and heapmax = 655360, the maximum possible. This *does not* allocate any memory; the memory for dynamic variables is allocated as needed, and the heap grows accordingly until all free memory is exhausted. Please see the chapter *Memory Issues* in the *Borland Pascal Language Guide* for details on this subject.

The next runtime check is {$I+} for "I/O Checking"; its purpose should be obvious. Note that it is common programming practise to turn this check off with {$I–} prior to an I/O call (like "Reset"), check the corresponding IOResult, take appropriate action, and finally turn I/O checking back on. In the {$I+} state, any I/O error will cause the program to halt, i.e, crash.

The final check is {$Q+} for "Overflow Checking". This checks arithmetic operations on integers for overflow. (It is Q+, not O+, which means "Overlay", something quite different.)

Any of these four runtime checkers can be turned on or off by compiler directives, or by using Options—Compiler.

A small warning: Borland's Pascal does not always report the correct error! Recently a colleague asked me to help him find the bug that was crashing his program with some kind of error message about an I/O error. He had I/O checking on, but not range checking. The first procedure in the program initialized some arrays, and the second procedure did some file management, so he naturally suspected an error in the second procedure. I suggested setting {$R+}. With this, the error message on crashing changed to "range value error", so the error was pinpointed to the first procedure.

Section 4.   Debugging Crashes

The first problem is to locate the crash. Suppose that you have a program of about 1000 lines, well organized into procedures and levels of subprocedures. At the outermost level the program body is

```
begin
statement_1;
statement_2;
. . . . . .
statement_19;
statement_20;
end.
```

Insert two lines halfway down:

```
    . . . . . .
    statement_10;
writeln('OK to here');
write('press <space>:  ');   readln;
    statement_11;
    . . . . . .
```

(The inserted diagnostic statements are flush left, for easily locating and deleting later. They are not formatted beautifully; that is a waste of time; "writeln" does the same as "WriteLn".) Run the program again. If you see the message on the screen, then you know that the crash is not during execution of the first 10 statements, so you might as well halt the program with <Ctrl–Break>. On the contrary, if you do not see the message, then the problem is certain with one of the first 10 statements. In either case, make a note of what you have learned (best with a comment right in the code at the halfway point) and select and copy the inserted diagnostic lines to about the center of the guilty lines. Assume that you will need to repeat these tests later; that's why I said "copy", not "move". Comment out with (* *) the diagnostics; only delete them when you are sure the problem(s) are fixed. (Just a *single* bug in a long program would be extraordinary indeed.)

In a few such steps you will pin the crash to one of the 20 statements. If that statement is a procedure call, go into the body of that procedure and do the same detective work. In a reasonable time you should pinpoint the error to a single statement.

This is good work, but usually not the end of the line. The statement that crashes the program probably deals with some variables that have been processed earlier, and possibly the earlier processing gave some variable a value that makes the suspicious statement crash the program. So now it is time to look at the values of individual variables just before the program fails. Insert some statements like

```
writeln('x = ', x);
writeln('ord(ch) = ', ord(ch));
for  k := 1  to  10  do
    writeln('a[k] = ', a[k]);
write('Press <Space>: ');  readln;
```

Eventually you will isolate the problem, but the process can be painstaking.

I wish I had kept careful records of every serious bug I have killed, but I didn't, so much wisdom is alas lost. However, I do have one rather hard to diagnose bug to show you.

Suppose a unit starts like this:

```
unit  vector_algebra;
interface
   uses  ABC;
   procedure  XYZ;
implementation
   var  v, w: vector;   { etc }
```

The implementation of procedure XYZ involves a call to a procedure BLAS, which is declared, along with the type vector, in the unit ABC:

```
unit  ABC;
interface
   type  vector = array[1..10500] of Integer;
   procedure  BLAS;
implementation
   procedure  BLAS;
      var  temp: vector;
      begin . . . end;   { etc }
```

This all compiles fine, but the main program, which uses the unit vector_algebra, crashes. If you compiled with the R+ and S+ options active, the crash is not too bad; there is a "range check error" message. Without the runtime options, you are in real trouble, and probably have to reboot.

It doesn't take too long to figure out that the crash occurs when BLAS is called, so you insert diagnostics into BLAS to isolate exactly where. After some experimentation, you find that BLAS never gets off the ground. Even more puzzling is your observation that XYZ runs without crashing if you work with smaller vectors, say 5,000 components instead of 10,500. So what is the problem?

This *is* a little subtle. There is a *stack segment* which contains all local variables. When a procedure is called, memory for its local variables is allocated in the stack, and when the procedure finishes, that memory is released. The maximum possible stack size is 64K, and that requires a {$M...} compiler directive before the main program, or it can be set in Options–Compiler. Anyhow, the vectors v, w, and temp altogether fill 31,500 words, or 63,000 bytes, almost the maximal stack size. Given that the program may have a few thousand other bytes allocated in the stack, it now is easy to see why there was a crash: the stack overflowed. Believe me, this kind of thing can be unnerving the first time you face it.

But this only explains the cause of the crash. What should you do if you really need vectors of length 10,500? The answer is to allocate them on the heap by means of pointers, "new", and "dispose".

Perhaps it is when you are debugging a serious error that you will most appreciate having your program modularized into small pieces. Sometimes debugging is so difficult that you will have to rewrite substantial portions of the program just so you can find its bugs. Always try to (* comment out *) portions of the program to see if they are causing the trouble. If you make all of your documentation comments { like this }, then you can surround portions of text, even those with { documentation }, with (* debugging *) comment brackets.

Extra lines that you add for debugging can be commented out wholesale by means of conditional compilation. First choose a key word, say "debug". Surround each series of lines that are strictly for debugging with the compiler directives:

```
{$ifdef debug }
. . . . . . . . . . . . . . .
{$endif }
```

The lines between the ifdef and endif are compiled only if the symbol debug is *defined*. It can be defined in two ways: either by a compiler directive at the beginning of the program:

```
{$define debug }
```

or by typing in "debug" when you choose (from the menu bar) Options, Compiler, Conditional defines. In the first choice, by putting a space between { and $, you convert the "define" to a comment, and all the debugging code is not compiled. For an example, see Program 7.6.5, "list_all_matrices".

You may wish to explore this family of directives for conditional compilation further. There are "ifndef" (if not defined) and "else" directives also:

```
{$ifdef xyz }...{$else }...{$endif }

{$ifopt N+ }...{$else}...{$endif }
```

The second example allows one piece of code to be compiled if compiling for the 80x87, and another piece of code otherwise. My MicroCalc program code is full of these directives as I have to make both math coprocessor and non-coprocessor versions available.

Remark Do not overlook a powerful debugging tool: your printer. It can be very useful at times to direct your diagnostic messages to the printer instead of to the screen. Simply reassign a "Text" variable from 'con' to 'prn'.

Section 5. Coprocessor Stack Overflow

Sometimes a program using the 80x87 (or coprocessor built into the 486DX, Pentium, etc.) will crash with Error 207: "Invalid floating point operation". There are various causes of this error message, such as Sqrt(negative), Round(too large), etc. There is another cause worth attention. The mathematical coprocessor has an evaluation stack of only 8 levels. A careless recursive calls might overflow this stack. For example:

```
program  crash2;    { 12/21/94 }

   type  Real = Extended;

   var   n: Word;

   function  power_of_2(n: Word): Real;

      begin
      if  n = 0    then   power_of_2 := 1.0
      else   power_of_2
              := power_of_2(n-1) + power_of_2(n-1);
      end;

   begin
   for  n := 0  to  8  do
      WriteLn(n, '  ', power_of_2(n):8:8);
   ReadLn;
   end.
```

This runs OK up to n = 7, and crashes at n = 8. Think about this example and how to fix it by adding a place holder variable to `power_of_2`, thereby moving what is held in the limited 80x87 stack to the Pascal stack in memory.

Section 6. Wrong Answers

So you finally have your program "debugged". It compiles without errors; it accepts input; it produces output; it never crashes. Now you must test it thoroughly for correctness. If at all possible, give it simple data for which you can compute, or at least estimate, the answer by hand. If the program is made to handle 50×50 matrices, change 50 to 2 for testing, and really give it a shakeout. Test extreme cases: sort an array of length one; check that it integrates correctly constant functions; check the cases of an empty file, or a file of only one record. Often these checks on minimal and extreme data can reveal major programming errors. If you have written your program to handle many different possibilities, be sure to check them all for correct output.

Don't be afraid to rewrite major portions of your code. After all, there are first drafts, second drafts, etc. of programs, not just of books about programming. (Of course, computer *manuals*, hardware and software, have *only* first drafts, and are the ultimate, modern example of "stream of consciousness" writing.)

Extensive testing turns up the sad result that your program does not always give correct answers. If it never produces correct answers, that is easier to fix than if it is sometimes right, sometimes wrong. Now you have to examine the logic of your program design very carefully, and test each subprogram. You may find yourself writing a spec sheet for each procedure, what goes in and what goes out. You may have to write quite a bit of extra code just for testing. When finished with testing code, don't toss it out, merely comment it out, as the likelihood of needing it again is usually higher than you think. When one subprogram seems fixed up, the next one may not be right, causing you to reexamine the previous one, etc.

Here is a point I cannot emphasize too strongly: Test extremes and test small. Test a program that reads a file both on an empty file and on a file with one record. Test a program that sorts a list on lists of one or two items. Test a program that deals with square matrices on 1×1 and 2×2 matrices. Etc, etc, etc.

As you gain experience in Pascal programming, you will find that most of your debugging time goes into making your programs give correct answers. Of course you will be tackling bigger projects, and errors in design probably grow at least as the square of program length, possibly exponentially. I cannot say too often how important it is to modularize your programs, breaking them into small units, small procedures. Eventually you should plan each part of a program with how you will test it in the back of your mind. Of course calling a procedure to do a little chore costs runtime. But only after you have everything thoroughly tested should you start to give real thought to fine polishing the speed, possibly by eliminating some procedures and putting their code "in line".

I have a little tip on test programs that use "Random". It is often important to test under exactly the same conditions after making a change to the code. If you comment out "Randomize" wherever it is called, then "Random" will produce the same sequence of pseudo-random numbers.

Section 7. Debugging Graphics

Errors in a program running on the graphics screen are a lot harder to find than errors in a program running on the text screen. When a program crashes while in graphics mode, you are back to DOS with the graphics screen active. There is no visible cursor, the screen responds slowly, and there is no obvious way to return to text mode. One remedy for this is to keep handy a little program that simply enters graphics mode with the procedure "InitGraph", and immediately returns to text mode with "CloseGraph", nothing more:

```
program   fix_text;

  uses   Graph, graphs;

  begin
  open_graph;   close_graph;
  end.
```

(The unit "graphs" was mentioned in Chapter 2.) Run "fix_text" to get out of trouble. Be prepared for the possibility that *nothing* whatsoever shows on the screen after the crash, so you must type blindfolded. Never rush to push the panic buttons: Alt-Ctrl-Del if you are only in modest trouble; Reset for deep trouble.

If you are testing a graphics program from the IDE and it crashes in this way, it is very possible that you will be back in the IDE with almost nothing visible. Then consider the sequence of key presses:

```
F10   F   D   cls <Enter>   exit <Enter>
<Alt-X>   cls <Enter>
```

or possibly a subset thereof. The idea is to shell out to DOS, clear the screen (which might look extremely peculiar), return to the IDE, exit, and clear the screen again. Then run "fix_text" if you are still in graphics mode. Alternatively, while shelled to DOS, change the screen mode to text mode by

```
MODE CO80 <Enter>
```

Sometimes you are not sure if you have exited the IDE and returned to DOS, or have shelled out from the IDE to DOS. The DOS "mem" call gives you a quick report on how much conventional memory is free. You should know the answers when nothing is running and when shelled out of the IDE. When running the IDE in protected mode, there is still almost 100K difference, probably more in real mode.

You can write diagnostic messages to the graphics screen by using the "Graph" unit procedure "OutTextXY". It takes three parameters, the coordinates (x, y) of the pixel where the message starts, and the message itself, a string. To write numbers to the graphics screen, there is nothing like "WriteLn" available, so you must first convert the numbers to strings with the "Str" procedure and then use "OutTextXY". Remember that text written to the graphics screen writes on top of whatever else is there, including previous diagnostic messages, without first clearing a little block, so it is easy for text to get lost. To fix this set a viewport for diagnostic text and use "ClearDevice" on that viewport to clear it. An alternative is to change the (x, y) for each message, for instance start at (10, 15) and increase y by 15 for each further message.

You can also write diagnostic messages to the printer.

Section 8.   Life Insurance

You install a new version of a highly regarded software product from an absolutely reliable house. Four days later, it wiped out your hard disk. Heard the story? You yourself write more and more complicated programs, binary trees of data with pointers in each node to other nodes, and highly complex procedures for such structures. You will make mistakes, and some day one of them may clobber your hard disk, your CMOS, your hidden system files, who knows what. What insurance policies will cover such unnatural disasters?

First policy. Have a tape drive in your computer, and back up the whole hard disk regularly. Have the hard disk partitioned into 40M, or maybe 80M pieces, so you easily can back up the partition that is used most. Keep all your programming files in one partition. Back up on diskette the programs you alter each day you do any programming. Keep in mind what I said in Appendix 1 about setting the archive bit to make it easy to back up only modified files. Most backup programs have options for this too. You can save time by archiving files with a utility like PKZip or LHArc before you copy them to diskette.

Warning If you have several files open in the IDE and you run a program which crashes, it is possible that one or more of the open files will be corrupted. I have had this happen many times. So after a crash, you should check all open files before saving them, thus overwriting good copies (except for their latest changes). Compiling is a fast, easy check. This again emphasizes the need for daily (at least) tape backups.

Second policy. Buy the *Norton Utilities* (Symantec Corp.) and create a "Rescue disk" for your computer. Put its "Image" command into your autoexec.bat file. Learn to use its "unerase" command if nothing else.

Third policy. Buy the *SpinRite* (Gibson Research Corp.) disk utility program. It will give you much information about your hard disk that is hard to get any other way, plus do marvelous things to the disk.

Index